ANDERSONIAN LIBRARY
WITHDRAWN
FROM
LIBRARY
STOCK
UNIVERSITY OF STRATHCLYDE

DEVELOPMENTS IN FOOD SCIENCE 31

MODERN METHODS IN FOOD MYCOLOGY

Proceedings of the
Second International Workshop on Standardisation of Methods for the Mycological Examination of Foods,
held in Baarn, The Netherlands, 20 - 24 August, 1990.

DEVELOPMENTS IN FOOD SCIENCE 31

MODERN METHODS IN FOOD MYCOLOGY

Edited by

ROBERT A. SAMSON
Centraalbureau voor Schimmelcultures
P.O. Box 273
3740 AG Baarn, The Netherlands

AILSA D. HOCKING, JOHN I. PITT
CSIRO Division of Food Processing
P.O. Box 52
North Ryde, NSW 2113, Australia

and

A. DOUGLAS KING
USDA Agricultural Research Service
Western Regional Research Center
Albany, CA 94710, USA

ELSEVIER
Amsterdam – London – New York – Tokyo 1992

ELSEVIER SCIENCE PUBLISHERS B.V.
Molenwerf 1
P.O. Box 211, 1000 AE Amsterdam, The Netherlands

Library of Congress Cataloging-in-Publication Data

Modern methods in food mycology / edited by Robert A. Samson ... [et al.].
p. cm. -- (Developments in food science ; 31)
Includes bibliographical references and index.
ISBN 0-444-88939-6 (acid-free)
1. Food--Microbiology--Methodology. 2. Mycology--Methodology.
I. Samson, Robert A. II. Series.
QR115.M58 1992
664'.07--dc20 92-27934
CIP

ISBN: 0-444-88939-6

This book is printed on acid-free paper.

Printed in The Netherlands

DEVELOPMENTS IN FOOD SCIENCE

Volume 1 J.G. Heathcote and J.R. Hibbert
Aflatoxins: Chemical and Biological Aspects

Volume 2 H. Chiba, M. Fujimaki, K. Iwai, H. Mitsuda and Y. Morita (Editors)
Proceedings of the Fifth International Congress of Food Science and Technology

Volume 3 I.D. Morton and A.J. MacLeod (Editors)
Food Flavours
Part A. Introduction
Part B. The Flavour of Beverages
Part C. The Flavour of Fruits

Volume 4 Y. Ueno (Editor)
Trichothecenes: Chemical, Biological and Toxicological Aspects

Volume 5 J. Holas and J. Kratochvil (Editors)
Progress in Cereal Chemistry and Technology. Proceedings of the VIIth World Cereal and Bread Congress, Prague, 28 June–2 July 1982

Volume 6 I. Kiss
Testing Methods in Food Microbiology

Volume 7 H. Kurata and Y. Ueno (Editors)
Toxigenic Fungi: Their Toxins and Health Hazard. Proceedings of the Mycotoxin Symposium, Tokyo, 30 August–3 September 1983

Volume 8 V. Betina (Editor)
Mycotoxins: Production, Isolation, Separation and Purification

Volume 9 J. Holló (Editor)
Food Industries and the Environment. Proceedings of the International Symposium, Budapest, Hungary, 9–11 September 1982

Volume 10 J. Adda (Editor)
Progress in Flavour Research 1984. Proceedings of the 4th Weurman Flavour Research Symposium, Dourdan, France, 9–11 May 1984

Volume 11 J. Holló (Editor)
Fat Science 1983. Proceedings of the 16th International Society for Fat Research Congress, Budapest, Hungary, 4–7 October 1983

Volume 12 G. Charalambous (Editor)
The Shelf Life of Foods and Beverages. Proceedings of the 4th International Flavor Conference, Rhodes, Greece, 23–26 July 1985

Volume 13 M. Fujimaki, M. Namiki and H. Kato (Editors)
Amino-Carbonyl Reactions in Food and Biological Systems. Proceedings of the 3rd International Symposium on the Maillard Reaction, Susuno, Shizuoka, Japan, 1–5 July 1985

Volume 14 J. Škoda and H. Škodová
Molecular Genetics. An Outline for Food Chemists and Biotechnologists

Volume 15 D.E. Kramer and J. Liston (Editors)
Seafood Quality Determination. Proceedings of the International Symposium, Anchorage, Alaska, U.S.A., 10–14 November 1986

Volume 16 R.C. Baker, P. Wong Hahn and K.R. Robbins
Fundamentals of New Food Product Development

Volume 17 G. Charalambous (Editor)
Frontiers of Flavor. Proceedings of the 5th International Flavor Conference, Porto Karras, Chalkidiki, Greece, 1–3 July 1987

Volume 18 B.M. Lawrence, B.D. Mookherjee and B.J. Willis (Editors)
Flavors and Fragrances: A World Perspective. Proceedings of the 10th International Congress of Essential Oils, Fragrances and Flavors, Washington, DC, U.S.A., 16–20 November 1986

Volume 19 G. Charalambous and G. Doxastakis (Editors)
Food Emulsifiers: Chemistry, Technology, Functional Properties and Applications

Volume 20 B.W. Berry and K.F. Leddy
Meat Freezing. A Source Book

Volume 21	J. Davídek, J. Velíšek and J. Pokorný (Editors) Chemical Changes during Food Processing
Volume 22	V. Kyzlink Principles of Food Preservation
Volume 23	H. Niewiadomski Rapeseed. Chemistry and Technology
Volume 24	G. Charalambous (Editor) Flavors and Off-Flavors '89. Proceedings of the 6th International Flavor Conference, Rethymnon, Crete, Greece, 5–7 July 1989
Volume 25	R. Rouseff (Editor) Bitterness in Foods and Beverages
Volume 26	J. Chelkowski (Editor) Cereal Grain. Mycotoxins, Fungi and Quality in Drying and Storage
Volume 27	M. Verzele and D. De Keukeleire Chemistry and Analysis of Hop and Beer Bitter Acids
Volume 28	G. Charalambous (Editor) Off-Flavors in Foods and Beverages
Volume 29	G. Charalambous (Editor) Food Science and Human Nutrition
Volume 30	H.H. Huss, M. Jakobsen and J. Liston (Editors) Quality Assurance in the Fish Industry. Proceedings of an International Conference, Copenhagen, Denmark, 26–30 August 1991
Volume 31	R.A. Samson, A.D. Hocking, J.I. Pitt and A.D. King (Editors) Modern Methods in Food Mycology

CONTENTS

LIST OF CONTRIBUTORS

AUSTRALIA

Dr S. Andrews, School of Chemical Technology, South Australian Institute of Technology, P.O. Box 1, INGLE FARM, SA 5098
Mr N.J. Charley, CSIRO Division of Food Processing, P.O. Box 52, NORTH RYDE, NSW 2113
Dr Ailsa D. Hocking, CSIRO Division of Food Processing, P.O. Box 52, NORTH RYDE, NSW 2113
Dr J.I. Pitt, CSIRO Division of Food Processing, P.O. Box 52, NORTH RYDE, NSW 2113

BELGIUM

A. Goris, Eco-Bio, Diagnostics Pasteur, Woudstraat 25, B-3600 GENK
L. Meulemans, Eco-Bio, Diagnostics Pasteur, Woudstraat 25, B-3600 GENK
Dr D. Stynen, Eco-Bio, Diagnostics Pasteur, Woudstraat 25, B-3600 GENK
N. Symons, Eco-Bio, Diagnostics Pasteur, Woudstraat 25, B-3600 GENK

DENMARK

Dr O. Filtenborg, Food Technology Laboratory, Technical University of Denmark, DK-2800 LYNGBY
Dr J.C. Frisvad, Food Technology Laboratory, Technical University of Denmark, DK-2800 LYNGBY
Ms Annette Lillie, Alfred Jørgensen Laboratories, Frydendalsvej 30, DK-1809 FREDERIKSBERG
F. Lund, Food Technology Laboratory, Technical University of Denmark, DK-2800 LYNGBY
Dr M. Jakobsen, Alfred Jørgensen Laboratories, Frydendalsvej 30, DK-1809 FREDERIKSBERG
Mr P. Nielsen, Food Technology Laboratory, Technical University of Denmark, DK-2800 LYNGBY
Dr U. Thrane, Food Technology Laboratory, Technical University of Denmark, DK-2800 LYNGBY

FRANCE

Mad. Helene Girardin, Unité de Mycologie, Institut Pasteur, 25 Rue de Dr. Roux, 75015 PARIS
Dr J.P. Latgé, Unité de Mycologie, Institut Pasteur, 25 Rue de Dr. Roux, 75015 PARIS

GERMANY

G. Cerny, Fraunhofer Institut für Lebensmitteltechnologie und Verpackung, Schragenhofstrasse 35, D-8000 MUNICH 50
Prof. J. Krämer, Institut für Mikrobiologie, Rhein. Friedr.-Wilh.-Universität, Meckenheimer Allee 168, D-5300 BONN
W. Röcken, Bundesforschunganstalt für Getreide und Kartoffelverarbeitung, Schutzenberg 12, D-4930 DETMOND

HUNGARY

Dr T. Deák, Department of Microbiology, University of Horticulture and Food Industry, Somloi ut 14-16, H-1118 BUDAPEST
G. Péter, Department of Microbiology, University of Horticulture and Food Industry, Somloi ut 14-16, H-1118 BUDAPEST
O. Reichart, Department of Microbiology, University of Horticulture and Food Industry, Somloi ut 14-16, H-1118 BUDAPEST

K. Szakmár, Department of Microbiology, University of Horticulture and Food Industry, Somloi ut 14-16, H-1118 BUDAPEST
V. Tabajdi-Pintér, Service for Veterinary and Food Examination, BUDAPEST

ISRAEL

Dr S. Steinman, Public Health Laboratory, HAIFA

NETHERLANDS

K.E. Dijkmann, Veterinary Faculty, University of Utrecht, P.O. Box 80175, 3508 TD UTRECHT-DE-UITHOF
B.J. Hartog, CIVO-TNO, P.O. Box 360, 3700 AJ ZEIST
T. Hoopman, Department of Food Science, Agricultural University, P.O. Box 8129, 6700 EV WAGENINGEN
Mr H. Kamphuis, National Institute of Public Health, P.O. Box 1, 3720 BA BILTHOVEN
Ir. Hetty Karman, Food Inspection Services, Nijennoord 6, 3552 AS UTRECHT
M.J.R. Nout, Department of Food Science, Agricultural University, P.O. Box 8129, 6700 EV WAGENINGEN
Ir. G.A. de Ruiter, Department of Food Science, Agricultural University, P.O. Box 8129, 6700 EV WAGENINGEN
Dr S. Notermans, National Institute of Public Health, P.O. Box 1, 3720 BA BILTHOVEN
Ms Marjolein van der Horst, Centraalbureau voor Schimmelcultures, P.O. Box 273, 3740 AG BAARN
A.W. van der Lugt, Department of Food Science, Agricultural University, P.O. Box 8129, 6700 EV WAGENINGEN
Drs Ellen S. van Reenen-Hoekstra, Centraalbureau voor Schimmelcultures, P.O. Box 273, 3740 AG BAARN
Dr R.A. Samson, Centraalbureau voor Schimmelcultures, P.O. Box 273, 3740 AG BAARN

NORWAY

S. Ewald, Department of Food Hygiene, Norwegian College of Veterinary Medicine, P.O. Box 8146, Dep. 0033 OSLO 1

SPAIN

V. Sanchis Almenar, Departament de Technicologia d'Aliments, Universitat Polytecnica de Catallunya, Av. Rovira Roure 177, 25006 LLEIDA

SWEDEN

Ms Kirsten Åkerstrand, Statens Livsmedelsverk, P.O. Box 622, S-751 26 UPPSALA
Mr P. Adamek, Swedish Institute for Food Research, Box 5401, S-402 29 GOTHENBERG
Mrs Birgitta Bergström, Swedish Institute for Food Research, P.O. Box 5401, S-402 29 GOTHENBERG
Mr T. Börjesson, Swedish Institute for Food Research, P.O. Box 5401, S-402 29 GOTHENBERG
Mr U. Stöllman, Swedish Institute for Food Research, P.O. Box 5401, S-402 29 GOTHENBERG

SWITZERLAND

Mad. Nadine Braendlin, Quality Assurance Department, Nestec Limited, Avenue Nestlé 55, CH-1800 VEVEY
Dr J.L. Cordier, Central Control Laboratory, Nestec Limited, CH-1814 LA-TOUR-DE-PEILZ
L. Cox, Quality Assurance Department, Nestec Limited, Avenue Nestlé 55, CH-1800 VEVEY

TURKEY

Dr Dilek Heperkan, Hikmet sok. Baris apt., 11/2, 81090 ERENKOY-ISTABUL

UNITED KINGDOM

Dr J.N. Banks, MAFF Central Science Laboratory, London Road, SLOUGH, Berks. SL3 7HJ
Dr J.H. Clarke, MAFF Central Science Laboratory, London Road, SLOUGH, Berks. SL3 7HJ
Dr Janet E.L. Corry, J. Sainsbury PLC, Stamford House, Stamford Street, LONDON SE1 9LL
Ms Sarah J. Cox, MAFF Central Science Laboratory, London Road, SLOUGH, Berks. SL3 7HJ
G. Edwards, RHM Research and Engineering, Lincoln Road, HIGH WYCOMBE, Bucks. HP12 3QR
Ms Josephine B. Head, J. Sainsbury PLC, Stamford Street, LONDON SE1 9LL
Mrs Judith Kinderlerer, Department of Biological Sciences, Sheffield City Polytechnic, Pond Street, SHEFFIELD S1 1WB
Dr M.O. Moss, Department of Microbiology, University of Surrey, GUILDFORD GU2 5XH
B.J. Northway, MAFF Central Science Laboratory, London Road, SLOUGH, Berks. SL3 7HJ
Ms Mary Phillips-Jones, Department of Molecular Biology and Biotechnology, University of Sheffield, SHEFFIELD, S10 2TN
Ms Kathleen Regan, J. Sainsbury PLC, Stamford Street, LONDON SE1 9LL
Dr D.A.L. Seiler, Flour Milling and Baking Research Association, CHORLEYWOOD, Herts. WD3 5SH
R.H. Shamsi, MAFF Central Science Laboratory, London Road, SLOUGH, Berks. SL3 7HJ

UNITED STATES OF AMERICA

Dr C.B. Anderson, Del Monte Corporation, 205 N. Wiget Lane, WALNUT CREEK, CA 94598
Dr L.R. Beuchat, Dept of Food Science, University of Georgia Experiment Station, GRIFFIN, GA 30223
R.E. Brackett, Dept of Food Science, University of Georgia Experiment Station, GRIFFIN, GA 30223
Dr L.B. Bullerman, Department of Food Science, University of Nebraska, LINCOLN, NE 68583
Dr L.S. Carlson, Pillsbury Company, Research and Development Laboratories, 311 Second Street Southeast, MINNEAPOLIS, MN 55414
J.J.Churey, NYS Agricultural Experiment Station, Cornell University, GENEVA, NY 14456
Dr D.E. Conner, Department of Poultry Sciences, Auburn University, AUBURN, AL 36849-5416
Dr Maribeth A. Cousin, Department of Food Science, Purdue University, WEST LAFAYETTE, IN 47907
M. Flores, USDA ARS Western Regional Research Center, 800 Buchanan Street, ALBANY, CA 94710
T.L. Fox, Statistical Data Services, 3M Pharmaceuticals, 3M Center, ST. PAUL, MN 55144-1000
Dr D.A. Golden, Dept of Food Science, University of Georgia Experiment Station, GRIFFIN, GA 30223
Dr A.D. King, USDA ARS Western Regional Research Center, 800 Buchanan St, ALBANY, CA 94710
Brenda V. Nail, Dept of Food Science, University of Georgia Experiment Station, GRIFFIN, GA 30223
Dr K.E. Olson, National Food Processors Association, 6363 Clark Avenue, DUBLIN, CA 94568-3097
M.E. Parish, Institute of Food and Agricultural Sciences, University of Florida Citrus Research and Education Center, LAKE ALFRED, FL 33850-2299
C. Royer, USDA ARS Western Regional Research Center, 800 Buchanan Street, ALBANY, CA 94710
Dr D.F. Splittstoesser, Department of Food Science and Technology, Cornell University, GENEVA, NY 14456-0462
Dr T.A. Török, USDA ARS Western Regional Research Center, 800 Buchanan Street, ALBANY, CA 94710
Ms Diane I. West, ConAgra Frozen Foods, Six ConAgra Drive, OMAHA, NE 68102-5006

PREFACE

This book represents the Proceedings of the Second International Workshop on Standardisation of Methods for the Mycological Examination of Foods, which was held in Baarn, the Netherlands, on 20 to 24 August, 1990. This Workshop was a follow up to the highly successful First International Workshop, held in Boston, USA in July 1984. The Proceedings of that Workshop were published as *Methods for the Mycological Examination of Food*, edited by A.D. King, J.I. Pitt, L.R. Beuchat and Janet E.L. Corry, Plenum Press, New York (NATO ASI Series, Series A: Life Sciences, Vol. 122, 1986).

Planning for the Second Workshop commenced in 1987, with a questionnaire seeking ideas for studies on media and techniques, and for potential participants. Suggestions from the survey were then developed into a series of collaborative studies among participants and other interested persons. Collaborative studies were set up and coordinated by Dr L.R. Beuchat. The editors wish to thank him, those who collaborated in interlaboratory studies and the other individuals listed on pages xi-xiii for their input which contributed so much to the success of this Workshop.

A major role of fungi in nature is to complete the carbon cycle by decomposition of organic matter. In this role, fungi become competitors with food technologists who wish to preserve food for later consumption by man or animals. Important reasons to study fungi are the economic losses associated with spoilage of foods and the production of mycotoxins in foods and feeds. Thus the main thrust of the workshop was to develop, test and promote better methodology for the detection and enumeration of fungi in foods. We wish to standardise methods, with resulting improvements in communications, interpretation of results between laboratories and improved facilitation of international commerce. Planning for the First Workshop involved a decision not to include any papers on mycotoxins, as this area is considered to be more closely related to chemistry and toxicology than to mycology. The Second Workshop also did not include mycotoxin research, but several studies have produced media for selectively isolating mycotoxigenic fungi.

The planning of the Workshop was in the hands of Larry R. Beuchat, Ailsa D. Hocking, A. Douglas King, John I. Pitt, Robert A. Samson, and David A.L. Seiler. Thirty five participants attended. Nine collaborative studies and 40 individual papers were presented over five consecutive days. Each day included sessions for discussion of the previous day's presentations. At a final plenary session, recommendations for methods to be adopted were prepared and agreed on. These are summarised at the end of this book, together with a section on the mycological methods and media considered, with current knowledge, to be the most satisfactory available. We recommend these methods to all who deal with the problems of fungi in foods.

The Second International Workshop was sponsored by the Dutch Royal Academy of Sciences, Amsterdam, Head Inspectorate for Health Protection, Rijswijk, Pharmachemie BV-Oxoid Nederland, Haarlem, Unilever Research, Vlaardingen, Avebe, Foxhol, Cehave, Veghel, Holland Biotechnologie BV, Leiden in the Netherlands, by Eco-Bio, Genk, in Belgium, and by Oxoid Australia. We wish to sincerely thank all of these sponsors. Local arrangements for the Workshop were very efficiently and smoothly organised with the assistance of Mrs Ans Spaapen-de Veer and Tineke van der Berg. We wish to thank each of them, as well as other members of staff of the Centraalbureau voor Schimmelcultures, Baarn.

The Editors

1

INTRODUCTION

INTRODUCTION AND SUMMARY OF THE FIRST INTERNATIONAL WORKSHOP ON STANDARDISATION OF METHODS FOR THE MYCOLOGICAL EXAMINATION OF FOODS

A.D. Hocking and J.I. Pitt

Food Research Laboratory
CSIRO Division of Food Processing
North Ryde, 2113 Australia

INTRODUCTION

The first International Workshop on "Standardization of Methods for Mycological Examination of Foods" (SMMEF I) was held in Boston, USA on July 11-13, 1984, just two years after it was conceived. The workshop was attended by twenty six people from Australia, Denmark, the United Kingdom, Hungary, the Netherlands, Turkey and the Unites States of America. The workshop proceedings were subsequently published (King *et al.*, 1986).

The aims of the workshop were to work towards standardisation of media and methods used in the mycological examination of foods. Attention was concentrated on dilution plating, a universally accepted procedure, and direct plating, a technique for particulate foods or commodities such as grains, where the particles are incubated directly on the media. Methods for estimation of fungi in foods by other methods, such as chitin or ATP assays, and consideration of the taxonomy of food spoilage fungi were also presented in review sessions.

The workshop was divided into sessions which considered particular aspects of enumeration of fungi in foods. Results of individual and collaborative studies were presented. The participants then formed working groups to discuss the results presented, and to formulate recommendations for development of standard procedures.

Broadly, the sessions were set out as follows.

1. Sample preparation (21 contributions)
2. General purpose media (18 contributions)
3. Selective media (11 contributions)
4. Media for yeasts and consideration of injured cells (6 contributions)
5. Baseline counts (11 contributions)
6. Unacceptable levels (6 contributions)
7. New techniques (5 contributions)
8. Taxonomic schemes (3 contributions).

The aims of the workshop were largely accomplished. Agreement was reached on the standardisation of a wide range of methods and media suitable for particular aspects of isolating and enumerating fungi in foods. However, it was recognised that in a number of important areas, there was insufficient information on which to make recommendations. A need for a second SMMEF workshop was clear. It took place in Baarn, the Netherlands, in August, 1991. The Proceedings of this workshop (SMMEF II) are published here.

Many of the methods used for collaborative studies presented at SMMEF II were based on recommendations from SMMEF I. The areas considered at SMMEF I, and the conclusions reached, are set out below.

1. SAMPLE PREPARATION

A wide variety of papers discussed all aspects of sample preparation and plating techniques. General agreement was reached on most aspects and a set of recommendations formulated, and agreed on. These are summarised below.

Plating systems

Two different plating techniques were discussed and recommended. ***Dilution plating*** techniques are similar to those used in bacteriology. Dilution plating is recommended for liquid and powdered raw materials and foods. ***Direct plating*** is a strictly mycological technique, and is recommended for all kinds of particulate foods. Sample preparation techniques for these two methods are quite different, as summarised below.

Sample preparation for dilution plating

Subsampling. The subsample size should be as large and representative as possible.

Diluent ratio. The ratio of sample to diluent should be 1:10 where practicable.

Homogenisers. The Colworth Stomacher and the Waring type Blender are preferable. Top drive macerators, or shaking by hand in a glass container with a few glass beads, are considered less acceptable. The recommended homogenising time is 2 minutes for Stomacher and blender, but should be less for top drive machines.

Settling time. Because fungal spores tend to settle out of diluents quickly, homogenised samples should not be allowed to stand more than 5 minutes before taking aliquots.

Presoaking. It was agreed that presoaking was a valuable procedure for dried food samples, but no recommendation was made, pending further studies.

Diluents. For general purpose dilution, 0.1% bacteriological peptone was recommended. However, it was recognised that for fungal spores, diluents had little effect, and 0.1 M phosphate buffer or 0.05% food grade detergent was also effective. Water used should be deionised or distilled.

For low a_w foods containing yeasts, more concentrated diluents are necessary, 20-40% (w/w) glucose being considered most satisfactory.

Plating. Spread plates are recommended. Aliquots should normally be 0.1 ml, but 0.3 ml can be used where small numbers of fungi must be enumerated. Aliquots should be spread on prepoured and set agar by means of bent narrow diameter glass rods or tubing.

Incubation. The standard incubation regime for general purpose enumerations is 25°C for 5 days. Plates should be incubated upright.

Counts. Although not a formal recommendation, participants agreed that satisfactory fungal counts lie in the range 10 to 100 colonies per plate, and preferably 10-50.

Results. Results are expressed as log count/g of original sample.

Sample preparation and direct plating

Surface sterilisation of particulate foods before direct plating is considered an essential for most purposes, to permit enumeration of fungi actually invading the food. An exception is to be made for cases where surface contaminants become part of the downstream mycoflora, e.g. wheat grains to be used in flour manufacture. In such cases, grains should not be surface sterilised.

Surface sterilisation. Surface sterilise food particles by immersion in 0.4% chlorine (household bleach, diluted 1:10) for 2 minutes. A minimum of 50 particles should be sterilised and plated. The chlorine should only be used once.

Rinse. After pouring off the chlorine, rinse once in sterile distilled or deionised water. **Note**: this step has not been shown to be essential, but is recommended.

Surface plate. As quickly as possible, transfer food particles with sterile forceps to previously poured and set plates, at the rate of 6-10 particles per plate.

Incubation. The standard incubation regime for general purpose enumerations is 25°C for 5 days. Plates should be incubated upright.

Results. Express results as per cent of particles infected by fungi. Differential counting of a variety of genera is possible if a stereomicroscope is used.

2. GENERAL PURPOSE ENUMERATION MEDIA

This proved to be one of the most contentious issues debated at the workshop. There is no single medium which can be used to enumerate all fungi from all foods. However, most fungi can be detected in most food samples using a well designed, high a_w medium. Ideally, a general purpose fungal enumeration medium should:

- suppress bacterial growth
- be nutritionally adequate
- induce compact colony formation
- suppress spreading fungi
- promote growth of relevant fungi.

In addition, desirable attributes of the medium include:

- easy preparation
- short counting time and minimal operator stress
- effective selective isolation from mixed cultures
- effective enumeration yeasts in the presence (or absence) of moulds
- ready identification of colonies.

Fifteen studies on media comparison were presented at SMMEF I. Seventeen different media were included, the most frequently used being Oxytetracycline Glucose Yeast Extract agar (OGYE) and some modifications of this, Potato Dextrose agar, antibiotic amended (PDAA), Rose Bengal Choramphenicol agar (RBC), Dichloran Rose Bengal Chloramphenicol agar (DRBC) and Dichloran 18% Glycerol agar (DG18). Other media used included Malt Salt agar (MSA), Malt Extract agar (MEA), acidified PDA and Plate Count Agar (PCA). The reference points for all enumeration studies were DRBC and/or DG18. Non-selective media such as PDAA and OGY were compared with selective media such as DRBC and RBC. Most studies found there was no significant difference (at the 95% confidence level) between **total** counts obtained on various media, but often counts were slightly higher on the more selective agars. Some studies found that rose bengal was inhibitory for yeasts, but this was not always the case.

The working group on general purpose enumeration media was unable to agree on recommendation of a single medium. They concluded that there was no overall best medium, but that media containing dichloran produced plates that were easier to count. There was some concern that media containing rose bengal can be inhibitory to yeasts.

Rose bengal is light sensitive, and exposure to light can cause formation of toxic products. Some people felt that colonies on DRBC were difficult to identify, but that otherwise, this was quite a good medium. The highest counts were often obtained on DG18.

3. SELECTIVE MEDIA AND PROCEDURES

Mycotoxigenic fungi

Few effective media have been developed for selective enumeration of mycotoxigenic fungi. However, two media were recommended for this purpose: Aspergillus flavus and parasiticus agar (AFPA) for isolation and enumeration of aflatoxigenic fungi, and DRYES for nephrotoxic *Penicillium* species. It was felt that there was no really satisfactory medium for selective enumeration of *Fusarium* from foods. SMMEF I recognised that much more work was needed in the development of selective and differential media for toxigenic fungi.

Xerophilic fungi

The use of reduced a_w media for low a_w foods (<0.90) is now widely accepted by food mycologists. SMMEF I was the first time this concept was fully aired and accepted, but the message has still not reached all laboratories enumerating fungi in intermediate and low moisture foods.

For enumeration of nonfastidious xerophilic fungi, e.g. *Wallemia sebi*, *Eurotium* species, *Aspergillus restrictus* series, other *Aspergillus* and *Penicillium* species, DG18 was recommended. For enumeration of fastidious xerophiles, e.g. *Xeromyces bisporus* and xerophilic *Chrysosporium* species, MY50G (or a medium of equivalent a_w) was recommended. In general, it was agreed that direct plating was the best method for detecting the presence of fastidious xerophilic fungi, but some success was reported with a dilution plating methods employing reduced a_w diluents and media.

For detection or enumeration of yeasts in concentrated samples like syrups, jams and fruit concentrates, the use of reduced a_w diluents (e.g. 20-40% glucose) is necessary to avoid osmotic shock to the cells. The diluents should be plated out onto media of similar a_w.

Heat resistant fungi

The methods outlined at the Boston workshop were accepted as being appropriate for detection of heat resistant fungi in raw materials and finished products. Samples should be heated at 70-80°C for 30 min. A sample size of at least 100 g was recommended, and the whole sample plated out after mixing with double strength plating medium. Plating media should be antibiotic amended, NOT acidified. Suitable media are MEA, PDA and OGY. As samples have been heat treated, fungal inhibitors designed to stop encroachment by Mucorales species are not considered necessary, and may be detrimental to heated ascospores. An incubation regime of 30°C for up to 4 weeks was recommended.

4. MEDIA AND METHODS FOR YEASTS and RECOVERY OF INJURED CELLS

Media and methods for yeasts

For enumeration of yeasts in foods or beverages where moulds are not expected to cause interference, antibiotic amended general mycological media (e.g. OGY, PDA, MEA,

TGY) are recommended. Media containing rose bengal are not generally recommended for enumeration of yeasts, as there is some evidence that rose bengal may be inhibitory to them. However, where yeasts need to be counted in the presence of moulds, DRBC was recommended, provided that the prepared medium is protected from the light.

For enumeration of yeasts in low a_w foods, both diluent and medium should be of reduced a_w, e.g. 40% glucose. Membrane filtration was recommended for enumeration of yeasts in wine and other clear beverages.

Recovery of injured cells

It was recognised that, although much work has been done on injury and recovery in bacteria, there is little published data on injury in yeasts and moulds. The SMMEF I workshop considered that, for recovery of fungi which may have suffered heating, freezing or desiccation injury, a plating medium containing minimal amounts of chemical inhibitors should be used. More work is needed to determine optimal diluent, media, sample preparation methods and incubation conditions.

5. BASELINE COUNTS

Variations reported in baseline counts from commodity to commodity and from region to region indicated that consensus on baseline counts, i.e. normal, commercially acceptable levels of fungi in commodities and other foods, is a long way off. The first need identified was for much more work to quantify variations in counts due to methodology. Quantitative information of this kind would then assist in an eventual understanding of variations due to agricultural practice.

If meaningful baseline levels for fungi in foods are eventually established it will no doubt be on a regional and commodity basis.

6. UNACCEPTABLE LEVELS

Certain fungi are unacceptable at any level in particular foods. Examples are *Zygosaccharomyces bailii* in acid, liquid, preserved products such as fruit juices, fruit based products, ciders, wines, tomato sauce, mayonnaise and salad dressings; *Xeromyces bisporus* in foods which rely on high carbohydrate contents and low a_w for stability; and heat resistant fungi in pasteurised fruit juices and fruit based products without preservatives.

7. NEW TECHNIQUES FOR ESTIMATING FUNGAL BIOMASS IN FOODS

A number of new techniques have been described for estimating fungal growth and biomass in foods. These include most probable number techniques, the use of impediometry, and the assay of chitin and ergosterol. It was agreed that further investigations were needed before these methods can be recommended in place of traditional procedures.

REFERENCE

KING, A.D., PITT, J.I., BEUCHAT, L.R. and CORRY, J.E.L. Eds. 1986. Methods for the Mycological Examination of Food. New York: Academic Press.

2

METHODS FOR ENUMERATION OF FUNGI IN FOODS

METHODOLOGY FOR ROUTINE MYCOLOGICAL EXAMINATION OF FOOD - A COLLABORATIVE STUDY

A. D. King

USDA Agricultural Research Service,
Western Regional Research Center,
Albany, CA 94710, USA

In collaboration with: P. Adamek, K. Åkerstrand, C.B. Anderson, S. Andrews, N. Aran, J. Banks, B. Bergström, N. Braendlin, L.C. Bullerman, D.E. Conner, T. Deák, J. Frisvad, A.D. Hocking, J.A. Magnuson, S. Notermans, J.I. Pitt, D.L. Splittstoesser, D. West, and L.C. Whitehand.

SUMMARY

A collaborative study to compare methods and media for enumeration of fungi in foods was carried out by 15 laboratories. Participants were requested to examine ten foods, including two samples of flour. A supplied reference method was compared with the method normally in use in their laboratory, which was preferably a published method. Eight standard methods were used, as well as those routinely used in individual laboratories. The major differences between the methods were the media used. Counts varied widely depending upon the food being examined, but the type of media had less influence on the counts obtained.

INTRODUCTION

This study was designed to gather data on methods used for routine enumeration of fungi from foods. The approach used was to compare a method in current use, especially an industry or government standard method, against a reference. All prospective workshop participants were invited to join in the study, and each was asked to give details of the methods they selected to include. With responses from a variety of places around the world, it was hoped that comparison of both media and methods would be possible. Collected data would also reflect the range of counts expected from particular classes of foods. As a kind of control, each participant was asked to include two domestic wheat flour samples.

METHODS

The following reference method and experimental plan was sent to workshop invitees. This reference dilution plating method for enumeration and fungal flora isolation from food was developed from the recommendations of the First International Workshop on Standardisation of Methods for the Mycological Examination of Foods (King *et al.*, 1986).

Sample preparation. Dilute samples (50 g) to 500 g (1/10) with sterile water. Homogenise sample in Colworth Stomacher (2 min) where possible, otherwise specify method of homogenising. Dilute as soon as possible; do not permit settling for more than 5 min at any stage. For liquid foods, omit the homogenising step.

Dilution. Dilute in 0.1% peptone water with 1:9 dilutions.

Plating. Spread plate appropriate dilutions (0.1 mL) in duplicate onto solidified agar (15-20 mL/plate).

Medium. For samples of greater than 0.9 a_w, use DRBC (Table 1) as the standard medium; for those less than 0.9 a_w, use DG18 (Table 1).

Incubation. Incubate plates upright for 5 days at 25°C or longer as necessary to achieve the maximum count and adequate development of colonies for identification. If necessary, transfer colonies to appropriate media for identification.

Counting. Count plates with between 10 and 100 colonies.

Collaborators were asked to identify mould colonies to genus or group, e.g. *Alternaria*, *Aspergillus*, *Cladosporium*, *Eurotium*, *Fusarium*, *Paecilomyces*, *Penicillium*, *Rhizopus* or *Mucor*, *Wallemia*, other dematiaceous Hyphomycetes, other hyaline Hyphomycetes and yeasts, and to identify *Aspergillus* species to series where possible. They were requested to clearly define the type of food being examined and the industry or governmental standard method being used. A minimum of 10 samples and two wheat flour samples was to be included in the comparison. Any deviations from this procedure were to be explained. Results were to be reported on standard data sheets, which were supplied.

Data were submitted from 16 laboratories, for a total of 378 food samples. Enumeration data were converted to $\log_{10}$ counts and analysed statistically. All but one laboratory used a Stomacher, so sample preparation methods were not analysed. Peptone water was used as diluent except for one laboratory in which tryptone saline was used. About half of the laboratories used pour plates as part of the standard method tested, and about half incubated plates upright. Incubation was 25 or 26°C for up to one week.

Table 1. Media used in survey

Acronym	Medium
AGY	Acidified glucose yeast extract agar, pH 4.5
CZCC	Czapek agar, chloramphenicol 50 mg/L, chlortetracyline 50 mg/L
CZCCP	Czapek agar, chloramphenicol 50 mg/L, chlortetracyline 50 mg/L, pour plates
CZA	Czapek Dox agar
DG18	Dichloran 18% glycerol agar
DG18P	Dichloran 18% glycerol agar, pour plates
DRBC	Dichloran rose bengal chloramphenicol agar
DRYS	Dichloran rose bengal yeast extract sucrose agar
GYC	Glucose yeast extract, chloramphenicol 100 mg/L
MACP	Malt agar, chloramphenicol 40 mg/L, pour plates
MEA	Malt extract agar
OGY	Oxytetracycline (100 mg/L) glucose yeast extract agar
PCAC	Plate count agar, chlortetracycline 40 mg/L
PDAA	Potato dextrose agar, pH 3.5 (tartaric acid)
PDAC	Potato dextrose agar, chlortetracycline 40 mg/L
PDACC	Potato dextrose agar, chloramphenicol 50 mg/L, chlortetracycline 50 mg/L
RBCA	Rose bengal chloramphenicol agar
SABCCP	Sabouraud dextrose agar, chloramphenicol 100 mg/L, chlortetracycline 100 mg/L, pour plates

Table 1 lists the various media used in the study: some media were in general use, while others were used by only one laboratory.

To assist in statistical analysis some data were pooled. Glucose yeast extract media, including OGY, were combined and designated GYC. Malt extract agar was used in three formulations: without antibiotics and with chloramphenicol with or without additional chlortetracycline. These three media were pooled for the analysis also. Count data for media with a common nutrient base, such as PDA, but with different antibiotics, were pooled under the extension -AN (e.g. PDAAN).

The samples examined were separated into seven food classes (Table 2). The number of samples in each class are as shown.

Table 2. Commodity class and number of samples

Class	Commodity description	No. of samples
1	Spices, herbs, dried vegetables, dry soup	44
2	Wheat flour	70
3	Cereals and products	82
4	Fresh and frozen fruits and vegetables	104
5	Dried fruits, sugar, etc.	11
6	Nutmeats, cocoa powder	31
7	Milk and milk products, dog food, meat	36

RESULTS

Fungal counts

Analysis of the enumeration data was complex because of the wide range of techniques, samples, and media used. However, the flour samples provided reasonably homogenous data, and it was possible to compare data from similar media against the reference media. In some cases, sufficient replications of the media used were not available.

Paired media counts. Results from the statistical analysis of counts on all samples are listed in Table 3. Samples that were analysed on two media were compared to determine the influence of the media on counts. A paired t-test was performed on each media pair for the yeast and mould counts obtained. The probabilities of significance for the t-tests are shown in Table 3 together with the number of samples compared.

Most of the paired media comparisons were not significantly different in probability values (Table 3), i.e. the general purpose media were reasonably comparable. However, several comparisons were significantly different at probability <0.05. DG18 produced a higher yeast count than DRBC for food samples but not for the flour samples. GYC showed higher mould counts on flour than DG18. This is surprising: moulds on dry products like flour would be expected to be xerophilic and grow well on DG18. The GYC plates showed higher numbers of dematiaceous and hyaline Hyphomycetes compared with DG18. PDAAN produced higher counts than DRBC for both mould and yeasts from food samples, and counts on PDAA were higher than those on DRBC or DG18 with food samples. MEA also produced higher counts for food samples for both yeasts and moulds than DRBC. These data are not surprising because both DRBC and DG18 are somewhat restrictive media by design while the other media in the comparisons are general purpose

Table 3. Significance probability from paired t-tests comparing media for mould and yeast log_{10} counts

Media Comparison	All Samples				Flour			
	Mould		Yeast		Mould		Yeast	
	No.	Prob.[1]	No.	Prob.	No.	Prob.	No.	Prob.
DRBC/DG18	179	0.39[a]	80	0.01[a]	28	0.16[a]	12	0.13[a]
DRBC/GYAN	138	0.20[a]	137	0.81[a]	20	0.66	20	0.37
DG18/GYAN	95	0.32[a]	94	0.15	18	0.01	18	0.08
DRBC/PDAAN	54	0.01	30	0.03	13	0.08	9	0.20
DG18/PDAAN	48	0.38	24	0.77	7	0.28[a]	3	0.42[a]
DRBC/PDAA	24	0.01			4	0.31[a]		
DG18/PDAA	24	0.01			4	0.37		
DRBC/AGY	22	0.05[a]	22	0.01[a]				
DG18/AGY	22	0.02[a]	22	0.03[a]				
DRBC/CZA	20	0.50	20	0.10[a]				
DG18/CZA	20	0.28[a]	20	0.69[a]				
DRBC/RBCA	8	0.27	8	0.33				
DG18/RBCA	17	0.92[a]	17	0.31				
DRBC/CZCCP	6	0.07	6	0.87[a]				
DG18/CZCCP	6	0.64	6	0.58[a]				
DRBC/DRYS	10	0.56	10	0.34[a]				
DG18/DRYS	10	0.05[a]						
DRBC/PCACC	6	0.06[a]	6	0.04[a]				
DG18/PCACC	4	0.45[a]	4	0.27				
DRBC/MEA	14	0.04	14	0.01				
DG18/MEA	12	0.29[a]	11	0.09				
AGY/GYAN	22	0.09	22	0.06				
CZA/DRYS	10	0.25						
CZCCP/DG18P	6	0.03[a]	6	0.25				
CZCCP/DG18P	6	0.72[a]	6	0.36[a]				
CZA/RBCA	6	0.99	6	0.33				
PDAA/PDAAN	24	0.08			4	0.55		
PDAAN/MAAN	26	0.21	25	0.62[a]	6	0.62	6	0.98

[1] Probability: t-test positive for media pair comparison marked [a]

media. In contrast DRBC and DG18 showed higher yeast and mould counts than AGY on food samples. DRBC counts were lower than either GYAN or DG18, but showed a smaller range in counts. Thus DRBC gave better performance on the wide range of samples tested.

Two comparisons of acidified and antibiotic media are listed in Table 3. Acidified PDA and GY agar were not significantly different from similar media containing antibiotics. These data are interesting because media containing antibiotics are now recommended to avoid the inhibition of injured cells by the low pH of acidified media. Table 5 shows similar results for a comparison of acidified and antibiotic amended PDA.

Food class counts. The mean counts for the three media most frequently used for different food classes are illustrated in Figure 1. The wide range of counts indicated by the error bar is due to some high counts from a very few samples and some samples with undetectable counts. The majority of counts for both moulds and yeasts were less than $\log_{10}$ 4. The highest average counts were in fresh and frozen fruit and vegetable samples. Dried fruit and nutmeats had low counts. The ratio of yeast to mould count varied with the food class.

Mean mould and yeast counts for the flour samples are listed in Table 4. The mould counts varied widely. It is not possible to determine from these data whether the flour samples were variable in contamination or whether the wide range of counts was due to differences between the various media used. The differences are probably related to both factors. In several cases, the minimum yeast count was undetectable; the maximum $\log_{10}$ 8.0. The mould counts in flour ranged from undetectable to $\log_{10}$ 6.7.

An extended study by one collaborator, D. L. Splittstoesser, evaluated the effect of standing on fungal counts as well as comparing four media, using 10 samples of frozen

Table 4. Counts and standard deviations on flour samples for all media[1]

Medium	No. of samples	Mould count ($\log_{10}$)	S.D.	Yeast count ($\log_{10}$)	S.D.
AGY	2	1.62	0.04	0	0
CZA	4	1.95	0.39	0	0
CZCCP	2	3.35	0.24	0	0
DG18	36	2.37	1.49	0.54	1.00
DG18P	2	3.44	0.11	0	0
DRBC	37	1.94	1.41	1.16	1.53
DRYS	2	2.41	0.07	0	0
GYAN	30	2.39	0.95	1.62	1.90
MAAN	12	1.45	2.14	1.66	2.27
PCACC	3	3.20	0.33	2.62	0.56
PDAA	4	2.94	0.35	---	---
PDAAN	16	2.13	1.77	1.23	1.96
RBCA	2	2.94	0.65	1.60	2.26

[1]For medium names, see Table 1. Suffix -AN, pooled data from a single medium type with different antibiotics.

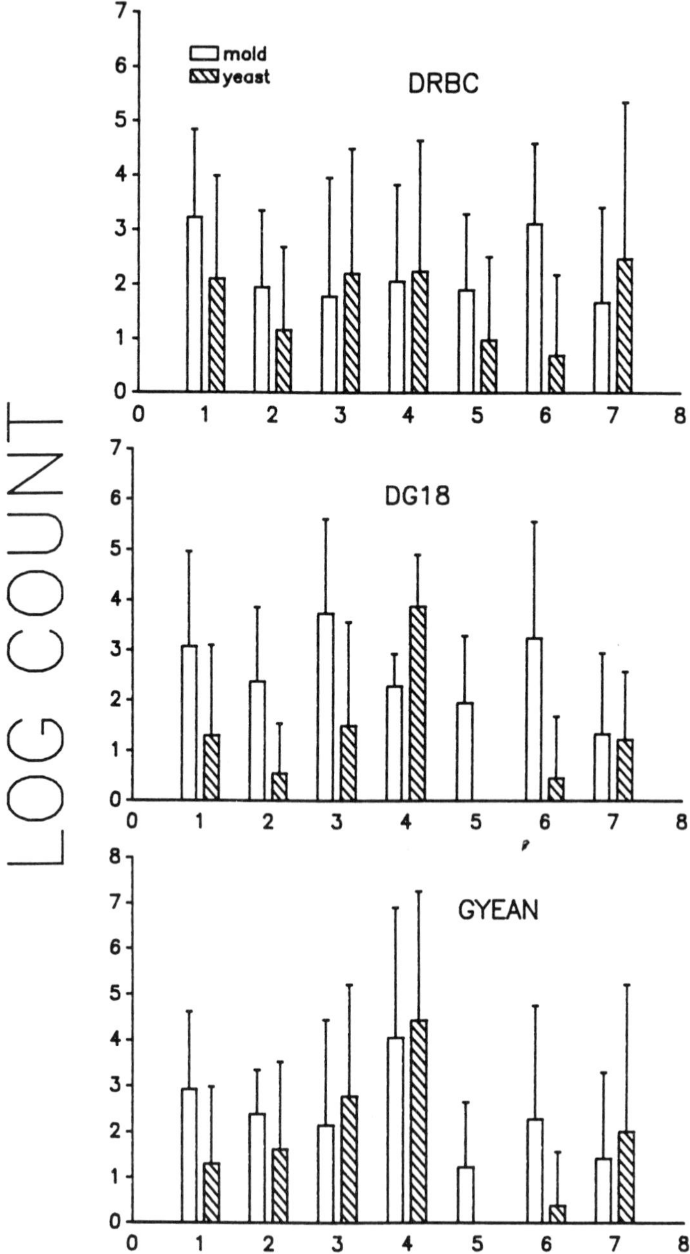

Figure 1. Mean $\log_{10}$ count and standard error bars for each food class on DRBC, DG18 and GYAN.

fruit and two of flour. Samples were homogenised in a Colworth Stomacher. One set of subsamples (5 mL) was plated immediately. A second set (5 mL) from the homogenised samples was allowed to settle for 1 min. After waiting a further 1 min., aliquots were plated. The media used were PDA plus rose bengal (8.4 mg/L) acidified to pH 3.5 with tartaric acid (PRDA); PDA plus rose bengal (8.4 mg/L) with chloramphenicol (100 mg/L) (PDRC); DRBC and DG18. Each sample was analysed 4 times; replicate counts for each sample were averaged before statistical analysis. Results are shown in Table 5.

Table 5. Means of mould counts ($\log_{10}$) of Splittstoesser data comparing four media and two sample preparation methods

Medium	Sample Preparation[1]	
	No standing	After standing
PDRA	3.15[a]	2.50[b]
PDRC	3.11[a]	2.71[a]
DRBC	2.95[b]	2.33[c]
DG18	2.78[c]	2.20[d]

[1] Means with the same letter are not significantly different, $p<0.05$

When samples were plated immediately, counts on PDRA and PDRC were not significantly different, but were higher on the non-acid medium after standing. Counts on DRBC and DG18 were lower. Fungal identifications made on each medium were similar. Subsamples plated immediately showed significantly higher counts than those that were allowed to stand before plating ($p<0.0001$). The lower counts were perhaps due to attachment of microorganisms to food or glassware surfaces during standing. This result is consistent with earlier recommendations (King *et al.*, 1986) that samples stand no more than five minutes before plating.

Identification of fungal flora

Table 6 lists the range of fungi identified from all samples and media reported in these studies. About 10 genera or groups of genera were recognised. The most frequently identified moulds were *Penicillium* and *Aspergillus* species. Yeasts were the next most frequent identification.

Influence of media on identifications. One purpose of this major collaborative study was to compare growth and ease of identifications on a range of media by different investigators. Table 7 provides a summary. The last column lists the wide range of average fungal identifications per test, arranged by the different media. The values for these averages reflect either the capacity of the medium to support growth of a range of fungi or the ability of the investigator to identify the fungi. Although it was clear that some investigators identified many more moulds than others, media differences were apparent also. As an example, studies with SABCCP, CZCCPP, and DG18PP were all carried out in the same laboratory by the same investigators. Note the wide range of identifications shown for these three media, clearly indicating large differences in the ability to allow growth

Table 6. Total numbers of identifications reported for major fungal groups

Fungi	Total
Penicillium spp.	402
Aspergillus niger	127
Aspergillus flavus	100
Aspergillus versicolor	62
Aspergillus ochraceus	41
Aspergillus restrictus	10
Other *Aspergillus* spp.	235
Eurotium spp.	143
Yeast spp.	323
Rhizopus/Mucor spp.	158
Cladosporium spp.	153
Fusarium spp.	49
Paecilomyces spp.	30
Alternaria spp.	21
Wallemia sebi	21
Dematiaceous Hyphomycetes	173
Hyaline Hyphomycetes	72
Total identifications	2120

and identification of a range of fungi. OGY was used in three laboratories; a considerably lower average identification rate is shown than for GYC or GYCH used in two or one laboratories, respectively. Perhaps the oxytetracycline is more inhibitory to some fungi than chloramphenicol or chlortetracycline. AGY, the acidified version of these media, gave similar results to the latter two media.

The difference in average identifications is quite small for the majority of media listed in Table 7. Because of the perceived variability between the different laboratories reporting data, statistical analysis of the data was not run. For this reason it is not possible to compare the range of species able to grow on a particular medium, or relative ease of identifications. However, SABCCP, MEA, MAAN, PDAAN and OGY were apparently much inferior to the other media.

Identifications from food classes. The richest source of moulds based upon the average identifications by food class listed in Table 8 were the spices. The fewest moulds identified per sample were from dried fruits. The protein rich foods, such as milk products, dog food and meats also had fewer moulds per sample. These classes of foods are processed and should have lower microbial counts.

Footnotes to Table 7

[1] See Table 1 for list of media

[2] Number of laboratories using the indicated medium

[3] Number of tests reported on each medium

[4] Total number of identifications reported for each medium

[5] Average number of identifications reported per test, i.e. Sum/No. of tests

Table 7. Fungal identifications reported from all media.[1]

	MEDIA GROUPS						
	SABCCP	CZCCP	DG18P	OGY	GYC	GYCH	AGY
No.of labs[2]	1	1	1	3	2	1	1
No.of tests[3]	6	12	6	69	46	42	22
Alternaria				3	1		1
Aspergillus flavus		4	3	9	10	7	2
A. niger		2	2	2	10	15	4
A. ochraceus			1	1	3	3	1
A. restrictus							
A. versicolor		2	3	6	4	3	
Other *Aspergillus* spp.		2	5	7	19	12	6
Cladosporium		8	6	15	9	16	7
Eurotium		1	6	9	11	11	7
Fusarium		4	4	4	4	2	
Paecilomyces					4	3	1
Penicillium		10	4	17	29	35	12
Rhizopus or *Mucor*		4	3	18	15	18	8
Wallemia sebi			1			1	
Demat. Hyphomycetes		4	3	13		6	
Hyaline Hyphomycetes		6	5		20	6	12
Yeasts	4	6	2	35	24	19	16
Sum[4]	4	53	50	139	163	157	77
Average[5]	0.7	4.4	8.3	2.0	3.5	3.7	3.5

	MEA	MAAN	PDAAN	PDAA	PCACC	CZA	DRYS	RBCA	DRBC	DG18
No.of labs	1	1	4	1	1	1	1	2	13	14
No.of tests	2	26	57	16	10	20	10	19	143	165
Alternaria		2	1		1	2			4	6
Aspergillus flavus		2	3	1		5	3	4	16	32
A. niger	1			1	4	8	3	7	29	40
A. ochraceus		1	1		1	1	1	2	13	12
A. restrictus		1								8
A. versicolor					1	4	2		12	25
Other Aspergilli		2	20	12	3		6	4	58	66
Cladosporium		8	3	1		13	1	1	38	39
Eurotium			1			1	4		28	61
Fusarium		1	4			4		1	8	13
Paecilomyces		2			2	3		1	5	10
Penicillium	1	7	25	12	5	2	10	11	99	106
Rhizopus or *Mucor*		3	10	1	6	17	2	8	29	30
Wallemia sebi						3	1		2	16
Demat. Hyphomycetes		4	1		2				16	23
Hyaline Hyphomycetes		3	16	11	4	1		4	51	34
Yeasts		11	22	11	9			7	101	56
Sum	2	47	108	48	38	64	34	50	509	577
Average	1.0	1.8	1.9	3.0	3.8	3.2	3.4	2.6	3.6	3.5

For footnotes see previous page

Table 8. Identifications by food class[1]

	Spices	Flour	Cereals	Fruit/veg.	Dry Fruit	Nuts	Protein
Count	128	139	223	128	30	74	74
Alternaria	4	4	9	1	1	0	2
Aspergillus flavus	32	27	33	0	0	8	0
A. niger	56	13	30	2	9	14	3
A. ochraceus	14	6	13	0	1	5	2
A. restrictus	5	2	3	0	0	1	0
A. versicolor	20	16	20	0	0	5	1
Other Aspergilli	54	58	64	43	1	12	3
Cladosporium	24	39	37	15	4	17	17
Eurotium	46	28	52	2	1	13	1
Fusarium	4	9	29	2	2	2	1
Paecilomyces	7	7	7	2	0	4	3
Penicillium	57	86	112	74	7	46	20
Rhizopus or *Mucor*	47	28	38	9	5	26	5
Wallemia sebi	4	2	8	0	0	4	3
Demat. Hyphomycetes	6	18	20	8	1	11	8
Hyaline Hyphomycetes	26	20	46	59	4	10	8
Yeasts	46	44	103	78	6	12	34
Sum	451	407	624	295	42	190	111
Average	3.52	2.92	2.80	2.30	1.40	2.57	1.50

[1]See Table 7 for abbreviations

CONCLUSIONS AND RECOMMENDATIONS

1. Most of the media produced comparable counts. Many of the media can be recommended for routine use.
2. GYAN and DG18 gave somewhat higher counts than DRBC for moulds from the food samples while MAAN gave lower counts. GYAN and MAAN had higher yeast counts than DRBC and DG18. DRBC was a more consistent media giving uniform mould and yeast counts for all seven classes of food.
3. Allowing samples to stand after stomaching gave lower counts.
4. *Penicillium* spp., *Aspergillus* spp. and yeasts were the most commonly identified fungi.
5. Average fungal identifications varied with the different media.
6. Average numbers of fungi identified varied between the classes of food. The types of fungi identified also differed between the classes. The widest range of fungi was identified from cereal products.

REFERENCE

KING, A.D., PITT, J.I., BEUCHAT, L.R. and CORRY, J.E.L., eds. 1986. Methods for the Mycological Examination of Food. New York: Plenum Press.

EVALUATION OF THE PETRIFILM^{TM} YEAST AND MOLD PLATE FOR ENUMERATION OF FUNGI IN FOODS

L. R. Beuchat,[1] Brenda V. Nail,[1] R. E. Brackett[1] and T. L. Fox[2]

[1]*Department of Food Science and Technology*
University of Georgia Agricultural Experiment Station
Griffin, GA 30223-1797 USA

[2]*Statistical Data Services*
3M Pharmaceuticals, 3M Center, 270-3A-01
St. Paul, MN 55144-1000, USA

SUMMARY

Acidified Potato Dextrose Agar (APDA) and Chloramphenicol supplemented Plate Count Agar (CPCA) were compared to Petrifilm^{TM} Yeast and Mold (YM) plates for their suitability to recover yeasts and moulds from a variety of dairy products and high acid foods. Correlation coefficients exceeded 0.99 for Petrifilm YM plates versus APDA and CPCA for recovering total yeasts and moulds from a composite of the eight test foods. Significantly higher populations were generally detected after 5 days compared to 3 days of incubation, regardless of the enumeration system. After 5 days of incubation, yeast or total fungal populations detected using Petrifilm YM plates were not significantly lower than using either pour- and spread plating techniques with either APDA or CPCA. The Petrifilm^{TM} YM plate offers an acceptable alternative to plate count methods for enumerating fungi in the dairy products and high acid foods evaluated in this study.

INTRODUCTION

Media traditionally used to enumerate fungi in food often do not provide good nutrient sources, may not inhibit bacteria, may inhibit resuscitation of injured cells and may cause precipitation of food particles due to low pH, thus interfering with colony detection. In addition, analysis of a large number of samples at any given time requires substantial space for incubation. In an attempt to eliminate or minimise these problems, the 3M Company, (Medical-Surgical Division, St. Paul, Minnesota, USA) has developed the Petrifilm^{TM} Yeast and Mold (YM) plate. The objective of this study was to compare Petrifilm YM plates to Acidified Potato Dextrose Agar (APDA) and Chloramphenicol supplemented Plate Count Agar (CPCA) for enumerating fungi in a wide range of dairy products, fruit juices, salad dressings, relishes and sauces.

MATERIALS AND METHODS

Products analysed

Two hundred and forty food samples, made up of 30 lots of each of eight food products (hard/soft cheese, cottage cheese, yogurt, sour cream, fruit juice, salad dressing, relishes and tomato based sauces) were analysed.

Enumeration systems evaluated

The Petrifilm YM plate consists of a dry culture film (75 x 95 mm) on which the diluted sample is applied and spread. This was compared with four other systems used for plate counts: Potato Dextrose Agar acidified to pH 3.5 with 10% tartaric acid (APDA) and Plate Count Agar supplemented with chloramphenicol (100 μg/mL) (CPCA, pH 6.8) to which appropriately diluted samples were applied using both pour and spread plating techniques.

Procedures for sample preparation and plating

Duplicate 10 g (or mL) samples from each lot were analysed. Each sample was combined with 90 ml of Butterfield's phosphate buffer (pH 7.0) and homogenised in a Colworth Stomacher for 30 sec. Samples were then serially diluted (1:10) and 1.0 mL aliquots placed in the centre of Petrifilm YM plates. This was carried out by lifting the cover film, applying the sample and rolling the cover back into place. Samples were then uniformly spread using a specially designed template. Serially diluted samples (1.0 mL) were pour plated with ca. 15 mL of molten (47°C) APDA or CPCA, and mixed. Serially diluted 0.1 mL samples were also surface plated on APDA and CPCA plates which had been dried at 21°C for 1 to 3 days after pouring.

Incubation conditions and enumeration of colonies

Inoculated Petrifilm YM and APDA and CPCA plates were incubated upright at 25°C. Colonies of yeasts and moulds were separately enumerated after 3 and 5 days. Plates containing 15 to 150 colonies were selected for counting, although occasionally plates containing numbers of colonies outside this range were included. Cells from randomly picked presumptive yeast colonies were examined microscopically to confirm the absence of bacteria, and occasionally to differentiate yeasts from moulds.

Statistical analysis

Colony counts were converted to $\log_{10}$ counts to more nearly match the underlying assumption of a normal distribution. Analysis of variance and Tukey's Honestly Significant Difference Test were used to compare the enumeration methods in this study. Paired t-tests were used to compare day 3 to day 5 counts. Standard regression techniques were used to calculate correlation coefficients, slopes and intercepts. Coefficients of variation were calculated by dividing the repeatability standard deviation by the mean count and expressing as a percentage.

RESULTS AND DISCUSSION

Observations on colony appearance

Most yeast colonies on Petrifilm YM plates were characterised by a blue colouration, more distinct after 5 days incubation. Occasionally white or cream colonies were also produced. Moulds produced variously coloured colonies but rarely the blue colour characteristic of yeasts. Yeast colonies were smaller than mould colonies on all media.

Comparison of enumeration systems

Populations of total yeasts and moulds detected on Petrifilm YM plates compared with APDA and CPCA (both pour and spread plated) after 5 days incubation are shown in Table 1. No significant ($P \leq 0.05$) differences in populations of fungi were detected in

Table 1. Total mean fungal counts ($\log_{10}$ CFU/g [mL]) obtained from 30 samples each of eight foods[1]

Product	Petrifilm YM plate		APDA Pour		APDA Spread		CPCA Pour		CPCA Spread	
Hard/soft cheese	6.76[ab]	(1.49)	6.69[c]	(1.47)	6.72[bc]	(1.48)	6.74[abc]	(1.45)	6.79[a]	(1.46)
Cottage cheese	5.74[abc]	(1.11)	5.67[c]	(1.11)	5.77[ab]	(1.12)	5.69[bc]	(1.10)	5.80[a]	(1.12)
Yoghurt	5.91	(1.20)	5.90	(1.23)	5.92	(1.21)	5.92	(1.21)	5.93	(1.21)
Sour cream	7.84[a]	(0.24)	7.83[ab]	(0.30)	7.80[ab]	(0.32)	7.82[ab]	(0.32)	7.77[b]	(0.28)
Fruit juice	5.61	(1.43)	5.61	(1.42)	5.65	(1.40)	5.58	(1.42)	5.62	(1.41)
Salad dressing	5.21[ab]	(1.46)	5.19[ab]	(1.43)	5.15[b]	(1.43)	5.23[a]	(1.43)	5.20[ab]	(1.41)
Relish	6.15[a]	(0.72)	6.02[b]	(0.79)	6.08[ab]	(0.76)	6.12[a]	(0.75)	6.12[a]	(0.72)
Sauces	5.63[a]	(0.69)	5.48[c]	(0.73)	5.55[b]	(0.70)	5.60[ab]	(0.72)	5.62[a]	(0.72)

[1] Values in the same row which are followed by the same letter are significantly different ($P \leq 0.05$). Values in parentheses indicate standard deviation

yoghurt and fruit juice. Overall, recovery using the pour plate technique was inferior to the spread plate and Petrifilm YM plate techniques.

Effect of time on colony formation

Table 2 compares the influence of 3 and 5 days incubation on yeasts, moulds and total fungal counts obtained from eight food products. No moulds were detected in sour cream and salad dressing; thus a total of six products is listed in the comparison of mould counts. In the majority of food products, yeast and total fungal counts after 5 days incubation were significantly higher ($P \leq 0.05$) than counts after 3 days. This observation was made for all enumeration systems.

Table 2. Yeast and mould populations detected after 5 days compared with 3 days of incubation

Fungus	Number of food products	Petrifilm YM plate	ADPA Pour	ADPA Spread	CPCA Pour	CPCA Spread
Yeast	8	6	7	5	5	4
Mould	6	2	2	3	2	2
Total fungi	8	6	6	4	4	6

Enumeration system[1]

[1]Number of times out of 6 (for moulds) or 8 (for others) counts were significantly higher ($P \leq 0.05$) after 5 days of incubation compared to 3 days.

Correlation coefficients, slopes and intercepts
Correlation coefficients, slope and intercept values derived from comparing $\log_{10}$ counts of total fungi obtained with the five enumeration systems are given in Table 3. Correlation coefficients of Petrifilm YM plates compared with APDA and CPCA (both pour and spread plated) were very high (>0.99), indicating comparable counts among all five systems. Slopes of the lines of best fit ranged from 0.984 to 1.008, with intercept values ranging from -0.051 to 0.149, again indicating close agreement between Petrifilm YM plates and the other techniques.

The coefficient of variation for total fungal counts on Petrifilm YM plates after 5 days of incubation was 1.0% compared with a range of 1.2 to 1.7% for the other systems. Except for yeast counts on Petrifilm YM plates, the coefficients of variation for counts on all test systems increased substantially after 5 days incubation.

Table 3. Correlation coefficients plus slopes and intercept values from lines of best fit for counts on Petrifilm™ YM plates versus APDA and CPCA for 240 food samples

		Petrifilm™ YM plate versus:			
	Incubation	APDA		CPCA	
Fungus	time (days)	Pour	Spread	Pour	Spread
Correlation coefficient					
Yeasts	3	0.992	0.992	0.993	0.993
	5	0.992	0.992	0.993	0.994
Moulds	3	0.964	0.996	0.959	0.968
	5	0.972	0.976	0.976	0.984
Total	3	0.994	0.992	0.993	0.994
	5	0.993	0.993	0.994	0.995
Slope value					
Yeasts	3	0.979	1.000	1.008	1.021
	5	0.974	0.995	1.000	1.015
Moulds	3	0.976	0.990	0.956	0.971
	5	0.981	0.998	0.961	0.978
Total	3	0.984	0.998	1.000	1.006
	5	0.984	0.997	0.997	1.008
Intercept value					
Yeasts	3	0.169	0.003	-0.036	-0.157
	5	0.222	0.061	0.019	-0.089
Mould	3	0.036	-0.135	0.063	-0.001
	5	0.065	-0.138	0.116	-0.023
Total	3	0.131	0.007	0.005	-0.067
	5	0.149	0.038	0.033	-0.051

Relative performance of Petrifilm YM plates

A comparison of Petrifilm YM plates to pour and spread plated APDA and CPCA for suitability to recover fungi from food products is summarised in Table 4. After 5 days incubation, yeast or total fungal counts in the eight food products using Petrifilm™ YM plates were not significantly lower than traditional pour and spread plating on APDA or CPCA. Petrifilm plates showed significantly higher total fungal counts than ADPA pour plates (3/8 food products), ADPA spread plates (1/8) and CPCA spread plates (1/8). Yeast counts were somctimes significantly higher on Petrifilm YM plates after 5 days incubation. On the other hand, recovery of moulds on APDA spread plates and CPCA was significantly better for two of six and one of six foods after 5 days incubation. This may be due to a more favourable oxygen tension associated with the spread plating technique.

Table 4. Comparison of Petrifilm YM plates with APDA and CPCA plate counts for fungi

			Number of times counts on Petrifilm™ YM plates were significantly higher/lower than on:			
			APDA		CPCA	
Fungus	Incubation time (days)	Number of food products	Pour	Spread	Pour	Spread
Yeasts	3	8	2/0	1/1	0/0	1/1
Moulds		6	0/1	0/2	0/1	0/2
Total		8	2/0	0/0	0/0	1/1
Yeasts	5	8	3/0	2/0	0/0	1/0
Moulds		6	0/0	0/2	0/0	0/1
Total		8	3/0	1/0	0/0	1/0

CONCLUSION

Counts on Petrifilm YM plates were comparable with plate counts on APDA and CPCA for enumerating fungi in the range of foods analysed in this study. Petrifilm YM plates also offer several inherent advantages over plate count systems. Among these are that Petrifilm YM plates are pre-prepared, less space is required for storing and incubation, less time is required for plating samples and secondary colony development is essentially eliminated. An autoclave is unnecessary unless sterilisation of Petrifilm YM plates is desired on disposal, nor is a water bath needed to keep molten agar for pour plating.

A disadvantage of Petrifilm YM plates is that a few yeasts do not develop blue colours. Occasionally, very small nonpigmented colonies are difficult to see, making the actual counting process potentially tedious to anyone not familiar with the technique. In addition, picking colonies of some fungal species from Petrifilm YM plates for identification may be more difficult than on traditional plates. None of the systems evaluated was exceptionally effective at controlling colony development by *Rhizopus* and *Mucor*. Notwithstanding these limitations, the Petrifilm YM plate can effectively enumerate fungi in the dairy and high acid foods selected for evaluation in this study. Additional work is underway to determine the performance of Petrifilm YM plates for enumerating fungi in a wide variety of other food products.

COMPARISON OF POUR AND SPREAD PLATING FOR ENUMERATION OF FUNGI IN FOODS

L. R. Beuchat,[1] Brenda V. Nail,[1] R. E. Brackett[1] and T. L. Fox[2]

[1]*Department of Food Science and Technology*
University of Georgia, Agricultural Experiment Station
Griffin, GA 30223-1797, USA

[2]*Statistical Data Services*
3M Pharmaceuticals, 3M Center,
270-3A-01, St. Paul, MN 55144-1000, USA

SUMMARY

Pour and spread plating techniques were used to compare Acidified Potato Dextrose Agar (APDA) and Chloramphenicol supplemented Plate Count Agar (CPCA) for their suitability to recover fungi from a variety of dairy products and high acid foods. Colonies of yeasts and moulds were separately enumerated after incubating plates upright for 3 and 5 days at 25°C.

Coefficients of variation for total fungal counts after 5 days on pour or spread plates of APDA or CPCA ranged from 1.2 to 1.7%. Yeasts and moulds and total fungal counts in the majority of food products were significantly higher after 5 days incubation, regardless of medium or application technique. Total fungal counts were significantly higher for two of eight food products when APDA was spread rather than pour plated, and for one of eight foods when CPCA was spread plated. Counts using pour plates were never significantly higher than on spread plates. CPCA spread plate counts were significantly higher than APDA spread plates for two of eight foods. Total counts on APDA spread plates were never significantly higher than CPCA spread plates. It is concluded that of the four plating technique/media combinations examined in this study for their ability to recover fungi from eight acid food products, CPCA using spread plates was clearly superior.

INTRODUCTION

Disagreement still exists among some food microbiologists concerning the relative merits of pour plates versus spread plates for enumerating fungi in foods. The pour plate technique is commonly used to enumerate bacteria and total microorganisms in foods, with little or no thought given to growth conditions, particularly oxygen requirements, necessary for fungal colony development.

The study reported here was designed to compare pour plate and spread plate techniques for their efficacy in recovering yeasts and moulds on Acidified Potato Dextrose Agar (APDA) and Chloramphenicol supplemented Plate Count Agar (CPCA), two media traditionally used in industrial quality assurance and food microbiology laboratories in North America. Foods selected for analysis were dairy products, fruit juices, salad dressings, relishes and sauces.

MATERIALS AND METHODS

The materials and methods used in this study are described in a companion paper (Beuchat *et al.*, 1992).

RESULTS AND DISCUSSION

Correlation coefficients, slopes and intercepts

Total yeast and mould populations detected in various food products using the five enumeration systems are reported in a companion paper (Beuchat *et al.*, 1992). Correlation coefficients, slopes of lines of best fit and intercept values for each medium and plating technique are listed in Table 1. Considerable differences exist in these values, reflecting the degree of performance of each system.

Table 1. Correlation coefficients, slopes of lines of best fit and intercept values for counts on APDA and CPCA using both pour and spread plates for 240 samples[1]

Fungus	Incubation time (days)	APDA spread vs: APDA pour	APDA spread vs: CPCA pour	APDA spread vs: CPCA spread	CPCA spread vs: APDA pour	CPCA spread vs: CPCA pour	APDA pour vs CPCA pour
Correlation Coefficients							
Yeasts	3	0.993	0.992	0.996	0.993	0.994	0.995
	5	0.993	0.993	0.996	0.992	0.994	0.995
Moulds	3	0.983	0.982	0.982	0.979	0.979	0.981
	5	0.980	0.984	0.983	0.977	0.981	0.978
Total	3	0.995	0.993	0.995	0.994	0.995	0.996
	5	0.995	0.994	0.996	0.994	0.996	0.996
Slopes of lines of best fit							
Yeasts	3	0.973	0.999	1.014	0.954	0.982	1.023
	5	0.973	0.997	1.013	0.955	0.981	1.020
Moulds	3	0.981	0.964	0.961	0.994	0.977	0.985
	5	0.978	0.958	0.956	1.000	0.979	0.954
Total	3	0.980	0.995	1.003	0.972	0.990	1.013
	5	0.981	0.993	1.004	0.972	0.986	1.009
Intercept of lines of best fit							
Yeasts	3	0.206	0.016	0.120	0.352	0.152	-0.170
	5	0.199	0.006	-0.112	0.338	0.134	-0.167
Moulds	3	0.189	0.202	0.228	0.085	0.095	0.096
	5	0.218	0.268	0.276	0.055	0.107	0.176
Total	3	0.157	0.041	-0.041	0.229	0.097	-0.105
	5	0.145	0.038	-0.049	0.222	0.101	-0.094

[1]Mean values of composite of eight food products.

The coefficients of variation of these enumeration systems are listed in Table 2. Coefficients varied only slightly and without substantial influence of type of medium or plating technique.

Comparison of plating technique within medium

In Table 3, the plating techniques and media used are ranked according to performance, taking into account only statistically significant differences. CPCA clearly performed better than APDA. These results are not surprising. Exposure of stressed fungal cells or spores to acid pH in recovery media can result in death. Likewise, exposure of stressed cells to elevated temperatures associated with pour plating can impose additional stress,

Table 2. Total fungal counts on APDA and CPCA, using both pour and spread plates, for 240 food samples

		Coefficient of variation (%)			
	Incubation	APDA		CPCA	
Fungus	time (days)	Pour	Surface	Pour	Surface
Yeasts	3	1.7	1.5	1.5	1.8
Moulds		1.8	2.4	1.5	1.8
Total		1.0	1.8	1.4	1.6
Yeasts	5	4.6	4.0	4.0	4.4
Moulds		4.7	5.1	4.2	5.0
Total		1.2	1.7	1.4	1.6

Table 3. Comparison of statistically significantly higher counts on APDA and CPCA, both pour and spread plates, for eight food types

			Number of times counts on APDA or CPCA were significantly higher or lower on pour or spread plates	
Fungus	Incubation time (days)	Number of food products	APDA (pour/spread)	CPCA (pour/spread)
Yeasts	3	8	0/2	0/2
Moulds		6	0/1	0/0
Total		8	0/2	0/2
Yeasts	5	8	0/1	0/1
Moulds		6	0/1	0/1
Total		8	0/2	0/1

perhaps resulting in death. The availability of oxygen to fungi is, of course, increased when using spread plates, thus enhancing colony development. In addition to the more favorable nutrient content in CPCA compared to APDA, the high pH (6.8) of CPCA is undoubtedly more favorable for resuscitating stressed cells.

CONCLUSIONS

Data obtained from this study in conjunction with observations reported in the companion study (Beuchat *et al.*, 1992) indicate that the spread plating technique is superior to pour plating for enumerating fungi in selected dairy products and high acid foods. Furthermore, spread plating using CPCA resulted in recovery of higher populations of fungi compared to spread plates of APDA. Based on the results of both studies reported here, spread plating is recommended in preference to pour plating for enumerating fungi in the dairy and high acid foods examined in this study.

REFERENCES

BEUCHAT, L.R., NAIL, B.V., BRACKETT, R.E. and FOX, T.L. 1992. Evaluation of the Petrifilm™ yeast and mold plate for enumeration of yeasts and moulds in foods. *In* Modern Methods in Food Mycology, eds R.A. Samson, A.D. Hocking, J.I. Pitt and A.D. King. pp. 21-25. Amsterdam: Elsevier.

MEDIA FOR ENUMERATING SPOILAGE YEASTS - A COLLABORATIVE STUDY

T. Deák

Department of Microbiology
University of Horticulture and Food Industry
Somloi ut 14-16
Budapest H-1118, Hungary

In collaboration with: N. Aran, L.B. Bullerman, D.E. Conner, J.E.L. Corry, A.D. King, J.I. Pitt, D.F. Splittstoesser and V. Tabajdi-Pintér

SUMMARY

Nine laboratories compared the following media: Dichloran Rose Bengal Chloramphenicol agar (DRBC), Acidified Tryptone Glucose Yeast extract agar (ATGY), Chloramphenicol Tryptone Glucose Yeast extract agar (CTGY) or Oxytetracycline Glucose Yeast extract agar (OGY), Dichloran 18% Glycerol agar (DG18), and one or more media of choice or preference. Conditions of media preparation and enumeration technique were standardised, and the experimental system arranged to enable statistical evaluation. Each participant investigated at least six types of food differing in composition, and consequently in the level of yeast contamination.

Samples investigated included a wide range of commodities, with yeast counts ranging from $\log_{10}0.6$ to $\log_{10}5.5$ per g, and, in some fermented or spoiled samples up to $\log_{10}8.2$ per g. Performance of media was unaffected by microbial load of the sample. Considering results from all laboratories, no unequivocal evidence was found for any medium being superior to the others. In some cases DRBC gave significantly higher counts. ATGY and Malt Extract Agar (MEA) were sometimes not selective enough for yeasts. In some cases, counts on DG18 were lower than on other media.

INTRODUCTION

At the first Workshop on Standardisation of Methods for the Mycological Examination of Food (King *et al.*, 1986), several recommendations were made on techniques and media for the enumeration and isolation of yeasts from food. It was recommended that general purpose media which inhibit bacterial growth and reduce fungal spreading be used for yeasts. The history and development of such media, of which there are many, has been well documented (King *et al.*, 1986; Jarvis and Williams, 1987; Fleet, 1990; Deák, 1991). Acidified media have been shown to be less suitable for the enumeration of yeasts than those supplemented with antibiotics (Beuchat, 1979; Thomson, 1984; Deák *et al.*, 1986). Chloramphenicol, oxytetracycline, chlortetracycline or gentamycin appeared to be equally effective, but the first is heat stable and more convenient to use (Anon., 1983; Beckers *et al.*, 1986). With some types of food, e.g. meats, a single antibiotic is not sufficient to control the growth of bacteria and the use of two antibiotics is recommended (Koburger and Rogers, 1978; Mossel *et al.*, 1980).

The inclusion of rose bengal and dichloran to retard mould growth and restrict colony size facilitates the enumeration of yeasts (Henson, 1981; Deák *et al.*, 1986), but rose bengal inhibits some yeasts (Banks *et al.*, 1985; Williams, 1986; Dijkman, 1986). DG18 also appears to be an appropriate medium for the enumeration of yeasts (Dijkman, 1986).

Comparative studies have indicated that none of the currently used media is effective for enumerating yeasts in all foods. Even when statistically significant differences can be found between media, these may not be of practical importance (Beckers *et al.*, 1986; Hastings *et al.*, 1986). This collaborative study was designed to test several media and to examine the influence of yeast load and food composition on their performance.

MATERIALS AND METHODS

Samples

Nine laboratories collaborated in the study. Each one was asked to include at least six types of food, such as wine or cider, fruit juice or other non-alcoholic beverage, fruit concentrate or puree, yoghurt or cheese, minced meat or hamburger, bread or flour, salads, acid preserved, fermented or salted food. In all, 119 samples of more than a dozen different kinds of food were examined (Table 1).

Media

The collaborating laboratories compared the following media: Dichloran Rose Bengal Chloramphenicol agar (DRBC), Acidified Tryptone Glucose Yeast extract Agar (ATGY), Tryptone Glucose Yeast extract agar with Chloramphenicol (CTGY) and Dichloran 18% Glycerol agar (DG18), as well as one or more media of preference. In all 12 different media were compared (Table 2).

Table 1. Yeast counts obtained by collaborating laboratories on various food samples

Food sample	No. of samples	Range of $\log_{10}$ counts
Fruits	6	0.6 - 4.1
Grapes	20	3.4 - 7.3
Wine	22	2.7 - 8.4
Fruit juices	12	2.7 - 6.9
Vegetables	2	3.6 - 6.0
Mayonnaise salads	9	2.5 - 5.1
Pickles	8	4.9 - 5.8
Cheese	7	0.8 - 3.2
Kefir	4	4.4 - 5.7
Cereals	10	0.7 - 1.1
Minced meat	7	1.8 - 4.3
Dried sausage	4	1.9 - 5.2
Others [1]	8	2.2 - 7.2
Total	119	0.6 - 8.4

[1] Dough, bread, poultry, shrimp

Table 2. Media used in the collaborative study

Medium	Abbreviation	No. of laboratories using
General		
Malt extract agar	MEA	2
Acidified		
Acidified tryptone glucose yeast extract agar	ATGY	7
Acidified glucose yeast extract agar	AGY	1
Acidified potato dextrose agar	APDA	1
Chloramphenicol		
Chloramphenicol tryptone glucose yeast extract agar	CTGY	6
Chloramphenicol glucose yeast extract agar	CGY	1
Chloramphenicol potato dextrose agar	CPDA	2
Other antibiotic		
Oxytetracycline glucose yeast extract agar	OGY	4
Gentamycin malt extract agar	GMEA	1
Rose bengal chlortetracycline agar	RBC	1
Multiple inhibitors		
Dichloran rose bengal chloramphenicol agar	DRBC	9
Dichloran 18% glycerol agar	DG18	9

Methods

The method of enumeration was standardised according to previous recommendations (King *et al.*, 1986). Solid products were homogenised in a Stomacher (1 to 10 ratio) in 1% peptone water diluent. After appropriate decimal dilutions with 1% peptone water, spread plates were prepared from subsamples (0.1 mL) on predried plates. Plates were incubated at room temperature (20 to 25°C), in the upright position, for 5 days. Plates with 10-100 yeast colonies were selected for counting. Data were transformed to $\log_{10}$ units and expressed per g or mL.

Each food was divided into two subsamples, and from each subsample two independent series of decimal dilutions were prepared. The same dilution was tested on at least 5 different media, by some but not all laboratories, in duplicate. Statistical evaluation was made by two factorial analysis of variance using a computer programme developed in this laboratory.

RESULTS AND DISCUSSION

Yeast counts in the samples investigated ranged from $\log_{10}0.6$ to $\log_{10}5.5$ per g, or up to $\log_{10}8.4$ in some fermented or spoiled samples (Table 1). Performance of the media was not dependent on load, nor was the standard error of the counts (Table 3). At high cell densities, ATGY produced significantly higher counts. This could probably be attributed to spoiled samples heavily contaminated with bacteria, which developed on ATGY.

Considering all results, no unequivocal evidence was found for any media producing consistently different results from the others (Table 4). In two laboratories, no statistically significant difference was found ($p = 0.05$) between the media compared when all samples

Table 3. Comparison of media at different yeast count levels[1]

Range of $\log_{10}$ count	$\log_{10}$ count per g on media				$SD_{95\%}$
	DRBC	DG18	ATGY	CTGY	
0 - 2	1.688[a]	1.505[a]	1.429[a]	1.501[a]	0.255
3 - 4	3.931[a]	3.838[a]	4.009[a]	4.018[a]	0.250
5 - 6	5.980[b]	6.032[ab]	6.209[a]	6.159[ab]	0.191
7 - 8	7.731[b]	7.697[b]	7.916[a]	7.703[b]	0.100

[1]Data represent averages of 5 randomly selected samples. Mean values followed by the same letter are not significantly different ($p < 0.05$).

Table 4. Summary of data from collaborative study[1]

Lab.	No. of samples	$\log_{10}$ yeast counts on media					$SD_{95\%}$
		DRBC	ATGY	CTGY	DG18	other	
1	12	3.603[a]	3.378[b]	3.172[c]	3.035[c]	3.193[c] (CPDA)	0.170
2	12	3.345[b]	3.792[a]	3.538[b]	3.113[c]	3.178[c] (RBC), 3.810[a] (MEA)	0.204
3	6	5.210[a]	5.290[a]	5.291[a]	5.180[a]	5.191[a] (MEA)	0.179
4	12	5.508[b]	6.115[a]	5.450[b]	5.039[c]	5.463[b] (APDT)	0.080
5	12	3.445[a]	3.518[a]	3.518[a]	3.406[b]	3.528[a] (OGY)	0.097
6	8	4.046[a]	4.082[a]	4.080[a]	4.198[a]	4.190[a] (OGY)	0.166
7	20	5.960[a]	5.817[b]	5.823[b]	5.728[c]	5.864[a] (CPDA), 5.856[a] (APDA)	0.118
	17	6.348[a]	6.373[a]	6.339[a]	6.310[a]	6.422[a] (CPDA), 6.239[a] (APDA)	0.156
8	15	4.526[a]	4.480[a]		4.499[a]	4.509[a] (OGY), 4.412[a] (GMEA)	0.118
9	9	4.762[b]	5.177[a]		4.975[a]	5.003[a] (OGY), 4.896[a] (GMEA)	0.275

[1] Mean values followed by the same letter are not significantly different ($p < 0.05$).

were evaluated. In other laboratories one or more media differed significantly from at least one other. In five cases, DG18 gave the lowest counts (in two cases compared to all media, and in one case each compared to ATGY, or DRBC, or both). ATGY resulted in the highest count in four cases, and DRBC in two cases.

Differences were found between media, however, when investigated by the same laboratory. In one laboratory significant differences were obtained between ATGY and DG18 when results from six samples were compared (Table 5). Closer analysis of the results revealed that this was mainly due to the great differences obtained with fresh ground beef samples. In orange juice concentrate and spoiled Mexican sauce only ATGY gave significantly higher counts. With spoiled cider DG18 and DRBC gave significatly lower counts, while in the case of frozen strawberry samples no differences were found between the media.

Table 5. Comparison of media within one laboratory[1]

Samples	No. of samples	log$_{10}$ counts per g on media					$SD_{95\%}$
		DRBC	DG18	PDAT	ATGY	CTGY	
Spoiled sauce	2	7.69^b	7.69^b	7.63^b	8.27^a	7.71^b	0.15
Orange juice	2	3.45^b	3.52^b	3.49^b	4.26^a	3.48^b	0.08
Strawberries	2	3.35	3.43	3.38	3.36	3.37	0.18
Ground beef	2	5.00^b	2.49^c	4.58^b	6.61^a	4.44^b	0.71
Spoiled salad	2	8.12^b	7.99^c	8.12^b	8.55^a	8.12^b	0.10
Spoiled cider	2	5.44^b	5.12^c	5.61^a	5.66^a	5.61^a	0.14
All samples	12	5.51^b	5.04^c	5.46^b	6.12^a	5.45^b	0.41

[1] Mean values followed by the same letter are not significantly different ($p < 0.05$).

In another laboratory 20 samples of grapes of various variety and degree of ripeness were examined; slight but significant differences were found between DRBC and DG18. However, no differences were found among the media when 17 samples of variously treated wines were studied.

In one laboratory countable plates were not obtained from all samples on all media, so these could only be evaluated separately. With two samples, DRBC produced significantly higher counts, but did not with another two samples, both hamburger mince (Table 6). With the rest of the samples studied, no differences were found between media, when the samples were evaluated either separately or together.

In some cases where the overall sample comparison did not show significant differences among media, separate evaluation of some samples showed significant differences

Table 6. Comparison of media within a second laboratory[1]

Food samples	No. of samples	log$_{10}$ counts per g on media					$SD_{95\%}$
		DRBC	ATGY	CTGY	DG18	MEA	
Ground mince	2	2.068^a	-	-	1.825^b	1.890^b	0.127
Hamburger mince 1	2	4.295^a	-	4.083^b	4.125^b	-	0.113
Hamburger mince 2	2	3.645	-	3.560	-	3.623	0.184
Mayonnaise salad	2	5.025	-	4.840	5.025	-	0.180
Coleslaw, orange juice, strawberries	6	5.210	5.290	5.291	5.180	5.191	0.179

[1] Mean values followed by the same letter are not significantly different ($p < 0.05$)

(Table 7). For example, on olives DG18 gave higher counts, while with mayonnaise salad DG18 gave lower counts than the other media. Overall, DG18 did not differ from other media in the average of counts on 15 samples. In kefyr samples, ATGY produced significantly higher counts than DRBC and DG18, but this was not the case when all samples were evaluated together.

Table 7. Outlying results for particular media found with individual samples

Food samples	No. of samples	Log_{10} count per g on media				$SD_{95\%}$
		DRBC	ATGY	DG18	OGY	
Olives	2	4.94[b]	4.91[b]	5.26[a]	5.06[b]	0.23
Mayonnaise salad	2	5.08[b]	5.15[b]	4.49[a]	5.10[b]	0.16
All samples	15	4.53	4.48	4.50	4.51	0.12
Kefyr	2	4.24[b]	4.92[a]	4.23[b]	4.74[a]	0.26
All samples	24	3.45	3.52	3.52	3.40	0.75
Sauerkraut	2	5.62[b]	5.85[b]	5.78[b]	4.96[a]	0.34
Kefyr	2	5.15[c]	6.18[a]	5.23[a]	5.69[b]	0.29
Sour cherries	2	5.24[b]	5.93[a]	6.07[a]	6.13[a]	0.37
All samples	9	4.76[b]	5.18[a]	5.00[a]	4.90[a]	0.27

[1] Mean values followed by the same letter are not significantly different ($p < 0.05$).

Table 8. Subjective evaluation of media: summary of comments by collaborators

Medium	Summary of comments
DRBC	Mould spreading is strongly inhibited, bacteria may grow, yeast colonies are smaller, but easily differentiated from bacteria, ingredients may differ from batch to batch
CTGY	Yeast colonies are greater, moulds can develop, bacteria may grow, easy to prepare
ATGY	Bacterial and yeast colonies could not be distinguished, but differences among yeast colonies can be recognised, lactic acid bacteria grow abundantly but form punctiliform colonies
OGY	Yeast colonies are larger, but wine yeasts developed poorly, moulds spread, bacteria may grow, difficult to differentiate bacterial from yeast colonies
DG18	Bacteria are suppressed, moulds are restricted, but the growth of yeasts may also be retarded, colonies are sometimes tiny
GMEA	Bacterial and yeast colonies could not be distinguished, red yeasts may be inhibited

These findings show that the different composition of the yeast flora may be reflected in the performance of the media. It has been demonstrated that of yeast species differ in sensitivity to inhibitory ingredients (rose bengal, dichloran, antibiotics, pH, a_w) of the media (Beckers *et al.*, 1986; Deák *et al.*, 1986). Clearly, yeast flora varies with different commodities (Fleet, 1990; Deák, 1991), resulting in differing counts on various media. In other words, no single medium will give satisfactory results for all types of food.

Media containing an antibiotic, or acidified, or of decreased a_w (e.g. DRBC, CTGY, OGY, ATGY, DG18) may be equally appropriate for recovery of yeasts from specific food samples. Objective evaluation of the performance of media in this collaborative study showed, in agreement with previous findings (e.g. King *et al.*, 1986), that ATGY is not selective enough against bacteria, and that DG18 is sometimes inhibitory to yeasts. As shown in the summary of collaborators' comments (Table 8), DRBC enhances recognition of yeast colonies.

REFERENCES

ANON. 1983. General guidance for detection and enumeration of molds and yeasts. International Standards Organisation /TC 34/ SC 9N 156.

BANKS J.G., BOARD, R.G., CARTER, J. and DODGE, A.D. 1985. The cyto-toxic and photodynamic inactivation of micro-organisms by rose bengal. *J. Appl. Bacteriol.* **58**: 391-400.

BECKERS, H.J., DE BOER, E., DIJKMAN, K.E., HARTOG, B.J., VAN KOOIJ, J.A., KUIK, D., MOL, N., NOOITGEDAGT, A.J., NORTHOLT, M.D. and SAMSON, R.A. 1986. Comparison of various media for the enumeration of yeasts and molds in food. *In* Methods for the Mycological Examination of Food, eds A.D. King, J.I. Pitt, L.R. Beuchat and J.E.L. Corry, pp. 66-76. New York: Plenum Press.

BEUCHAT, L.R. 1979. Comparison of acidified antibiotic-supplemented potato dextrose agar from three manufacturers for its capacity to recover fungi from foods. *J. Food Prot.* **42**: 427-428.

DEAK, T. 1991. Foodborne yeasts. *Adv. Appl. Microbiol.* **36**: 179-278.

DEAK, T., TÖRÖK T., LEHOCZKI, J., REICHART, O., TABAJDI-PINTER, V. and FABRI, I. 1986. Comparison of general purpose media for the enumeration of molds and yeasts. *In* Methods for the Mycological Examination of Food, eds A.D. King, J.I. Pitt, L.R. Beuchat and J.E.L. Corry, pp. 114-119. New York: Plenum Press.

DIJKMAN, K.E. 1986. Merits and shortcoming of DG18, OGY, OGGY and DRBC media for the examination of raw meats. *In* Methods for the Mycological Examination of Food, eds A.D. King, J.I. Pitt, L.R. Beuchat and J.E.L. Corry, pp. 120-123. New York: Plenum Press.

FLEET, G.H. 1990. Food spoilage yeasts. *In* Yeast Technology, eds J.F.T. Spencer and D.M. Spencer, pp. 124-166. Berlin: Springer Verlag.

HASTINGS, J.W., TSAI, W.Y.J. and BULLERMAN, L.B. 1986. Comparison of general purpose mycological medium with a selective mycological medium for mold enumeration. *In* Methods for the Mycological Examination of Food, eds A.D. King, J.I. Pitt, L.R. Beuchat and J.E.L. Corry, pp. 94-96. New York: Plenum Press.

HENSON, O.E. 1981. Dichloran as an inhibitor of mold spreading in fungal plating media: effect on colony diameter and enumeration. *Appl. Environ. Microbiol.* **42**: 656-660.

HOCKING, A.D. and PITT, J.I. 1980. Dichloran-glycerol medium for enumeration of xerophilic fungi from low-moisture foods. *Appl. Environ. Microbiol.* **39**: 488-492.

JARVIS, B. and WILLIAMS, A.P. 1987. Methods for detecting fungi in foods and beverages. *In* Food and Beverage Mycology, ed. L.R. Beuchat, 2nd edn, pp. 599-636. New York: AVI Van Nostrand Reinhold.

KING, A.D., PITT, J.I., BEUCHAT, L.R. and CORRY, J.E.L., eds. 1986. Methods for the Mycological Examination of Food. New York: Plenum Press.

KOBURGER, J.A. and RODGERS, M.F. 1978. Single or multiple antibiotic-amended media to enumerate yeasts and molds. *J. Food Prot.* **41**: 367-369.

MOSSEL, D.A.A., DIJKMAN, K.E. and KOOPMANS, M. 1980. Experience with methods for the enumeration and identification of yeasts occurring in foods. *In* Biology and Activities of Yeasts, eds F.A. Skinner, S.M. Passmore and R.R. Davenport, pp.279-288. London: Academic Press.

SEILER, D.A.L. 1986. A collaborative exercise comparing media for enumerating fungi in foods. *In* Methods for the Mycological Examination of Food, eds A.D. King, J.I. Pitt, L.R. Beuchat and J.E.L. Corry, pp. 83-85. New York: Plenum Press.

THOMSON, G.H. 1984. Enumeration of yeasts and moulds - media trial. *Food Microbiol.* 1: 223-227.

WILLIAMS, A.P. 1986. A comparison of DRBC, OGY and RBC media for the enumeration of yeasts and molds in foods. *In* Methods for the Mycological Examination of Food, eds A.D. King, J.I. Pitt, L.R. Beuchat and J.E.L. Corry, pp. 89-91. New York: Plenum Press.

COMPARISON OF YEAST IDENTIFICATION METHODS

A. D. King and T. Török

USDA Agricultural Research Service,
Western Regional Research Center,
Albany, CA 94710, USA

SUMMARY

Using standard yeast identification texts, API commercial identification kits and the simplified system described by Deák in 1986, 239 yeast strains were identified. The test results were interpreted by conventional means and by computer assisted keys and databases. The identified yeast strains from fruits and vegetables represented 36 species from 19 genera. The simplified scheme proved correct in over 80% of cases, and was significantly better than all other methods.

INTRODUCTION

Deák (1986) proposed a simplified identification key for yeast species associated with food. The scheme was improved by Deák and Beuchat (1987) then tested and compared with commercial systems (Deák and Beuchat, 1988). Rohm and Lechner (1990) strongly criticised the scheme as unreliable. They claimed that Deák and Beuchat (1987) did not consider the variability of some characters, and the simplified scheme failed to identify certain reference strains and over 50% of species they isolated from cultured milk products. Rohm and Lechner (1990) examined the species in all 16 subgroups of the simplified scheme and concluded that only 19 out of some 200 species listed could be unequivocally identified by the scheme.

The purpose of this investigation was to compare conventional yeast identification methods (Barnett *et al.*, 1983; Kreger-van Rij, 1984), commercially available identification kits (API 20C and API YEAST-IDENT™), and Deák and Beuchat's simplified identification scheme (Deák and Beuchat, 1987), by using a wide range of yeast strains.

MATERIALS AND METHODS

Pure cultures of yeasts were grown on Yeast extract Peptone Dextrose agar (YPDA) (yeast extract 1%, peptone 1%, dextrose 1% and agar 2%). Preliminary morphological screening was used to select 239 isolates for identification. We selected 127 yeast isolates from fruits or fruit products, 14 of which had caused spoilage, and 112 yeast isolates from partially and fully processed vegetables. The wide variety of food types and products from which yeasts were isolated ensured that many types of yeasts were studied. This allowed a general comparison of the identification methods.

The yeast isolates were characterised according to Barnett *et al.*, (1983) and Kreger-van Rij (1984), and the instructions of the commercial identification kits (API 20C, API

Yeast-IDENT™). Characterisation of isolates by Deák and Beuchat's system followed the steps as described (Deák and Beuchat, 1987) with the exception that the nitrate assimilation test was modified according to Pincus *et al.* (1988).

The yeasts were identified using the dichotomous keys in Kreger-van Rij (1984). Yeast identification following Barnett *et al.* (1983) was achieved using the computer program designed by the same authors (Barnett *et al.*, 1985). The identification of yeasts within the simplified scheme (Deák and Beuchat, 1987) was carried out using the master key and the detailed identification keys described.

To evaluate the test results from the commercial kit API 20C, we used the "API Analytical Profile Index" (API, 1988). The API Yeast-IDENT results were evaluated using the "API Yeast-IDENT Directory" (API, 1986).

Computer assisted identification software packages tested were "Yeasts" by the American Society for Microbiology Computer Users Group (American Society for Microbiology, 1989), "Yeast-ID" (Deák, 1990) and "Yeast Identification Program" (Reichart, 1990). As no strain characterisation methods had been described for any of these programmes, we used the test results obtained in the other identification schemes.

RESULTS AND DISCUSSION

The fruit and vegetable samples from which yeasts were isolated are listed in Table 1. A total of 624 isolates was obtained: 239 were identified as summarised in Table 2. A larger number of ascomycetous species could be isolated from fruits than from vegetables, but more basidiomycetous species were isolated from vegetables than from fruits.

Table 1. Fruit and vegetable samples included in this study

Sample	Number of samples
Apple juice	1
Apricot pouch	1
Blueberry yoghurt	1
Cabbage, fresh or partially processed	4
Cauliflower, fresh or partially processed	3
Carrot, fresh or partially processed	6
Cherry frozen dessert with yoghurt	2
Dried apricots	1
Fruit product processing equipment	28
Grapefruit juice	1
Lettuce, fresh or partially processed	77
Orange juice	5
Peach concentrate	2
Pear	20
Pineapple	6
Pineapple concentrate	15
Pineapple juice	12
Pineapple and orange juice	10
Raspberry puree	1
Red grape concentrate	1
Unknown origin	3

Table 2. Summary of yeast identifications

	Yeasts isolated from		
	Fruits	Vegetables	Both
Identified isolates	127	112	
Ascomycetous	106	77	
Basidiomycetous	21	35	
Number of genera	18	17	
Number of species	37	35	
Number of genera in common			12
Number of species in common			18

Although the food samples were selected randomly, the number of yeast genera and species were similar for both (Table 2). A total of 54 species was identified, from 23 genera. This is an indication of the relatively limited number of yeasts described as associated with food (Deák and Beuchat, 1987).

Identifications of the yeasts from fruits and vegetables are listed in Table 3. *Candida tropicalis* and *Saccharomyces cerevisiae* were most frequently isolated from fruits, and weakly or non fermenting species, *Candida lambica* (the anamorph of *Pichia fermentans*), *Cryptococcus albidus* and *Trichosporon cutaneum*, were predominant among isolates from vegetable origin. Overall, *Candida lambica* and *Saccharomyces cerevisiae* were the most frequently isolated species. This is remarkable, because while *Candida lambica* can hardly be misidentified, due to its overwhelming number of negative characteristics, *Saccharomyces cerevisiae* has a large number of variable reactions in standard discriptions (Barnett *et al.*, 1983; Kreger-van Rij, 1984) and can be very difficult to identify.

We based our comparison of species names, resulting from the various identification schemes, mainly on the work of Kreger-van Rij (1984). "Black yeasts" could not be identified with that system, and we turned to other sources (de Hoog and Hermanides-Nijhof, 1977; Hermanides-Nijhof, 1977). The comparisons of species names are shown in Figure 1 for yeasts of fruit origin, and in Figure 2 for those of vegetable origin. The mean values of correct identifications were calculated for all schemes. If an identification scheme performed within the limits of the calculated standard deviation ($SD_{95\%}$) we have recomended it for identification of food-borne yeasts.

The simplified identification scheme of Deák and Beuchat (1987) provided significantly more correct identifications than any other system in this study when compared with the standard identification scheme of Kreger-van Rij (1984). The major advantage of the simplified scheme is that it usually requires only two Petri dishes, three test tubes and a microscope to achieve identifications, in contrast with the 80-100 morphological and physiological tests demanded by conventional schemes (Barnett *et al.*, 1983; Kreger-van Rij, 1984). Occasionally the simplified scheme needs up to eight additional tests for species identification.

Deák and Beuchat (1987) used the standard descriptions of yeasts provided by Barnett *et al.* (1983). However, Barnett *et al.* (1983) did not specify in detail how data were accumulated. Deák and Beuchat (1987) is difficult to use when variable, delayed or weak test results were encountered, or when information was lacking, e.g. on sexual reproduction. Thus, some important food-borne yeasts either cannot be identified or will be continuously misidentified in certain sub-groups of Deák and Beuchat's scheme. Some specific examples from our experience are listed below:

Table 3. Frequency of isolation and identification of yeast species from fruit and vegetables

Species	Fruit	Vegetables
Arthroascus javanesis	-	1
Aureobasidium pullulans	1	1
Candida ciferrii	-	1
C. colliculosa	5	6
C. diversa	1	-
C. famata	5	3
C. haemulonii	-	2
C. humicola	3	-
C. intermedia	-	1
C. krusei	-	1
C. lambica	8	22
C. magnoliae	3	-
C. parapsilosis	3	1
C. sake	2	6
C. tropicalis	10	3
C. versatilis	1	-
Clavispora lusitaniae	1	-
Cryptococcus albidus	3	10
Cr. curvatus	-	1
Cr. flavus	-	3
Cr. hungaricus	-	1
Cr. laurentii	2	7
Cr. luteolus	-	2
Cr. macerans	1	-
Cr. neoformans	-	1
Debaryomyces castellii	1	-
D. hansenii	6	2
D. marama	-	1
Hanseniaspora guillermondii	1	-
H. valbeyensis	1	1
Hansenula anomala	4	-
Issatchenkia orientalis	1	-
Kloeckera apis	2	-
K. japonica	-	1
Kluyveromyces marxianus var. *lactis*	-	1
Metschnikowia pulcherrima	-	1
M. reukaufii	5	-
Pichia membranaefaciens	1	1
Rhodosporidium infirmo-miniatum	-	2
Rhodotorula glutinis	5	6
R. minuta	3	1
R. rubra	4	1
Saccharomyces cerevisiae	18	7
Stephanoascus ciferrii	-	1
Torulaspora delbrueckii	3	5
Trichosporon adeninovorans	1	-
Tr. cutaneum	4	7
Tr. pullulans	4	-
Trichosporonoides sp.	-	1
Wickerhamiella domercqii	1	-
Yarrowia lipolytica	2	-
Zygosaccharomyces bailii	4	-
Z. rouxii	6	-
Unidentified isolate	1	-
Total	127	112

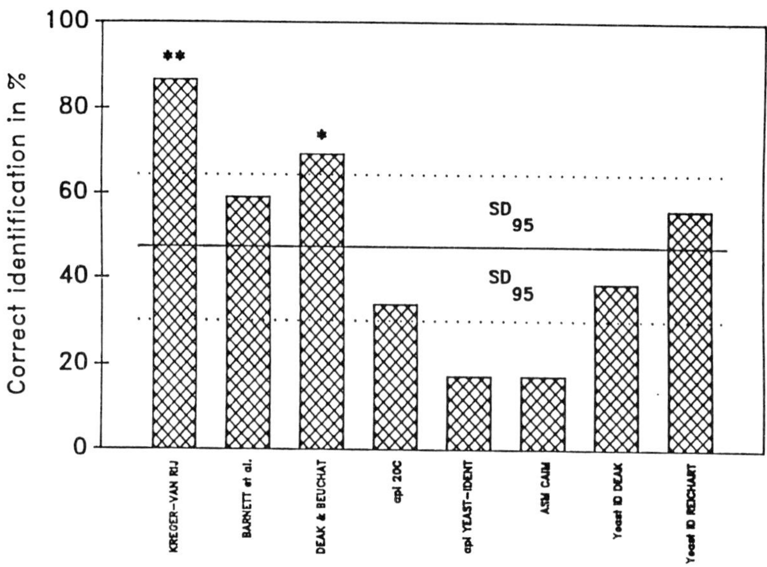

Figure 1. Comparison of identification for yeasts isolated from fruits. (**: significant at p = 0.99; *: significant at p = 0.95; SD_{95}: standard deviation at p = 0.95).

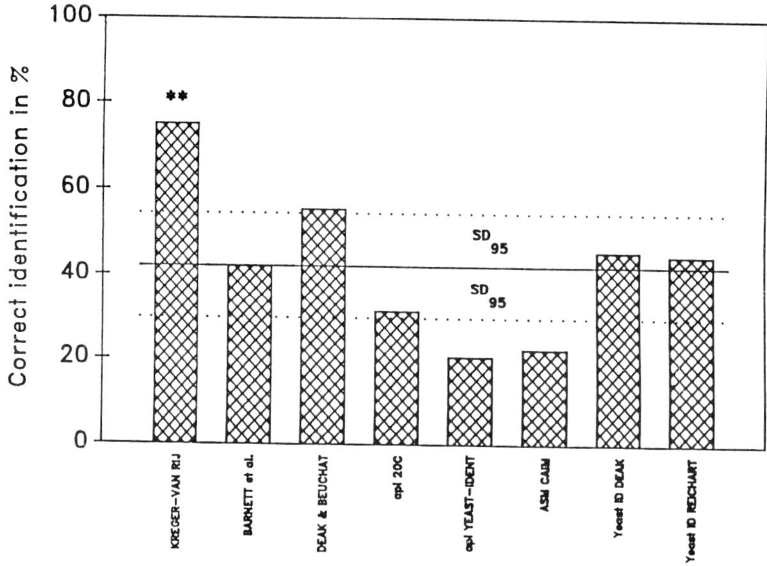

Figure 2. Comparison of identification for yeasts isolated from vegetables. (**: significant at p = 0.99; SD_{95}: standard deviation at p = 0.95).

1. In the standard descriptions (Barnett *et al.*, 1983; Kreger-van Rij, 1984) the following food-borne species vary (±) in certain characters which are important in the master key of Deák and Beuchat (1987):

Candida magnoliae: growth on galactose and cellobiose;
Candida tropicalis: growth on cellobiose;
Candida versatilis: growth on cellobiose;
Cryptococcus albidus: growth on galactose, melibiose, raffinose and erythritol;
Debaryomyces hansenii: fermentation of glucose;
Torulaspora delbrueckii: growth on galactose, maltose, raffinose and xylose.

2. In other cases the standard descriptions of Barnett *et al.* (1983) and Kreger-van Rij (1984) disagree about particular characters such as:

Growth of *Candida castelli* on trehalose;
Growth of *Cryptococcus gastricus* on lactose;
Growth of *Sporobolomyces roseus* on nitrate;
Growth of some *Cryptococcus* species on inositol.

In all these and similar cases yeasts may be misidentified with the simplified identification scheme because Deák and Beuchat (1987) failed to include both positive and negative test results in their subgroups.

The identification scheme by Barnett *et al.* (1983), with the help of the computer assisted key (Barnett *et al.* 1985), can be recommended for food-borne yeasts. The software uses the same set of tests as the manual. The database has been updated for new or amended taxa and now contains information on 497 yeast species. Its main disadvantage lies in the large number of tests required.

The commercial API kits were easy to use, and the biochemical tests proved to be highly reproducible. Nevertheless there were major problems. With the Yeast-IDENT kit, the last 11 tests out of a total of 20 were almost always positive, and so provided no selection power. Also, due to the limited database, API 20C and API Yeast-IDENT may result in the same, but incorrect species identification, e.g. *Candida guilliermondii* instead of *Debaryomyces hansenii*. Thus, API kits cannot be recommended for effective yeast identification of food-borne yeasts, especially because some diagnoses may seem reasonable to the taxonomically inexperienced microbiologist. These limitations arise because the kits were developed to identify medically important yeasts. However, the use of quick and reliable identification kits would be of great advantage to the food industry as well. Attempts to combine API 20C with appropriate microscopic morphological characteristics were only successful for a limited number of genera (Lin and Fung, 1987). Perhaps the simplified identification scheme of Deák and Beuchat (1987) could meet a market demand if it were to be made available commercially.

The "Yeast" program of the American Society for Microbiology Computer Users Group (1989) uses a probabilistic approach for yeast identification. It needs to be updated with approved names of yeast species (Barnett *et al.*, 1983, 1985; Kreger-van Rij, 1984; Moore and Moore, 1989). Also, the continual reloading of the database required during program operation causes some frustration. The program's vulnerability lies in the use of probability assessment for identifying microorganisms.

The "Yeast-ID" software by Deák (1990) was based on the same concepts as Deák and Beuchat (1987). However, some tests have been changed or added (e.g. assimilation of erythritol). Twenty five test results must be entered into the program (Table 4). The program calculates the frequency and the modal, making possible a comparison of the unknown's results with a single species or with all species entered into the database. The user will decide whether to rely more on the frequency, on the modal, or on both. The program is not yet finished; only two levels are available (all known food-borne yeasts and commonly occurring food-borne yeasts). Deák (pers. comm.) intends to extend the database to include all known yeast species. The weakness of the program remains that it adopts literature descriptions of species without considering how those characters were gathered.

Reichart's yeast identification program (Reichart, 1990) uses the same database as Deák's software. The user has the option of setting confidence values for the diagnosis, providing an intelligent challenge for the experienced zymologist. This interactive program offers additional tests from the database for better separation of species.

Table 4. Typical printout comparing an unknown with a known yeast in the computer key by Deák (1990).

Test	*Sacch. kluyveri*[1]	Unknown
Ure	1[2]	-
Ery	1	+
Nit	1	-
Ce	1	+
Mt	50	+
M	80	+
R	80	+
G	99	+
Tre	80	+
Mz	1	-
Me	99	+
L	1	-
Rm	1	+
Xyl	10	+
It	1	-
d	99	+
cyc	1	+
vit	60	+
37	99	+
50	50	-
art	1	-
hy	1	-
psm	1	-
pell	1	-
pink	1	-

[1] *Sacch. kluyeri* was randomly chosen for comparison. Computed values for the unknown compared with *Sacch. kluyveri*: frequency = 0.07902718; modal = 1.145122×10^{-7}.

[2] Percent positive reaction cited in the literature. For computing purposes, 1 was assigned to negative reactions instead of 0.

RECOMMENDATIONS

1. For taxonomic purposes or for definitive "identification" (Kurtzman, 1988), standard yeast identification methods must be used. We prefer the work of Kreger-van Rij (1984).
2. For routine or "presumptive" yeast identification (Kurtzman, 1988) in the food industry, we recommend the simplified identification scheme developed by Deák and Beuchat (1987). Adapting the tests from Deák's newly released software and revising it

using the available standard works will make both the simplified identification key and the computer program more accurate and user friendly.

3. Computer assisted identification keys make identification of yeast species easier. Their use is to be encouraged. However, we must caution against the assumption that high scores always mean correct identifications.

ACKNOWLEDGMENTS

We thank T. Deák and O. Reichart for permission to use their then unpublished software. Material from this paper has been published in *Appl. Environ. Microbiol.* **57**: 1207-1212, 1991.

REFERENCES

AMERICAN SOCIETY FOR MICROBIOLOGY Computer User Group. 1989. Computer-assisted identification of microorganisms (CAIM) - Yeasts. Diskette No. 1106. Washington, D.C.: American Society for Microbiology.

API. 1986. Yeast-IDENT™ directory for the biochemical identification of yeast and yeast-like organisms. Product No. 8886-436016. API, Bio Mérieux SA, 69280 Marcy-L'Etoile, France.

API 20C. 1988. Analytical profile index yeast. 43-103.

BARNETT, J.A., PAYNE, R.W. and YARROW, D. 1983. Yeasts: characteristics and identification. Cambridge: Cambridge University Press.

BARNETT, J.A., PAYNE, R.W. and YARROW, D. 1985. Yeast program. Cambridge: Cambridge Micro Software (Cambridge University Press).

DEAK, T. 1986. A simplified scheme for the identification of yeasts. *In* Methods for the Mycological Examination of Food, eds A.D. King, J.I. Pitt, L.R. Beuchat and J.E.L. Corry, pp. 278-293. New York: Plenum Press.

DEAK, T. 1990. Yeast-ID. Software available upon request. Department of Microbiology, University of Horticulture and Food Industry, Somlui ut 14-16, Budapest H-1118, Hungary.

DEAK, T., and BEUCHAT, L.R. 1987. Identification of foodborne yeasts. *J. Food Prot.* **50**: 243-264.

DEAK, T., and BEUCHAT, L.R. 1988. Evaluation of simplified and commercial systems for identification of foodborne yeasts. *J. Food Microbiol.* **7**: 135-145.

DE HOOG, G.S., and HERMANIDES-NIJHOF, E.J. 1977. Survey of the black yeasts and allied fungi. *Stud. Mycol., Baarn* **15**: 178-223.

HERMANIDES-NIJHOF, E.J. 1977. *Aureobasidium* and allied genera, *Stud. Mycol., Baarn* **15**: 141-177.

KREGER-VAN RIJ, N.J.W., ed. 1984. The Yeasts, a Taxonomic Study, 3rd edn. Amsterdam: Elsevier.

KURTZMAN, C.P. 1988. Identification and taxonomy. *In* Living Resources for Biotechnology: Yeasts, eds B.E. Kirshop, and C.P. Kurtzman, pp. 99-140. Cambridge: Cambridge University Press.

LIN, C.C.S., and FUNG, D.Y.C. 1987. Conventional and rapid methods for yeast identification. *CRC Crit. Rev. Microbiol.* **14**: 273-289.

MOORE, W.E.C. and MOORE, L.V.H. eds. 1989. Index of the Bacterial and Yeast Nomenclatural Changes. Washington, D.C.: American Society for Microbiology.

PINCUS, D.H., SALKIN, I.F., HURD, N.J., LEVY, I.L. and KEMNA, M.A. 1988. Modification of potassium nitrate assimilation test for identification of clinically important yeasts. *J. Clin. Microbiol.* **26**: 366-368.

REICHART, O. 1990. Yeast identification program. Software available upon request.

ROHM, H. and LECHNER, F. 1990. Evaluation and reliability of a simplified method for identification of food-borne yeasts. *Appl. Environ. Microbiol.* **56**: 1290-1295.

EXPERIENCES WITH, AND FURTHER IMPROVEMENT TO, THE DEAK AND BEUCHAT SIMPLIFIED IDENTIFICATION SCHEME FOR FOOD-BORNE YEASTS

T. Deák

Department of Microbiology
University of Horticulture and Food Industry
Somloi ut 14-16, Budapest H-1118
Hungary

SUMMARY

At the first International Workshop on Standardisation of Methods for the Mycological Examination of Foods, a simplified scheme was presented for the identification of 100 common food-borne yeasts. A revised version appeared later. This paper reports further improvements to this simplified identification method. The database has been reduced to 76 species most frequently occurring in foods, and the number of physiological and morphological tests has been limited to less than 20. The revised method is outlined here, and some experiments checking its performance are reported.

INTRODUCTION

At the First International Workshop on Standardisation of Methods for the Mycological Examination of Foods (King *et al.*, 1986) a simplified scheme for identification of foodborne yeasts (SIM) was presented. After discussions at that Workshop, the scheme was subsequently modified to restrict it to the 76 species most frequently occurring in foods, using less than 20 physiological and morphological tests for their identification.

Only two Petri dishes and three test tubes are used to examine each strain for ability to assimilate 10 carbon sources, fermentation of glucose, assimilation of nitrate, and splitting of urea. These biochemical tests are supplemented by morphological observations.

In its present form the SIM consists of the basic set of tests that form the basis of a master key leading to subgroups. The master key is arranged dichotomously, such that each subgroup is split by a single specific property. Eventually 5 tests are used for separating 6 subgroups, each containing on average 13 species. Within each subgroup a dichotomous key leads to species identification using the remaining tests and the characteristics established.

The SIM has been tested by using it to identify more than 210 yeast isolates from a range of foods. The results of tests were compared with those obtained by API 20C kits, and by extended investigation of traditional identification procedures. Assignment of species was evaluated by referring to the standard descriptions.

MATERIALS AND METHODS

The identification procedure and methods in SIM have been described by Deák and Beuchat (1987). Only two Petri dishes and three test tubes are used per strain. The following characters are studied: assimilation of maltose, raffinose, galactose, cellobiose and trehalose (plate 1); mannitol, erythritol, inositol, melibiose and xylose (plate 2); fermentation of glucose, assimilation of nitrate, and splitting of urea (Figure 1). These biochemical tests are supplemented by morphological observations for the occurrence of arthroconidia, true mycelium and pseudomycelium, formation of pellicle, colour of colony, and microscopic appearance of cells.

These tests form the basis of a master key leading to groups. The master key is arranged so that a group is defined by a single specific test, then from the remaining species the next group is defined by another test (Figure 2). Five tests separate six groups, each containing an average of 13 species. Within each group a dichotomous key leads to species based on the results of the others tests and characters (Table 1).

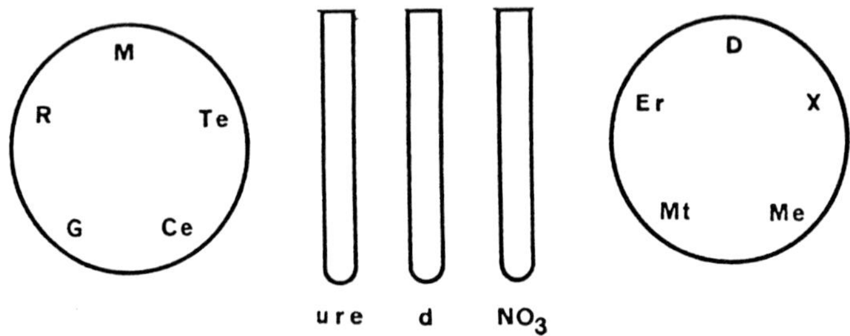

Figure 1. Standard tests for identification

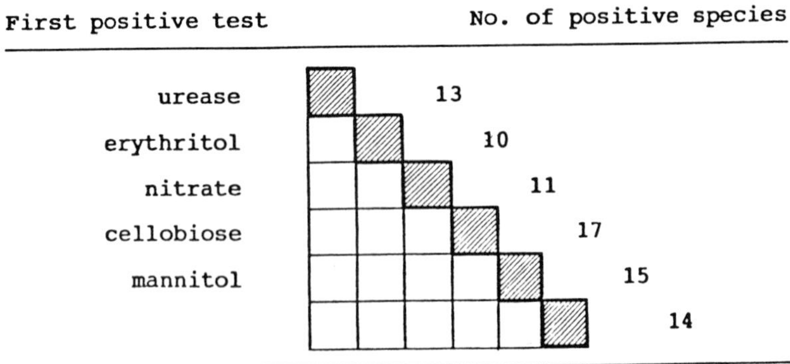

Figure 2. Master key for identification of common foodborne yeasts (76 species)

Table 1. Keys for simplified identification of common foodborne yeast species

Character[1]	Positive	Negative
Master key		
1. Urease	Group 1	2
2. Erythritol	Group 2	3
3. Nitrate	Group 3	4
4. Cellobiose	Group 4	5
5. Mannitol	Group 5	Group 6
Group 1: urease +		
1. Red colony	2	6
2. Nitrate	3	5
3. Inositol	*Rp. infirmominiatum*	4
4. Galactose	*Rh. glutinis*	*Sp. roseus*
5. Raffinose	*Rh. mucilaginosa*	*Rh. minuta*
6. Maltose	7	*Ya. lipolytica*
7. Arthroconidia	8	11
8. True hyphae	9	10
9. Nitrate	*Tr. pullulans*	*Tr. cutaneum*
10. Raffinose	*Sc. pombe*	*Sc. octosporus*
11. Nitrate	*Cr. albidus*	12
12. Pseudohyphae	*Cr. humicolus*	*Cr. laurentii*
Group 2: erythritol +		
1. Nitrate	2	4
2. Maltose	3	*Ca. boidinii*
3. Galactose	*Pi. anomala*	*Pi. subpelliculosa*
4. Maltose	5	9
5. True hyphae	6	7
6. Galactose	*Hy. burtonii*	*End. fibuliger*
7. Raffinose	8	*Ca. diddensiae*
8. Glucose fermentation	*De. polymorphus*	*De. hansenii*
9. Galactose	*Pi. farinosa*	*Ca. cantarelli*
Group 3: nitrate +		
1. Acetate produced	2	3
2. True hyphae	*Dk. anomala*	*Dk. intermedia*
3. Glucose fermentation	4	5
4. Maltose	*Pi. canadensis*	*Wc. domercquiae*
5. Maltose	6	9
6. Galactose	7	8
7. Trehalose	*Ca. versatilis*	*Ca. etchellsii*
8. Cellobiose	*Pi. jadinii*	*Ci. matritensis*
9. Mannitol	10	*Ca. lactis-condensii*
10. Galactose	*Ca. magnoliae*	*Ca. norvegica*

[1] Inclusion of names of the following substrates indicate assimilation of them: cellobiose, erythritol, galactose, inositol, maltose, mannitol, melibiose, nitrate, raffinose, trehalose and xylose; urease means hydrolysis of urea.

Table 1 (continued)

Character[1]	Positive	Negative
Group 4: cellobiose +		
1. Acetate produced	2	3
2. True hyphae	*Dk. anomala*	*Dk. intermedia*
3. Bipolar budding	4	6
4. Large cells	*S'codes ludwigii*	5
5. Maltose	*Hp. osmophila*	*Hp. uvarum*
6. Raffinose	7	14
7. Maltose	8	13
8. Hyphae	9	12
9. Pellicle	10	11
10. Growth at 37°C	*Pi. ohmeri*	*Ca. intermedia*
11. True hyphae	*Za. hellenicus*	*Pi. guillermondii*
12. Glucose fermentation	*Zy. fermentati*	*De. hansenii*
13. Trehalose	*Kl. lactis*	*Kl. marxianus*
14. Growth at 37°C	15	16
15. Pseudohyphae	*Ca. tropicalis*	*Pi. etchellsii*
16. Reddish colony	*Me. pulcherrima*	*Ca. sake*
Group 5: mannitol +		
1. Glucose strongly fermented	2	11
2. Conjugation	3	6
3. Raffinose	*Zy. microellipsoideus*	4
4. Maltose	*Zy. rouxii*	5
5. Large cells	*Zy. bailii*	*Zy. bisporus*
6. Pseudohyphae	7	10
7. Growth at 37°C	8	*Ca. sake*
8. Raffinose	*Kl. marxianus*	9
9. Germ tubes	*Ca. albicans*	*Lo. elongisporus*
10. Trehalose	11	*Ca. apicola*
11. Maltose	*Kl. thermotolerans*	*Tp. delbrueckii*
12. Arthroconidia	*Ge. candidum*	13
13. Growth at 37°C	14	15
14. Maltose	*Ca. catenulata*	*Ca. rugosa*
15. Trehalose	*Ca. zeylanoides*	*Ca. vini*
Group 6: mannitol -		
1. Large cells (>5 um)	2	4
2. Galactose	3	*Sa. bayanus*
3. Melibiose	*Sa. pastorianus*	*Sa. cerevisiae*
4. Conjugation	5	6
5. Raffinose	*Zy. microellipsoideus*	*Zy. bisporus*
6. Raffinose	7	9
7. Trehalose	8	*Ca. stellata*
8. Maltose	*Sa. kluyveri*	*Sa. exiguus*
9. Pseudohyphae	10	13
10. Growth at 37°C	11	*Pi. membranaefaciens*
11. Xylose	*Pi. fermentans*	12
12. Growth without vitamin	*Is. orientalis*	*Is. terricola*
13. Trehalose	*Ca. glabrata*	*Ca. inconspicua*

The performance of SIM has been tested by using it to identify yeasts isolated from various foods, including bran, bread crumbs, meat salad, frozen raspberries, raisins, wine, sauerkraut, pickled cucumbers and kefyr. Enumeration and isolation of yeasts were made on selective media, including DRBC, CTGY, ATGY and others (Deák, 1992). Results from SIM were compared with those obtained by API 20C kits, and traditional identification procedures using standard descriptions (Barnett *et al.*, 1983; Kreger-van Rij, 1984).

The probability of accurate identification, expressed as a percentage, was calculated according to Lapage *et al.* (1973) based on our probabilistic data matrix and computerised identification system (not yet published).

RESULTS AND DISCUSSION

From the samples investigated 30 strains were isolated, from which, 21 species were identified (Table 2). The frequency of occurrence of each species was assessed as the proportion of the total yeast count. Apart from the meat salad, in which a broad spectrum of species was found, samples were populated with only a few yeast species, often dominated by one or two types. The range of counts was also small, from 2.0×10^0 to 4.9×10^2 per

Table 2 Origin and distribution of food isolates

Sample	Yeast count ($\log_{10}$/g)	Species	Frequency of occurrence (%)	Representative isolates
Bran, bread crumbs	0.7-1.1	*Ca. catenulata*	5	A3
		De. hansenii	33	A1, A2
		Pi. anomala	48	D1
		Rh. rubra	14	D2
Frozen raspberry	0.62	*Cr. laurentii*	17	G3
		Me. pulcherrima	66	G1
		Rh. glutinis	17	G2
Raisins	0.30	*Kl. thermotolerans*	66	F2
		Sc. pombe	34	F1
Meat salad	2.55	*Ca. parapsilosis*	11	I1
		Ca. sake	4	I10
		Ca. zeylanoides	8	I7
		De. hansenii	11	I3
		Ge. candidum	16	I2
		Kl. marxianus	8	I11
		Sa. dairensis	23	I5, I8
		Ya. lipolytica	19	I4, I9
Wine	2.69	*Pi. membranaefaciens*	6	K1
		Sa. cerevisiae	94	K2, K3, K4
Sauerkraut	5.3-5.9	*Pi. membranaefaciens*	53	T1
		Sa. exiguus	20	M1
		Tp. delbrueckii	27	T2
Kefyr	4.4-4.8	*Ge. candidum*	10	03
		Is. orientalis	10	01
		Tp. delbrueckii	80	02

Table 3 Species correctly identified by SIM[1]

Strain	Reactions	Group	Identity
D2	ure+ red+ NO3- R+	1	*Rh. rubra*
G2	ure+ red+ NO3+ It- G+	1	*Rh. glutinis*
F1	ure+ red- M+ arthro- R+	1	*Sc. pombe* [2]
G3	ure+ red- M+ arthro- NO3-	1	*Cr. laurentii*
D1	Er+ NO3+ M+ R+	2	*Pi. anomala* [3]
G1	Ce+ R- 37- reddish+	4	*Me. pulcherrima* [4]
I1	Mt+ d+ psh+ 37+	5	*Ca. parapsilosis*
K1, T1	Mt- R- psh+ 37-	6	*Pi. membranaefaciens* [5]
01	Mt- R- psh+ 37+ X-	6	*Is. orientalis* [6]

[1] Abbreviations: assimilation tests Ce, cellobiose; d, glucose fermentation; Er, erythritol; G, galactose; It, inositol; M, maltose; Mt, mannitol; R, raffinose; X, xylose; NO3, nitrate; arthro, arthroconidia; red, reddish colony colour; psh, pseudohyphae; ure, urea hydrolysis; 37, growth at 37°C
[2] growth on M and R very poor, growth in 1% acetate
[3] hat-shaped ascospores
[4] large cells with lipid globules
[5] pellicle, growth in 1% acetate
[6] pellicle, growth without vitamins

gram or mL, except for fermented foods which ranged from 2.5×10^4 to 8×10^5 per gram.

From the test results, the SIM gave straightforward identifications of 9 species (Table 3). Certain microscopic and macroscopic morphological properties were very useful in the identification, e.g. *Rhodotorula* and *Metschnikowia* species developed reddish colonies, *Schizosaccharomyces pombe* showed splitting cells, *Pichia membranaefaciens* and *Issatchenkia orientalis* produced a pellicle in broth culture, and well developed pseudomycelium.

Due to minor variation in test results, five species were identified with 65 to 85 % probability (Table 4). These species all fell into Group 5 because, except for one strain of *Geotrichum candidum*, their first positive master key reaction was the assimilation of mannitol. The isolate of *Kluyveromyces thermotolerans* assimilated trehalose slowly, and test results were rather similar to *Torulaspora delbrueckii*, which sometimes also assimilates trehalose and maltose. However, *K. thermotolerans* grew on cadaverine, on which *T. delbrueckii* is negative. *Candida zeylanoides* assimilated trehalose in two days but galactose only after 10 days, and was therefore negative by SIM where tests are read at 7 days. *Candida sake* fermented glucose rather slowly, but was positive after 7 days. *Candida catenulata* did not assimilate maltose, but differed from *C. rugosa* in assimilating trehalose and fermenting glucose slowly. From their morphological properties the identity of isolates I2 and 03 as *Geotrichum* sp. was unequivocal. The first isolate, however, failed to assimilate mannitol and grew at 37°C.

One species, *Saccharomyces dairensis*, represented by two isolates (I5 and I8) is not included in the SIM database. Application of the key would lead to *Candida inconspicua* in Group 6. This possible misidentification is, however, revealed by the strong fermentation of glucose and the formation of ascospores.

At this point it is worth noting that biochemical test results may be identical for widely different species. Examples are given in a recent publication by Rohm and

Table 4. Species identified with high probability[1]

Strain	d	M	X	psh	pell	37	Identity	Probability of correct ident. (%)	Discrepancy
	Distinguishing properties								
F2	+	+	-	-	-	+	*Kl. thermotolerans*	65	slow Te
I7	(+)	-	-	+	-	(+)	*Ca. zeylanoides*	83	slow G
I10	(+)	+	+	+	+	-	*Ca. sake*	78	slow d
A3	(+)	-	+	+	+	+	*Ca. catenulata*	85	M-
I2	-	-	+	-	+	+	*Ge. candidum*	66	Mt- 37+
O3	-	-	+	-	+	-	*Ge. candidum*	100	-

[1] Abbreviations as in Table 3, plus: pell, pellicle; Te, assimilation of trehalose.

Table 5. Species with uncertain identity or misidentified[1]

Isolate	Species	Uncertain reactions leading to misidentification	Possible misidentification
I4, I9	*Ya. lipolytica*	ure-	*Hy. burtonii* *E'ces fibuliger*
I11	*Kl. marxianus*	Mt-	*Ca. stellata*
I3,A1,A2	*De. hansenii*	Er- R- d+	*Ca. diddensiae* *Ca. sake* *Zy. fermentati*
T2,02	*Tp. delbrueckii*	Mt- lys- 37+	*Sa. kluyveri* *Sa. exiguus*
M1	*Sa. exiguus*	lys+	*Tp. delbrueckii*
K2,K3,K4	*Sa. cerevisiae*	Te- G- R-,+ lys-,+	*Tp. delbrueckii* *Sa. exiguus*

[1] Abbreviations as in Table 3, plus: lys, lysine

Lechner (1990), who pointed out that because of similar biochemical data *Hanseniaspora guilliermondii* could be mistaken for *Saccharomycodes ludwigii*, *Rhodotorula minuta* for *Schizosaccharomyces octosporus*, *Saccharomyces cerevisiae* for *Endomycopsella vini*, etc. However, these species are readily recognisable by their sharply differing morphology. Rohm and Lechner (1990) also evaluated the reliability of the version of SIM which covers a larger number of species (Deák and Beuchat, 1987) by comparing it to the computerised method of Barnett *et al.* (1985). Their criticism, however, is based on a different database, considering a number of variations which were omitted from SIM in order to simplify it.

Certainly, SIM may give unsatisfactory results, especially in the case of species with several variable properties. These are exemplified by four species in this study which gave incorrect results leading to false or uncertain identifications (Table 5).

Isolates of *Yarrowia lipolytica* gave correct reactions to all tests except they failed to hydrolyse urea. It is suspected that this was due to an error in methodology. This possibility is being currently investigated.

Kluyveromyces marxianus did not assimilate mannitol and hence might be misidentified as *C. stellata* in Group 6. Approximately 20% of isolates of *K. marxianus* may be negative in this property, and this variability is not considered in the SIM database. The three isolates of *Debaryomyces hansenii* varied in the assimilation of erythritol and raffinose and the fermentation of glucose. Because in SIM only the variability of erythritol assimilation is considered, raffinose negative and glucose fermenting strains would be misidentified.

A group of isolates identified in Group 6 assimilated only a few compounds and fermented glucose strongly. They mainly resembled *Saccharomyces cerevisiae*. Three isolates, however, were found to grow on lysine (a test not routinely applied in SIM), and were identified as *Saccharomyces exiguus* and *Torulaspora delbrueckii*, although they failed to assimilate mannitol, again an infrequent variation of the latter, not considered in the SIM database.

The results of identification by SIM were compared to those obtained by the API 20C kit, a conventional identification method. The API 20C kit gave correct results in most tests, however, correct identification of species was achieved only in 6 cases: *D. hansenii* (as its anamorph *Ca. famata*), *C. zeylanoides, Cr. laurentii, K. marxianus* (as its anamorph *C. kefyr*), and *S. cerevisiae* (isolates K2 and K3). With another five species, data in the API Profile Index either differed from the standard description (e.g. cellobiose assimilation by *R. rubra* is stated to be negative), or the kit did not perform important identification reactions (e.g. nitrate assimilation, glucose fermentation). Hence the data obtained resulted in misidentifications: *R. rubra* for *R. minuta*; *R. glutinis* for *R. rubra*; *D. hansenii* for *C. guilliermondii*; *S. cerevisiae* for *C. glabrata*; and *G. candidum* for *G. penicillatum*. The rest of the species are not included in the Profile Index and their identification was incorrect or could not be achieved at all. The present database of the API 20C system has been focused on clinically important yeasts, and this limits its applicability for the identification of food-borne yeasts. We have previously reported this (Deák and Beuchat, 1988).

REFERENCES

BARNETT, J.A., PAYNE, R,W, and YARROW, D. 1983. Yeasts: characte-ristics and identification. Cambridge: Cambridge University Press.

BARNETT, J.A., PAYNE, R.W. and YARROW, D. 1985. Yeast identification program. Cambridge: Cambridge University Press.

DEAK T. 1986. A simplified scheme for the identification of yeasts. *In* Methods for the Mycological Examination of Food, eds A.D. King, J.I. Pitt, L.R. Beuchat and J.E.L. Corry, pp.278-293. New York: Plenum Press.

DEAK, T. 1992. Media for enumerating of spoilage yeasts- a collaborative study. *In* Modern Methods in Food Mycology, eds R.A. Samson, A.D. Hocking, J.I. Pitt and A.D. King. pp. 31-38. Amsterdam: Elsevier.

DEAK T. and BEUCHAT, L.R. 1987. Identification of foodborne yeasts. *J. Food Prot.* **50**: 243-264.

DEAK T. and BEUCHAT, L.R. 1988. Evaluation of simplified and commercial systems for identification of foodborne yeasts. *Int. J. Food Microbiol.* **7**: 135-145.

KING, A.D., PITT, J.I., BEUCHAT, L.R. and CORRY, J.E.L., eds. 1986. Methods for the Mycological Examination of Food. New York: Plenum Press.

KREGER-VAN RIJ, N.J.W., ed. 1984. The Yeasts, a Taxonomic Study. 3rd edn. Amsterdam: Elsevier.

LAPAGE, S.P., BASCOMB, S., WILLCOX, W.R. and CURTIS, M.A. 1973. Identification of bacteria by computer: general aspects and perspectives. *J. Gen. Microbiol.* **77**: 273-290.

ROHM, H. and LECHNER, F. 1990. Evaluation and reliability of a simplified method for identification of food-borne yeasts. *Appl. Environ. Microbiol.* **56**: 1290-1295.

SPOILAGE YEASTS OF UNUSUAL SORBATE RESISTANCE

T. Deák, O. Reichart, K. Szakmár and G. Péter

Department of Microbiology
University of Horticulture and Food Industry
Somloi ut 14-16, Budapest, H-1118,
Hungary

SUMMARY

Four strains of yeast were isolated from spoiled citrus drink, pH 2.4, containing 900 mg/L sorbic acid. They were identified as *Zygosaccharomyces bailii*, *Pichia membranaefaciens*, *Candida parapsilosis*, and *Trichosporon pullulans*. They survived in the presence of 1000 mg/L sorbic acid (pH 3.0) for 30 min. *T. pullulans* did not grow in the presence of 4 g/L sorbic acid, and was considered an incidental contaminant. However, *Z. bailii* grew in the presence of 6 g/L sorbic acid, and *C. parapsilosis* at 10g/L, while *P. membranaefaciens* grew in 20 g/L sorbic acid (pH 4.0). Decrease in sorbic acid concentration was observed only at the lowest concentration (5g/L) in aerobic cultures of *P. membranaefaciens*.

INTRODUCTION

It is well known that sorbic acid (SA) possesses a specific fungistatic effect (Liewen and Marth, 1985). As its efficiency is largely dependent on the presence of undissociated acid, its inhibitory effect on growth is pH dependent. At pH 4 the growth of most yeasts is inhibited by 100 to 200 mg/L SA (Warth, 1985). Certain yeasts, such as *Issatchenkia orientalis* (anamorph *Candida krusei)* and *Zygosaccharomyces bailii* are resistant to SA and capable of growing in the presence of 800 mg/L (Pitt and Richardson, 1973; Warth, 1986).

Resistance to SA has been attributed to the induction of an active transport system that pumps undissociated acid molecules out of cells (Warth, 1977). During aerobic growth some yeasts are also able to decompose sorbic acid (Deák *et al.*, 1970). SA can be assimilated by some yeasts, such as *Candida albicans, C. pseudotropicalis* and *C. utilis* (Deák and Novák, 1972). By these or other means resistance is built up after repeated incubation in the presence of SA (Bills *et al.*, 1982).

Some yeasts recently isolated in our laboratory have shown unusual sorbate resistance. They are described in this paper.

MATERIALS AND METHODS

Isolation and identification of yeasts

Yeasts were isolated from spoiled citrus drink (pH 2.44) preserved with 900 mg/L SA, and from processing line samples in the manufacturing plant. The isolation medium was Tryptone Glucose Yeast extract (TGY) medium supplemented with 100 mg/L chloramphenicol. Yeasts were identified by conventional fermentation and assimilation tests according to the simplified method of Deák and Beuchat (1987). Additional tests, growth in 50% glucose, in vitamin free medium, in the presence of 0.01% cycloheximide and 1% acetic acid were also carried out.

Growth and survival tests

Growth and survival tests were performed in TGY broth (pH adjusted to 3.0 and 4.0) with 1 to 20 g/L of SA added. Growth was measured spectrophotometrically by absorbance at 550 nm. Cell suspensions were inoculated into TGY broth, pH 3, containing 1000 mg/L SA, and samples were taken after 10 and 30 min to determine the number of survivors. Number of survivors were determined by plate counts.

Sorbic acid concentration

Concentration of sorbic acid was determined spectrophotometrically at 254 nm. The usual steam distillation extraction was omitted as Deák *et al.* (1970) showed that direct samples gave identical results .

RESULTS AND DISCUSSION

Yeasts

Four preservative resistant yeast species were identified, *Zygosaccharomyces bailii*, *Candida valida*, *C. parapsilosis* and *Trichosporon pullulans*. Distinguishing characteristics of representative isolates are shown in Table 1. Identification of strain 641 as

Table 1. Characteristics distinguishing yeast isolates

Test	*Z. bailii* 641	*C. valida* 649	*C. parapsilosis* 643	*T. pullulans* 650
Glucose fermentation	+	+	+	-
Assimilation:				
sorbose	-	+	+	+
xylose	-	+	+	+
maltose	-	-	+	+
trehalose	-	-	+	-
melezitose	-	-	+	+
mannitol	+	-	+	-
inositol	-	-	-	+
nitrate	-	-	-	+
Hydrolysis of urea	-	-	-	+
Growth in				
50% glucose	+	-	-	+
without vitamins	-	+	-	+
1% acetic acid	+	+	-	-
Formation of				
true hyphae	-	-	-	+
pseudohyphae	-	+	+	+
conjugating cells	+	-	-	-
pellicle	-	+	-	+

Zygosaccharomyces bailii was clear-cut; microscopically cells were observed showing conjugation and producing ascospores. This species occurred in all spoiled samples of citrus drink investigated.

Strain 649 gave very few positive test results. As ascospores were not observed, it was identified as *Candida valida*, the anamorph of *Pichia membranaefaciens*. Its ability to grow in 1 % acetic acid was confirmatory. It was isolated from a line sample of diluted citrus concentrate. Strain 643 fitted the description of *Candida parapsilosis,* except for the assimilation of starch and rhamnose. The single isolate came from sugar syrup.

Strain 650 produced arthroconidia, pseudomycelium and true hyphae, and hydrolised urea. It was identified as *Trichosporon pullulans* although it failed to assimilate erythritol.

Inhibition by sorbic acid

As the citrus drink was preserved by 900 mg/L SA, the fungistatic or fungicidal effect of SA was tested in TGY broth plus 1000 mg/L SA. As shown in Table 2, no decrease in cell counts was observed.

Table 2 Survival of yeasts in the presence of sorbic acid[1]

	Yeast count/mL			
Time (min)	*Z. bailii* 641	*C. valida* 649	*C. parapsilosis* 649	*Tr. pullulans* 650
0	7.6×10^5	1.5×10^6	1.3×10^7	4.7×10^5
10	8.0×10^5	1.5×10^6	1.0×10^7	4.5×10^5
30	7.3×10^5	1.7×10^6	1.2×10^7	4.4×10^5

[1] TGY broth (pH 3.0) with 1 g/L sorbic acid, incubation at 23°C. Data are means of three determinations.

The fungistatic effect of SA was therefore studied in higher concentrations up to 20 g/L. As shown in Table 3, *T. pullulans* did not grow in the presence of 4 g/L SA at pH 4, and it was considered as an incidental contaminant in citrus drink. *Z. bailii* isolated from spoiled bottled drink grew in up to 6 g/L SA at pH 4. Strains of *C. parapsilosis* and *C. valida* showed even higher sorbate resistance, and developed in the presence of 10 g/L and 20 g/L SA at pH 4, respectively (Table 3).

Table 3. Growth inhibition by sorbic acid[1]

	Growth in the presence of SA (g/L)				
Yeast	4	6	8	10	20
Z. bailii 641	+	+	-	-	-
C. valida 649	++	++	++	+	(+)
C. parapsilosis 643	++	++	+	(+)	-
T. pullulans 650	-	-	-	-	-

[1] ++, strong; +, good; (+), poor; -, no growth after 4 days incubation at 28°C in TGY broth, pH 4.0.

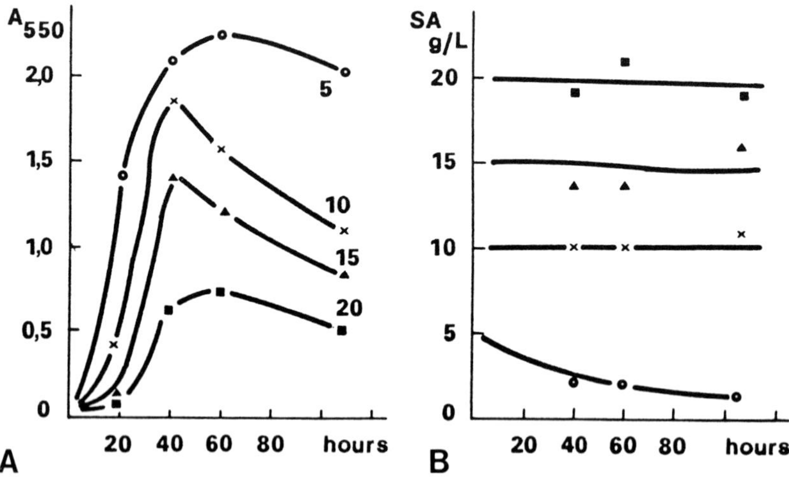

Figure 1. (A) Effect of sorbic acid on the growth of *Pichia membranefaciens*; (B) change in sorbic acid concentration during growth of *P. membranaefaciens*

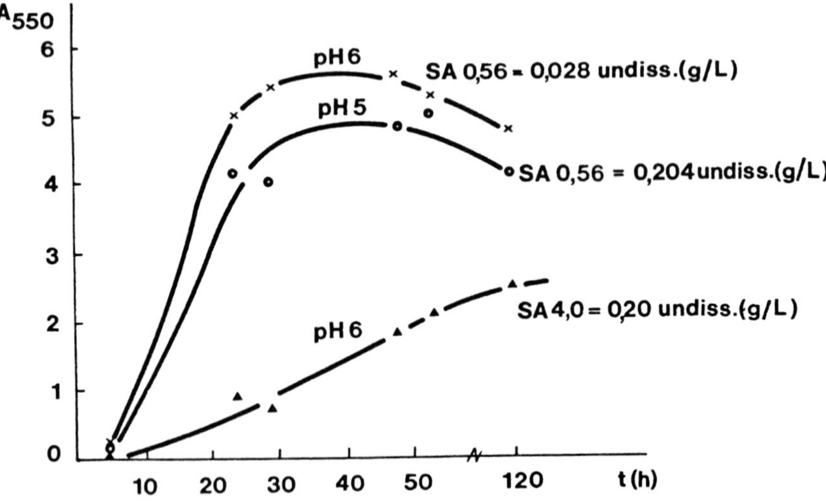

Figure 2. pH dependence of inhibition by sorbic acid on growth of *Pichia membranaefaciens*

Growth in the presence of sorbic acid

Increasing SA concentrations extended the lag phase of *C. valida,* and shortened the length of the exponential growth phase, but did not decrease the exponential growth rate (Figure la). Decrease of SA concentration was observed only at the lowest concentration (5 g/L) in aerobic cultures of *C. valida* (Figure 1b). It appears that this yeast possesses unusually high tolerance of SA. The ablity to grow in up to 20 g/L SA makes it a real risk to the stability of soft drinks preserved by SA.

Although it is generally believed that the antimicrobial effect of weak acid preservatives depends on the undissociated molecule (Liewen and Marth 1985), it has also been shown that the anionic species may contribute to some extent (Eklund 1983). This may well be the case in the high concentrations of SA used in the present experiments. When the degree of growth inhibition of *C. valida* was studied at different pH levels (Figure 2), it was found that approximately the same concentration of undissociated SA together with a high ratio of sorbate anion was more inhibitory than when SA was present mainly in the undissociated form.

REFERENCES

BILLS, S., RESTAINO, L. and LENOVICH, L.M. 1982. Growth response of an osmotolerant sorbate-resistant yeast, *Saccharomyces rouxii*, at different sucrose and sorbate levels. *J. Food Prot.* **45**: 1120-1124.

DEAK, T. and BEUCHAT, L.R. 1987. Identification of food-borne yeasts. *J. Food Prot.* **50**: 243-264.

DEAK, T. and NOVAK, E.K. 1972. Assimilation of sorbic acid by yeasts. *Acta Aliment.* **1**: 87-104.

DEAK, T., TÜSKE, M. and NOVAK, E.K. 1970. Effect of sorbic acid on the growth of some species of yeasts. *Acta Microbiol. Acad. Sci. Hung.* **17**: 237-256.

EKLUND, T. 1983. The microbiological effect of dissociated and undissociated sorbic acid at different pH levels. *J. Appl. Microbiol.* **54**: 383-389.

LIEWEN, M.B. and MARTH, E.H. 1985. Growth and inhibition of microorganisms in the presence of sorbic acid: a review. *J. Food Prot.* **48**: 364-375.

PITT, J.I. and RICHARDSON, K.C. 1973. Spoilage by preservative-resistant yeasts. *CSIRO Food Res. Q.* **33**: 80-85.

WARTH, A.D. 1977. Mechanism of resistance of *Saccharomyces bailii* to benzoic, sorbic and other weak acid used as food preservatives. *J. Appl. Bacteriol.* **43**: 215-230.

WARTH, A.D. 1985. Resistance of yeast species to benzoic and sorbic acids and to sulfur dioxide. *J. Food Prot.* **48**: 564-569.

WARTH, A.D. 1986. Preservative resistance of *Zygosaccharomyces bailii* and other yeasts. *CSIRO Food Res. Q.* **46**: 1-8.

COMPARATIVE STUDY OF WL NUTRIENT AGAR WITH DRBC AND OGY FOR YEAST ENUMERATION IN FOODS

S. Andrews

School of Chemical Technology
South Australian Institute of Technology
Ingle Farm, S.A. 5098, Australia

SUMMARY

WL Nutrient agar was originally developed for detection of yeasts, moulds and bacteria in brewery products. Today it is widely used in the wine and fruit juice industries also. WL agar permits the development of yeasts from any of these acidic products, but more importantly, it also supports the growth of acid tolerant bacteria such as *Gluconobacter*, *Acetobacter*, *Zymomonas*, *Lactobacillus* and *Pediococcus*. This study compared WL Nutrient agar with DRBC and OGY for enumeration and identification of yeasts in fruit juice, beer and wine. Where products contained a mixture of yeasts, these could usually be differentiated on the basis of colony morphology on all three media, although this was often more distinctive on WL. These results indicate that the brewing, wine and fruit juice industries could readily adopt DRBC or OGY for enumeration of yeasts in their products.

INTRODUCTION

Wallerstein Laboratories' Nutrient Agar (WL) was originally developed for the growth of yeasts and bacteria from brewery products (Green and Gray, 1950). This medium (Table 1) is available commercially in a dehydrated form from Oxoid Ltd (Basingstoke, UK) and Difco (Detroit, USA).

WL was developed to detect the presence of bacteria in brewery products. Many beer spoilage bacteria grew indifferently on wort agar, so WL was developed and shown to be more satisfactory for their detection. Actidione was added to WL to suppress the large number of yeasts present. Aerobic or anaerobic incubation was used to detect the different types of bacteria.

Table 1. The composition of Wallerstein Laboratories' Nutrient (WL) agar (Green and Gray, 1950)

Glucose	50.0 grams	$MgSO_4.7H_2O$	0.125
Casein hydrolysate	5.0	$FeCl_3.6H_2O$	0.0025
Yeast extract	4.0	$MnSO_4.4H_2O$	0.0025
KH_2PO_4	0.55	Bromcresol Green	0.022
KCl	0.425	Agar	20.0
$CaCl_2.2H_2O$	0.125	Distilled water to 1L	

Adjust pH to 5.5 with HCl

At the same time, WL was shown to be equally as effective as wort agar for enumerating yeasts, and that wild yeasts could be differentiated from brewery yeasts by careful examination of colony morphology. In many cases, the brewery yeasts produced dark green colonies, in contrast to pale green, bluish or white colonies produced by wild yeasts (Hall, 1971). This was considered advantageous because Lysine Medium (Walters and Thiselton, 1953) used to detect wild yeasts fails to detect *Saccharomyces* strains unable to use lysine as sole nitrogen source.

The use of WL for the detection of yeasts and bacteria from other acidic beverages and foods was demonstrated by Green and Gray (1951). Widespread use in the wine and fruit juice industries followed. In Australia over 80% of laboratories involved in quality control of beer, wine and fruit juice still use WL for the detection of contaminants (Andrews 1992). WL continues to be used because it effectively recovers yeasts and spoilage bacteria, including lactic acid and acetic acid bacteria.

As Oxytetracycline Yeast extract Glucose agar (OGY) and Dichloran Rose Bengal Chloramphenicol agar (DRBC) are now widely used for yeast and mould counts, a comparative evaluation with WL for the enumeration of yeasts from acidic beverages seemed pertinent.

Products tested

A selection of acidic beverages was tested by spreading 0.1 mL aliquots in duplicate onto WL, OGY and DRBC. As most of the commercially packaged samples tested were sterile or nearly so, unfinished or alternative products were used. Hence the "wine" samples were samples of grape juice undergoing spontaneous natural fermentations, the beer samples had completed maturation but were yet to be fined and pasteurised, and the fruit juices were commercial products that had been incubated at 25°C for 48 hours to enhance the level of contamination.

RESULTS

Analysis of wine samples

Counts on 6 wine samples were high enough to be included in the comparative study (Table 2). As counts on individual samples varied widely, counts on the three media were ranked on a scale of 1 (lowest count) to 3 (highest count). The highest cumulative

Table 2. Yeast counts (per mL) detected in wine samples cultured on three media

Wine sample	DRBC	OGY	WL
1	4.8×10^6	3.6×10^6	3.2×10^6
2	1.2×10^6	2.0×10^6	1.5×10^6
3	5.1×10^4	4.8×10^4	5.9×10^4
4	1.4×10^4	3.3×10^4	2.7×10^4
5	1.7×10^5	1.6×10^5	2.6×10^5
6	3.0×10^4	5.9×10^4	6.5×10^4
Cumulative ranking score	10	12	14

Ranking Score indicates the medium with the recovery of the highest yeast counts overall. Using this system it is obvious that no one medium is superior or inferior to the others. Microscopic examination showed that the higher counts observed on WL for some samples were not due to bacterial contaminants. Some of the yeasts isolated from these samples were purified and identified as *Saccharomyces cerevisiae, Hanseniaspora guilliermondii, H. uvarum,* and species of *Candida, Rhodotorula* and *Aureobasidium.*

Analysis of beer samples

Counts for 7 beer samples (Table 3) also are similar on the three media. OGY and WL rate similarly, with DRBC lower. Most yeasts cultured from the samples were brewery yeasts, *Saccharomyces cerevisiae,* that had failed to settle out of the beer during storage.

Table 3. Yeast counts (per mL) detected in beer samples cultured on three media

Beer sample	DRBC	OGY	WL
1	2.0×10^3	3.9×10^3	4.1×10^3
2	2.3×10^3	2.4×10^3	1.5×10^3
3	9.0×10^2	3.0×10^3	2.8×10^3
4	1.1×10^3	2.3×10^3	3.0×10^3
5	9.0×10^3	2.3×10^3	7.0×10^3
6	7.3×10^3	1.0×10^4	1.4×10^4
7	3.7×10^3	3.9×10^3	3.4×10^3
Cumulative ranking score	11	16	15

Analysis of fruit juice samples

Initially, fruit juices were enumerated on all three media, but some WL plates could not be counted due to the high level of bacterial contamination. Hence all samples were retested after chloramphenicol, 200 mg/L, was added to WL.

As shown in Table 4, similar counts were observed on the three media, WL and OGY outranked DRBC. Most juice samples contained a mixture of yeasts including species of *Candida, Hansenula, Saccharomyces, Rhodotorula* and others still to be identified.

Table 4. Yeast counts (per mL) detected in fruit juices cultured on three media

Fruit juice sample	DRBC	OGY	WL
1	5.8×10^3	7.5×10^3	6.1×10^3
2	8.5×10^3	8.7×10^3	8.7×10^3
3	1.6×10^3	3.3×10^3	4.9×10^3
4	6.0×10^3	6.0×10^3	4.8×10^3
5	2.8×10^3	4.1×10^3	4.4×10^3
6	1.0×10^3	0.6×10^3	1.9×10^3
7	3.9×10^3	overgrown *Rhizopus*	overgrown *Rhizopus*
Cumulative ranking score	8	12	14

One sample of juice contained a few spores of *Rhizopus stolonifer,* which grew vigorously over the surface of OGY and WL plates, which could not be counted. However, yeasts could still be readily enumerated on the DRBC plates. This highlights the use of DRBC when yeast enumeration is required in the presence of spreading fungi which grow rapidly on OGY and WL.

DISCUSSION

This study has shown that yeasts were equally recovered on OGY and WL. On WL, the number of different yeast species present was readily detected by the different colours and sizes of the colonies. For OGY, this was less obvious, and size, shape and outline of the colonies were required to quantify species numbers.

The results of this limited study show that OGY could be successfully substituted for WL for enumerating contaminant yeasts in beer, wine and fruit juices. However, WL is also used for enumerating spoilage bacteria, so before recommending substitution of OGY for WL in these industries, evaluation of the glucose yeast extract base in OGY in necessary. In our experience, for this purpose OGY without oxytetracycline is effective for lactic acid bacteria, provided the plates are incubated in a CO_2 incubator or anaerobic jar, and acetic acid bacteria. However, growth of some fastidious *Lactobacillus* and *Leuconostoc* species still needs to be evaluated.

The results with DRBC suggest it may be slightly inhibitory to some yeasts. This has been reported with other foodstuffs (Dijkmann, 1986; Williams, 1986; S. Andrews, unpublished) as well as pure yeast cultures (Williams, 1986). This limits its usefulness for sterility testing or testing of products with low levels of contamination.

For the purposes described here, WL has the ability to recover yeasts with a similar efficiency to OGY, and marginally more effectively than DRBC.

RECOMMENDATIONS

It is recommended that extensive work should be performed on the relative recovery of bacteria and yeasts from acidic beverages on OGY and WL. Conditional on satisfactory results, the use of OGY for the detection of yeasts and bacteria in wine, beer and fruit juices can be recommended. Where both yeast and bacterial contaminants are to be detected, oxytetracycline should be omitted from the medium, but where yeasts need to be detected in the presence of high numbers of bacteria, the antibiotic should be incorporated. The addition of actidione to the GY base is recommended for the detection of bacteria in the presence of high numbers of yeasts.

For the enumeration of yeasts in products contaminated with spreading fungi, DRBC is recommended, as the addition of Rose Bengal to OGY has been found to be unsatisfactory in our experience.

REFERENCES

ANDREWS, S. 1992. Specifications for yeasts in Australian beer, wine and fruit juice products. *In* Modern Methods in Food Mycology, eds R.A. Samson, A.D. Hocking, J.I. Pitt and A.D. King. pp. 111-118. Amsterdam: Elsevier.

DIJKMANN, K.E. 1986. Merits and shortcomings of DG18, OGY, OGGY and DRBC media for the examination of raw meats. *In* Methods for the Mycological Examination of Food, eds A.D. King, J.I. Pitt, L.R. Beuchat and J.E.L. Corry, pp. 120-123. New York: Plenum Press.

GREEN, S.R. and GREY, P.P. 1950. A differential procedure applicable to bacteriological investigation in brewing. *Wallerstein Commun.*, **13:** 357-366.

GREEN, S.R. and GREY, P.P. 1951. A differential procedure for bacteriological studies useful in fermentation industries. *Proc. Am. Soc. Brew. Chemists* (1950), 19-32.

HALL, J.F. 1971. Detection of wild yeasts in the brewery. *J. Inst. Brewing* **77:** 513-516.

WALTERS L.S. and THISELTON, M.R. 1953. Utilisation of lysine by yeasts. *J. Inst. Brewing* **59**: 401-404.

WILLIAMS, A.P. 1986. A comparison of DRBC, RBC and MEA media for the enumeration of molds and yeasts in pure culture and in foods. *In* Methods for the Mycological Examination of Food, eds A.D. King, J.I. Pitt, L.R. Beuchat and J.E.L. Corry, pp. 85-89. New York: Plenum Press.

COLLABORATIVE STUDY TO COMPARE FUNGAL COUNTS IN FLOUR SAMPLES

D. A. L. Seiler

Flour Milling and Baking Research Association
Chorleywood, Herts., WD3 5SH, U.K.

In collaboration with: C.B. Anderson, S. Andrews, H.J. Beckers, L.R. Beuchat, L.B. Bullerman, J.E.L. Corry, M.A. Cousin, T. Deák, D. Eke, R.B. Ferguson, J.C. Frisvad, B.J. Hartog, A.D. Hocking, M. Howell, A.D. King, J.A. Koberger, L.M. Lenovich, D.A.A. Mossel, K.E. Olson, D. Richard-Molard, N. Shapton, D.F. Splittstoesser and A.P. Williams.

SUMMARY

Variations in fungal counts on flour reported at the workshop on Methods for the Mycological Examination of Foods in Boston in July 1984 led to the collaborative study reported here. Twenty four laboratories were each sent two samples of flour and asked to carry out counts of moulds and yeasts in duplicate subsamples using the method normally adopted by that laboratory. Participants were also asked to test a sample of locally acquired domestic white flour. A wide variety of media and methods was used, but despite this, mould counts in the reference samples were reasonably uniform (coefficient of variation 46-50%). However, yeast counts varied widely. The results suggest that analysts had difficulty in separating yeasts from bacteria. Some media were more effective at suppressing bacteria than others. The spread plate method appears to give better recovery of yeasts. Low counts of moulds and yeasts (< 1000 per g) were obtained from the majority of the domestic flours.

INTRODUCTION

At the workshop on Methods for the Mycological Examination of Foods held in Boston in July 1984 (King *et al.*, 1986), a number of papers were presented which showed widely different counts for fungi in flour. There was doubt whether the results were a true reflection of the range of counts which can be expected in this commodity or whether the variations were due, at least in part, to the different methods of enumeration employed.

In an attempt to clarify the situation, a collaborative study was organised involving the majority of participants at the current workshop. Each participant was sent samples of the same two flours and asked to carry out counts of moulds and yeasts on duplicate subsamples using the method normally adopted in his or her laboratory. In addition, participants were also asked to test a sample of domestic white flour obtained from a local source. Twenty four laboratories took part in the exercise.

METHODOLOGY

Details of the media and methods used by the different laboratories are given in Table 1. Comments on these methods follow.

More laboratories used the spread plate method (13) than the pour plate method (9), while two laboratories compared both. Usually 0.1 mL samples were spread plated, but a few American laboratories distributed 1 mL of the 10^{-1} dilution of sample over the surface of two or three plates. The reason for this procedure is not clear.

Table 1. Media and methods used by collaborating laboratories

Laboratory[1] and region		Plating method	Diluent[2]	Primary dilution	Mixing[3]	Medium[4]	Incubation[5]
1	EUR	Spread	Saline peptone	40/310	St. 2.0	OGY	5/27
2	EUR	Spread	Peptone Ringers	10/90	St. 2.0	RBC, DRBC, & DG18	5/25
3	EUR	Spread	Peptone Ringers	1/10	St. 1.0	OMEA	5/25
4	EUR	Pour	Ringers	10/90	St. 1.0	RBC	4/25
5	EUR	Pour	Saline peptone	20/180	Shake 30	DG18, MSSA, & Malt	5-7/25
6	EUR	Spread	Saline peptone	40/360	St. 2.0	DRBC	5/25
7	EUR	Pour	Saline peptone	20/180	St. 0.5	OGY	5/24
8	EUR	Pour	Saline peptone	20/180	St. 1.0	OGY	5/24
9	EUR	Spread	Buffered peptone	40/360	Soak St. 2.0	OGGA	5/24
10	EUR	Spread	Saline peptone	8/72	St. 2.0	DRBC	5/25
11	EUR	Spread	0.05% Tween	Various	St. 3.0	MEA	5/25
12	EUR	Spread	Peptone	10/90	Shake?	OGY, DRBC, GYC & GYA	7/RT
13	AUS	Spread	Peptone	11/99	St. 2.0	DRBC	5/25
14	AUS	Spread	Peptone	20/180	St. 0.5	DRBC	5/25
15	USA	Pour	Peptone	20/180	St. 1.0	APDA	4/25
16	USA	Pour	Distilled water	11/99	Shake 2.0	PDA	5/27
17	USA	Pour	Peptone	50/450	Shake?	PDACC, APDA	5/25
18	USA	Pour	Buffer	25/225	Blender 2.0	SMACC	6/25
19	USA	Spread	Peptone	50/450	St. 2.0	DRBC	5/25
20	USA	Pour	Buffer	25/225	Blender 2.0	PDAC	5/25
21	USA	Pour & spread	Buffer	22/198	St. 0.5	MycA, OGY	5/25
22	USA	Pour & spread	Peptone	30/270	St. 2.0	PDACC	5/25
23	USA	Spread	Buffer	11/99	St. 3.0	DRBC	5/25
24	USA	Spread	Buffer Tween	40/360	St. 2.0	OGY, RBC, DRBC, APDA, & PCACC	5/25

[1] EUR, Europe; AUS, Australia.
[2] Peptone, usually 0.1%; Saline peptone, usually 0.85-0.9% NaCl and 0.1% peptone; Buffer, usually Butterfields; Peptone Ringers, usually 0.1% peptone and Ringers solution.
[3] St., Stomacher; figure, time in min.
[4] OGY, oxytetracycline glucose yeast extract agar; RBC, rose bengal chloramphenicol agar; DRBC, dichloran rose bengal chloramphenicol agar; DG18, dichloran 18% glycerol agar; OMEA, oxytetracycline malt extract agar; MSSA, malt salt sucrose agar; Malt, 2% malt agar; OGGA, oxytetracycline gentamycin glucose yeast extract agar; GYC, glucose yeast extract chloramphenicol agar; GYA, acidifed glucose yeast extract agar; APDA, acidified potato dextrose agar, pH 3.5; PDA, potato dextrose agar; PDACC, PDA plus chlortetracycline and chloramphenicol; PDAC, PDA with chlortetracycline; SMACC, standard methods agar (BBL) with chlortetracycline and chloramphenicol; MycA, mycological agar (Difco), pH 4.0; PCACC, plate count agar with chlortetracycline and chloramphenicol.
[5] Time in days/temperature in °C; RT, room temperature

Eight different diluents were used, of which 0.1% peptone and saline/peptone (0.85-0.9% NaCl, 0.1% peptone) were most common. The ratio of sample to diluent varied greatly, eleven different ratios being used.

A large majority of laboratories (18) used the Colworth Stomacher for preparing the primary dilution. Mixing times varied from 0.5 to 3.0 min, with 2 min being used most often. Two laboratories used a Waring blender, three shook samples by hand and one used a mechanical mixer for 30 min. Only one laboratory reported soaking the sample in diluent prior to mixing.

No less than 18 different media were employed, of which DRBC and OGY were most popular. Six laboratories compared counts on two or more media.

In general, plates were counted after 5 days at 24-27°C (18 laboratories), but two laboratories counted after four days.

RESULTS AND DISCUSSION

Counts in reference samples

The average counts obtained by the different laboratories for moulds and yeasts in the duplicate subsamples of the two reference flours are given in Table 2. Results from comparison of different media and methods are included.

With sample A the mean mould count for all laboratories was 1.13×10^3/g with a standard deviation of 520 and coefficient of variation of 46%. Sample B had a higher mean count of 2.13×10^3/g with a standard deviation of 1.07×10^3 and coefficient of variation of 50%. While a variability of 46-50% is not particularly satisfactory, it is probably no more than to be expected when so many different media and methods were employed.

The low mould counts obtained by laboratories 4 and 15 may possibly be explained by the fact that the plates were examined after only 4 days incubation. Comparisons made by laboratories 17, 21 and 24 and the results from laboratory 15 suggest that reduced mould counts will be obtained on acidified media using the pour plate method.

Yeast counts, especially for sample B, varied considerably. After removal of obvious outliers, and ignoring "<" signs, the mean for 37 counts on sample A was 155/g and sample B 236/g. Some laboratories found hundreds and even thousands per g to be present whereas others detected very few if any. It is clear that many workers have difficulty in distinguishing these microorganisms. Three laboratories commented that bacteria were present on DRBC plates. Confusion between yeasts and bacteria may explain some of the higher counts obtained. Contamination with bacteria could also explain the high yeast counts quoted by laboratories 11 and 16 who used media without acidification or antibiotic supplement. Yeasts counts were generally higher using a spread plate than a pour plate method.

Counts in domestic flours

The counts of fungi in the samples of domestic flour tested by the different laboratories are listed in Table 3. With few exceptions, the laboratories obtained lower counts in the domestic flour than in the reference flours. The average mould count for all samples was 520/g. In general, counts were higher in samples examined by European than by American or Australian laboratories (average 850/g and 410/g respectively). Yeast counts were very variable but mostly they were low, under 100/g.

Table 2. Counts/g in reference flour samples

			Sample A		Sample B	
Lab.	Method	Medium[1]	Moulds	Yeasts	Moulds	Yeasts
1	Spread	OGY	860	240	2600	500
2	Spread	RBC	6700	100	3000	400
	"	DRBC	1000	<50	1100	50
	"	DG18	1800	<50	3000	230
3	Spread	OMEA	1500	25	2800	100
4	Pour	RBC	210	320	320	180
5	Pour	DG18	370	<20	1700	20
	"	MSSA	360	<20	2300	20
	"	Malt	430	<20	2500	20
6	Spread	DRBC	1500	280	1700	800
7	Pour	OGY	580	30	2200	60
8	Pour	OGY	1700	20	4900	35
9	Spread	OGGA	1300	250	1500	700
10	Spread	DRBC	1200	230	2900	50
11	Spread	MEA	6200	94,000	5000	240,000
12	Spread	OGY	1200	100	1200	100
	"	DRBC	700	100	750	100
	"	GYC	100	100	4600	100
	"	GYA	1400	100	750	100
13	Spread	DRBC	1400	280	2400	980
14	Spread	DRBC	2300	480	3400	300
15	Pour	APDA	470	190	590	500
16	Pour	PDA	1900	1200	3300	2300
17	Pour	PADCC	690	<10	1100	10
	"	APDA	450	<10	320	10
18	Pour	SMACC	1800	<100	3200	650
19	Spread	DRBC	1500	130	2200	500
20	Pour	PDAC	1700	100	4400	100
21	Spread	MycA	1300	120	2500	90
	Pour	MycA	750	10	1400	10
	Spread	OGY	1300	180	2400	150
	Pour	OGY	950	10	1500	10
22	Spread	PDACC	1100	200	1400	360
	Pour	PDACC	960	30	1100	70
23	Spread	DRBC	480	70	580	85
24	Spread	OGY	1300	400	2800	430
	"	RBC	1400	880	2200	250
	"	DRBC	1200	1400	1700	2400
	"	APDA	1300	130	1000	230
	"	PDACC	1900	350	2500	430
MEAN			**1100**	**155**	**2100**	**236**

[1] For medium abbreviations, see Table 1.

Table 3. Counts/g in locally purchased flour samples

Laboratory	Sample	Moulds	Yeasts
1	A	110	100
	B	500	100
	C	300	<100
	D	800	400
	E	1400	<100
	F	1000	800
	G	600	100
	H	900	100
	I	800	100
	J	4000	200
	K	800	<100
	L	500	<100
	M	1900	200
2		730	250
4		320	30
5		530	<20
6		320	<100
7	A	300	110
	B	400	120
8		1000	10
9	A	1000	400
	B	700	600
10		680	50
12		300	<100
13	A	150	<100
	B	200	100
14		650	350
15	A	660	480
	B	410	280
16	A	210	30
	B	180	10
	C	220	20
17		1000	<10
18		150	<100
19		950	150
20	A	200	10
	B	150	10
21	A	240	20
	B	370	30
	C	120	10
	D	150	10
	E	300	140
	F	220	90
22	A	760	50
	B	730	80
23	A	700	20
	B	700	70

Other observations

Counts on duplicate samples of the same flour and on duplicate plates at each dilution were similar at the majority of laboratories. Calculated counts from plates inoculated with consecutive tenfold dilutions of sample were similar for eight of the 20 laboratories who provided this information. With the other laboratories the calculated counts from plates inoculated with the higher dilution of sample were 30-210% higher than the calculated counts from plates inoculated with the next lower dilution. This dilution error tended to be more pronounced with laboratories using the pour plate method.

CONCLUSIONS

This collaborative study highlighted the diversity of media and methods currently used for enumerating fungi in foods. Bearing in mind the differences in methodology, the mould counts in the reference samples are reasonably uniform. However, yeast counts varied widely and this is a matter for further investigation.

The spread plate method appears to be preferable to the pour plate method as it gave better recoveries of yeasts and lower dilution errors. Acidified media with pour plates and enumeration after a short incubation period also gave reduced counts. The results suggest that problems of bacterial contamination occur with media containing no acid or antibiotic and also, on occasion, with DRBC.

The low counts of moulds and yeasts quoted for flour by most participants at the workshop in Boston (King *et al.*, 1986) are fully justified on the basis of the results from the domestic flours tested by the different laboratories. Based on the results reported here, it is unlikely that differences in methodology can account for the large variations in count reported. The variations can probably be explained by differences in the climatic conditions for growing and harvesting the wheat from which the flour was derived. Differences in post harvest storage time and conditions prior to testing also probably contributed. Flours from freshly harvested wheat may contain relatively high numbers of field fungi and yeasts. Many of these do not survive for long under the dry conditions exisiting in most flours.

REFERENCES

KING, A.D., PITT, J.I., BEUCHAT, L.R. and CORRY, J.E.L. eds. 1986. Methods for the Mycological Examination of Food. New York: Plenum Press.

MONITORING MYCOLOGICAL MEDIA

D. A. L. Seiler

Flour Milling and Baking Research Association
Chorleywood, Herts., WD3 5SH, U.K.

SUMMARY

Few microbiology laboratories monitor the performance of batches of media used. However, with laboratory accreditation becoming a requirement in many countries, laboratories increasingly will need to consider this. Evidence in this paper indicates that monitoring is not only desirable but is necessary if accurate estimations of the numbers of fungi in foods are to be obtained. Batches of mycological medium from the same supplier may vary in performance as can the same medium from different suppliers. This is particularly the case with selective media where small differences in inhibitory ingredients, such as rose bengal, dichloran or antibiotics can markedly affect properties.

No standard method exists for assessing the performance of batches of mycological media. A method whereby plates of the medium are stab inoculated with a range of test moulds and yeasts and the colony diameter measured with incubation time is suggested. Although differences between batches are clearly defined using this technique, the extent to which the differences noted will affect recovery of fungi from foods is still uncertain. The possibility of monitoring media by carrying out counts on a standardised, freeze-dried mixed culture of test microorganisms is discussed.

INTRODUCTION

At the present time it seems certain that few laboratories routinely carry out tests to monitor the performance of new batches of dehydrated media before use. In a busy, cost conscious microbiology laboratory it can be argued that there is neither the staff nor time available to carry out tests to ensure each batch of the large number of different media used comes up to standard. However, from time to time most microbiologists meet problems with poor recovery, lack of sensitivity or contamination with unwanted microorganisms which suggest that the medium is at fault. Usually it is found that a mistake has been made in medium preparation, sterilisation or storage, but occasionally good evidence indicates a faulty batch of dehydrated medium. Whatever the cause, such incidents can result in considerable loss in time and money, or cause embarrassment. The effort required for monitoring may seem well spent when such incidents occur.

A second urgent reason for development of suitable methods for assessing the performance of microbiological media is the growing importance of laboratory accreditation in many countries. Quality control of media is usually a necessity for accreditation and the methods used must be accurately defined.

Most problems of poor performance occur with selective media, where slight variations in concentration of highly active selective agents can have a deleterious effect on microbial growth. For mycological media, it is important that levels of antibiotics and antispreading agents, such as rose bengal and dichloran, are correct. Exposure to light will markedly increase the inhibitory properties of rose bengal and recovery of fungi can be seriously impaired if either dehydrated media or poured plates are exposed to light. Dichloran is a powerful inhibitor used at a concentration of 2 mg/L: only a small deviation from this concentration can result in an incorrect level of inhibition.

Performance of selective bacteriological media performance can be monitored by comparing the recovery of desired and unwanted species against a general purpose medium. However, the selective principle with a mycological medium is different: a major function is to restrict the growth of spreading moulds which cannot be judged by recovery.

The best approach for monitoring mycological media is to measure colony diameters of specific fungal strains following a standardised incubation. One methods is to use a stab inoculation procedure and measure the rate of increase in colony diameter with incubation time. The work which has been undertaken to standardise such a technique and assess its usefulness for comparing the performance of different batches of mycological media is discussed in this paper.

METHODS

Stab inoculation

The stab inoculation method used here simply involves inserting a straight wire loaded with spores of the test fungus into the agar medium. To minimise scattering of spores around the inoculation site it is preferable to turn the plate upside down and inoculate from below. After preliminary experiments, a standard method was used. Plates were poured with 20 mL of medium, the surface dried with lids removed in an incubator either at 40°C for 2h or 50°C for 1h. Stab inoculation was from below. Colony diameters were measured daily during incubation at 25°C and growth rate calculated from the plot of the average colony diameter against time.

The most effective inoculation technique for moulds was a wire heavily loaded with dry spores (or mycelium) taken from a 7-14 day old culture. Less variation in diameter between replicate colonies was also observed. The maximum number of effective inoculation sites per plate varied. With rapid spreading species, one central position per plate was best, and for less rapid and slowly growing moulds 3 and 5 positions were satisfactory. Yeasts and bacteria could be inoculated in up to 8 positions per plate. At least 3 colonies of each test species were used on separate plates to obtain an average growth rate.

To assess the value of the stab inoculation technique for monitoring purposes two series of tests were carried out.

(i) Comparison of batches of different media from the same manufacturer. Samples of different production batches of five dehydrated mycological media from a single manufacturer were prepared according to manufacturer's instructions and poured (20 mL) into Petri dishes. After drying, triplicate plates were stab inoculated in three positions with a wire heavily loaded with spores or cells of the microorganisms listed in Table 1.

(ii) Comparison of batches of the same medium from different manufacturers. Samples of Rose Bengal Chloramphenicol agar (RBC; Jarvis, 1973) received from three media manufacturers were used to prepare plates as described above. Stab inoculations were carried out with a range of microorganisms. As an additional test, spread plate counts were also carried out with two samples of cereals.

The stab inoculation procedure was used to compare the growth rates of a variety of test microorganisms on the same batches of ready prepared mycological media made strictly according to the manufacturers instructions at weekly intervals over a period of 5 weeks. Plates which had been stored in a refrigerator for 5 weeks were also tested.

Reference microorganisms

For monitoring mycological media, microorganisms exhibiting stable characteristics should be used. Both rapidly and slowly growing moulds, plus yeasts and bacteria should be used. The microorganisms used in these investigations, and which may be suitable reference species are listed in Table 1. Use of all of them on each medium is not necessary. Those used should be related to the type of food to be examined and the nature of the medium. For example, with low or intermediate moisture foods, where a medium of reduced a_w such as Dichloran 18% Glycerol agar (DG18) should be used (Hocking and Pitt, 1980), a range of xerophiles which commonly spoil these foods, such as *Aspergillus flavus*, *Eurotium repens* and *Zygosaccharomyces rouxii*, should be included. For high a_w media and moist foods a range of less xerophilic fungi would be more suitable.

RESULTS

Preliminary tests showed that the inoculation method influenced the percentage of successful (viable) inoculation points and the rate of colony growth. However, using the method outlined, and with the media and test microorganisms employed, the growth rates rarely varied by more than 10% from the mean. Thus, providing the procedure is carefully followed, either growth rate or average colony diameter in a given time is a reliable indication of the performance of a medium.

Table 1: Suggested reference species for monitoring mycological media

Spreading moulds	*Mucor racemosus* *Rhizopus stolonifer* *Fusarium culmorum*
Rapidly growing moulds	*Aspergillus flavus* *Penicillium aurantiogriseum* *Alternaria alternata*
Slowly growing moulds	*Eurotium repens* *Cladosporium herbarum* *Verticillium alboatrum*
Yeasts	*Hansenula anomala* *Saccharomyces cereviseae* *Pichia burtonii* *Zygosaccharomyces rouxii*
Bacteria	*Bacillus subtilis* *Pseudomonas aeruginosa* *Staphylococcus aureus* *Escherichia coli*

Comparison of batches of different media from the same manufacturer

The average increase in colony diameter at 25°C of three rapidly spreading moulds and six others of various growth rates on different batches of five media are shown in Fig. 1. Significant batch to batch variations in growth rate were observed with Oxytetracycline Glucose Yeast extract agar (OGY; Mossel *et al.*, 1962) and Dichloran Rose Bengal Chloramphenicol agar (DRBC; King *et al.*, 1979). Small but significant differences were also observed between batches of DG18.

Growth rates of some yeast species were seriously reduced on DG18 and DRBC and recovery in mixed populations in food may be low. The oxytetracycline in OGY prevented growth of the four test bacteria, but the chloramphenicol in the other three selective media failed to prevent the growth of *Pseudomonas aeruginosa*. An alternative antibiotic or combination of antibiotics may be required if *Ps. aeruginosa* is present in foods.

Comparison of batches of the same medium from different manufacturers

No growth of the test bacteria occurred on any of the batches in 7 days at 25°C. The growth rate of the test yeasts was essentially similar on all media. The mean mould counts from the cereal samples, and the mean growth rates and colony diameters after 4 days at 25°C for stab inoculated plates are given in Table 2. Significantly lower mould counts in the cereal samples were observed with the new batch of medium from manufacturer A and the medium from manufacturer B. Significant differences were also observed in mean growth rates and average colony diameters. However, no relationship was observed between reduced counts and growth rates. Observation suggests that higher mould counts on manufacturer A old stock medium may have been due to the reduced colony size which enabled more colonies to be counted.

In addition to the differences noted above, marked differences in appearance and colour of the moulds was seen on the various batches of RBC. On the medium from manufacturer B it was very difficult to make identifications of commonly occuring moulds on the basis of appearance.

Based on this study, the stab inoculation procedure may prove suitable for monitoring purposes. A certain expertise is necessary when stabbing the medium to ensure that mould spores are restricted to the inoculation site but otherwise the method is straightforward. Moreover, it need not be particularly time consuming if it proves possible to assess performance on the basis of average colony diameter at a specific incubation time rather than on growth rate.

Table 2. Comparison of mould counts from two cereal samples with growth rates and colony diameters of reference cultures on four batches of RBC medium

	Old stock	New batch A	New batch B	New batch C
Mean count per g (x 10^5)	2.1	1.5	1.6	2.0
Mean growth rate (mm/day at 25°C)	3.26	3.74	3.97	4.33
Mean colony diameter (4 days at 25°C)	11.93	14.35	14.78	16.07

From the tests described and general observations in the laboratory, a real need is perceived for a simple and reliable method for assessing the performance of selective media. This need is becoming more urgent with the growing requirement for laboratory accreditation. Stab incoculation is one approach, but further work is required to assess its potential. Further collaborative exercises appear warranted.

ACKNOWLEDGEMENT

The author is grateful to the UK Home Grown Cereals Authority who funded some of this study.

REFERENCES

HOCKING, A.D. and PITT, J.I. 1980. Dichloran-glycerol medium for enumeration of xerophilic fungi from low moisture foods. *Appl. Environ. Microbiol.* **39**: 488-492.

JARVIS, B. 1973. Comparison of an improved rose bengal-chlortetracycline agar with other media for the selective isolation and enumeration of moulds and yeasts in foods. *J. Appl. Bacteriol.* **36**: 723-727.

KING, A.D., HOCKING, A.D. and PITT, J.I. 1979. Dichloran-rose bengal medium for enumeration and isolation of molds from foods. *Appl. Environ. Microbiol.* **37**: 959-964.

MOSSEL, D.A.A., VISSER, M. and MENGERINK, W.H.J. 1962. A comparison of media for the enumeration of moulds and yeasts in foods and beverages. *Lab. Pract.* **11**: 109-112.

REPORT ON A COLLABORATIVE STUDY ON THE EFFECT OF PRESOAKING AND MIXING TIME ON THE RECOVERY OF FUNGI FROM FOODS

D. A. L. Seiler

Flour Milling and Baking Research Association
Chorleywood, Herts., WD3 5SH, U.K.

In collaboration with : C.B. Anderson, S. Andrews, L.R. Beuchat, G. Cerny, J. Clarke, J.E.L. Corry, M.A. Cousin, T. Deák, K.E. Dijkmann, G. Edwards, J.C. Frisvad, D. Heperkan, A.D. Hocking, A.D. King, W. Röcken and R.A. Samson.

SUMMARY

This paper presents results from a collaborative study designed to determine the effects of presoaking and mixing times on the recovery of fungi from cereals and cereal derived products. Results of the study indicated that, without presoaking, increasing the mixing time from 2 to 10 min led to an increase in recovery of moulds and yeasts in the cereal samples examined. The level of count and variability between laboratories after presoaking for 30 min and mixing for 2 min was similar to that obtained by mixing for 10 min without soaking. Soaking for 30 min followed by the shorter mixing time was preferred because it was more convenient. There was remarkably good agreement between laboratories on the composition of the mould flora in the different samples, but serious dilution errors were encountered which were larger with some samples than others and varied with laboratory. The best method of calculating the count needs further consideration. Diluting 1:5 would probably improve the accuracy of mould counts.

INTRODUCTION

Seiler (1986) reported that soaking cereals, cereal products and cereal based foods in diluent for 30 or 60 min before mixing usually significantly increased fungal counts. Some evidence also suggested that counts increased with increasing mixing time in the Stomacher. This paper reports on a collaborative study designed to assess the significance of these results. It involved 17 laboratories, four each in the UK and USA and two each in Denmark, Hungary and Turkey. Six cereal samples, selected for varied consistency and fungal population were distributed in 250 g lots to each laboratory. Sufficient Dichloran 18% Glycerol agar (DG18) (Hocking and Pitt, 1980) and antibiotic supplement for the study were also supplied.

METHODOLOGY

Each laboratory was asked to carry out the following protocol. For each sample, six primary dilutions of 40 g in 360 mL of saline peptone diluent were prepared and duplicates soaked for 0, 30 and 60 min. After each soaking time, subsamples were mixed using the Stomacher 400 for 2 min and again after a further 8 min (Table 1). Mould and

Table 1. Protocol

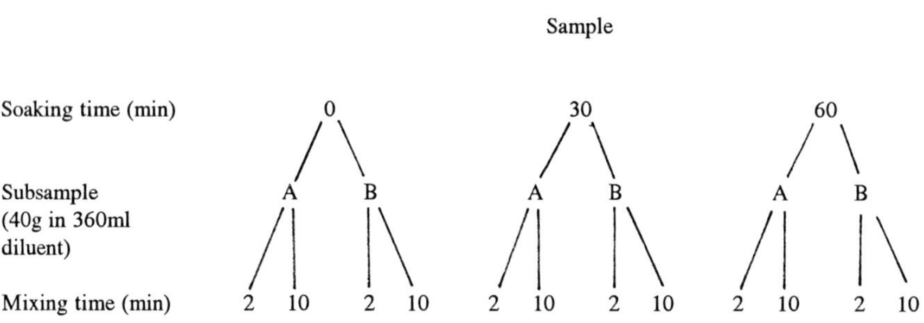

yeast counts were determined in duplicate using a 0.1 mL spread plate technique and DG18. Specific instructions were given on the method of preparing dilutions, drying and spreading plates, and preparing the medium. Laboratories were asked to record separately the numbers of mould, yeast and total colonies on the duplicate plates at all countable dilutions after 7 days incubation at 25°C, and to note any increase in count between 5 and 7 days. Some of the laboratories also agreed to make a broad identification of the moulds present on the countable plates from the subsamples which had been soaked for 60 min and mixed for 10 min. The samples tested are given in Table 2.

RESULTS

Counts

The means of all viable counts obtained for each sample are given in Table 2. Only four of the six samples proved to have sufficiently high counts to allow effective comparisons between treatments. Low counts in the flour sample prevented comparison of the effect of presoaking and mixing in homogeneous and non-homogeneous materials.

Table 2: Samples used in the collaborative exercise

Code	Description of sample	Mean count per g
A	Winter wheat (Slejpner)	3.1×10^5
B	Winter wheat (Avalon)	2.1×10^2
C	Spring oats	4.6×10^5
D	Spring barley (Corniche)	5.2×10^4
E	Breadmaking flour	2.9×10^2
F	Mixed animal feed	3.8×10^4

Table 3. Average $\log_{10}$ counts/g for moulds, all laboratories and treatments

	Sample			
Laboratory code	A	C	D	F
1	5.18	5.67	4.17	4.41
2	4.74	5.65	3.72	4.36
3	5.72	5.54	-	-
4	-	-	-	-
5	5.53	5.78	4.95	4.65
6	5.28	5.58	4.76	4.59
7	5.53	5.77	5.00	4.61
8	5.87	5.60	4.98	4.36
9	5.05	5.69	4.13	4.53
10	4.98	5.61	4.65	4.42
11	-	5.59	4.49	4.43
12	5.91	5.68	-	-
13	5.48	-	4.20	4.37
14	4.86	5.50	3.67	4.26
15	-	-	-	4.44
16	5.38	5.71	4.77	4.57
17	-	-	-	4.45
Mean	5.35	5.64	4.46	4.46

Bacterial contamination

Only one of the 17 laboratories reported the presence of bacteria on plates from one sample (F). Perhaps antibiotics were not added to that batch of medium. Taken as a whole the results indicate that the combined effects of low a_w and chloramphenicol in DG18 are adequate to prevent bacterial growth in 7 days at 25°C.

Incubation time

Most laboratories noted an increase in count from 5 to 7 days incubation. Usually the increase was small and rarely more than 10%. With samples A and C, which contained large numbers of *Wallemia sebi*, a number of laboratories commented that counts may have increased because of secondary colonies. This point, and the small increases in 7 day counts suggest that 5 days incubation at 25°C is preferable.

Mould counts: between laboratory variability

The mean $\log_{10}$ count of moulds obtained for all treatments with samples A, C, D and F are shown in Table 3.

The variation in counts between laboratories was lower for samples C and F than for samples A and D. The reason for this is not clear but may be linked with type of fungi present and ease of counting colonies on the plates.

Mould counts: within laboratory variability

The coefficient of variation in mould count between subsamples for all soaking and mixing treatments, but eliminating the effect of treatment, are shown for each laboratory for samples A,C,D and F in Table 4.

Table 4. Moulds: Within laboratory variability (coefficients of variation, %)[1]

Laboratory code	Sample A	C	D	F
1	23.62	21.71	35.33	30.50
2	68.73	19.15	39.93	14.26
3	84.33	67.46	-	-
4	-[2]	-	-	-
5	19.62	8.89	13.74	25.97
6	16.68	26.75	5.09	52.03
7	54.81	18.11	13.89	24.22
8	20.56	13.66	17.35	27.35
9	19.15	9.95	30.29	33.11
10	36.30	8.42	19.07	7.72
11	-	15.29	61.66	52.06
12	86.51	24.54	-	-
13	51.74	-	16.55	17.60
14	104.13	35.08	60.84	68.07
15	-	-	-	41.45
16	8.52	20.48	30.87	15.40
17	-	-	-	12.77

[1] Coefficients of variation above 50% are underlined
[2] Sample not tested or only a single subsample tested.

The within laboratory variability was higher with sample A than with the other samples, 6 laboratories showing coefficients of variation of over 50%. This suggests that the mould contamination was less evenly distributed in the wheat sample than in the others.

Yeast counts: between and within laboratory variability

Significant numbers of yeasts were only encountered in samples D and F (Table 5). The between laboratory variability for these samples was acceptable, less than one log cycle for all but a single figure.

The within laboratory variability is greater than that encountered with moulds (Table 2) with more than half the laboratories showing coefficients of variation greater than 50%. It is not clear whether laboratories have difficulty in identifying and counting yeasts, or samples are inherently more variable than for moulds.

Dilution error

It is usually found that the mould count calculated from the average number of colonies per plate at the highest countable dilution is higher than the count calculated from plates at the next lower tenfold dilution, i.e. there is a dilution error. Table 6 shows the number of times that the count at the higher dilution was more than that at the next lower dilution for the various laboratories and samples examined in this collaborative exercise.

It is apparent that the extent of the dilution error varies considerably between samples and laboratories. With samples D and F most laboratories obtained counts at the higher dilution which were no more than 50% greater than those from the next lower dilution.

Table 5: Yeasts: between and within laboratory variability[1]

Laboratory code	Between laboratory variability (ave. $\log_{10}$ count/g)		Within laboratory variability (coefficient of variation %)[2]	
	Sample D	Sample F	Sample D	Sample F
1	3.81	3.33	111.2	63.4
2	3.70	3.80	91.2	52.4
3	-	-	-	-
4	-	-	-	-
5	3.89	3.98	105.5	72.0
6	3.87	3.67	87.8	39.7
7	3.82	3.64	30.5	77.9
8	-	3.87	-	143.2
9	3.56	-	27.9	-
10	3.32	-	63.8	-
11	3.76	4.07	17.2	97.6
12	3.58	3.59	124.4	36.8
13	-	3.34	-	91.9
14	4.51	3.59	50.7	36.5
15	-	3.96	-	31.0
16	-	3.83	-	31.5
17	-	4.03	-	26.1

[1] Missing values: either sample not tested or only a single subsample tested
[2] coefficients of variation above 50% are underlined

Table 6. Dilution errors with mould counts at two countable dilutions[1]

Laboratory code	Sample				Mean
	A	C	D	F	
1	3.5	2.0	1.4	1.3	2.1
2	1.4		2.6	0.8	1.6
3	5.6			2.0	3.8
4	1.3	1.7	1.3	1.1	1.4
5	1.8		1.0	1.6	1.5
6					
7	2.8	2.4	1.1	1.4	1.9
8	3.9	2.4	1.7	1.6	2.4
9	2.4		0.9	0.8	1.4
10	1.9		1.1	1.1	1.4
11		3.1	1.8	1.5	2.0
12					
13	8.0		8.5	10.1	8.9
14	1.7	2.4	5.3	0.9	2.6
15	2.0	3.8	1.1	1.5	2.1
16	2.9	2.4		1.2	2.2
17				1.1	1.1

[1] Ratio of the average mould count at the highest countable dilution over the average mould count at the lower tenfold dilution. A large figure indicates a higher dilution error.

With samples A and C, on the other hand, the majority of counts were more than twice as great at the higher dilution than at the lower dilution. i.e. a dilution error of 2.0 or more. With the latter samples the larger errors are probably due to inaccuracies in counting the large numbers of colonies present on plates at the lower dilution. Some laboratories encountered higher dilution errors than others. Taken over all samples, 8 of the 15 laboratories who counted at two dilutions showed a high mean dilution error of over 2.0.

Calculating counts in samples
In view of the errors described above it is not desirable to average counts at two dilutions when calculating the count per g of sample. Certainly, the use of the Farmiloe equation (Farmiloe *et al.*, 1954), where the greatest weight is placed on the countable plates with the largest number of colonies, cannot be justified. It can be argued that it is best to count plates at the highest countable dilution even if the plates contain 10 or less colonies. However, it must be considered dubious practise to rely on counts obtained from such low numbers per plate unless a number of replicate subsamples are compared. A better assessment may be achieved by counting plates containing 10-50 colonies, but it will be difficult to obtain such counts on all occasions using a tenfold dilution series. Perhaps there is a case for using a 5 fold dilution series where accurate mould counts are required.

Effects of presoaking and mixing
The results for presoaking and mixing with samples A, C, D and F have each been analysed statistically. Comparisons of the differences in mould count between a number of the soaking and mixing treatments are given for each laboratory as a histogram in Figure 1.

When the mixing time was increased from 2 to 10 min without soaking the majority of laboratories (79%) obtained an increase in count. With the exception of sample C, the within laboratory variability was such that most of the increases were not statistically significant. However, the results present good evidence to indicate that the additional mixing will improve recovery. When the counts obtained from samples mixed for 10 min without soaking are compared with those when the samples were soaked for 30 min and then mixed for 2 min, only half the laboratories noted an increase and only in four cases was the difference significant. In most laboratories (67%) mould counts increased when the mixing time was increased from 2 to 10 min after soaking for 30 min. However, the differences in count were small and only five were statistically significant. By increasing the soaking time from 30 to 60 min followed by mixing for 2 min, less than half the laboratories obtained an increase in count and only on three occasions was the increase significant. An important general finding was that the variability in count between laboratories was reduced when the samples were soaked or when mixed for 10 min.

The results indicate that equivalent mould counts can be expected either by mixing the sample for 10 min or by soaking for 30 min and mixing for 2 min. There is some evidence to suggest that increasing the mixing time to 10 min after soaking will increase recovery, but the effect is small. No additional increase in count occurred by soaking for 60 min rather than 30 min. On grounds of convenience, a soaking time of 30 min followed by mixing using the Stomacher for 2 min is recommended for mould counts in cereals and other foods where the fungal contamination is likely to be deep seated.

Identification
Ten laboratories took part in tests to identify the moulds present in the samples. Identifi-

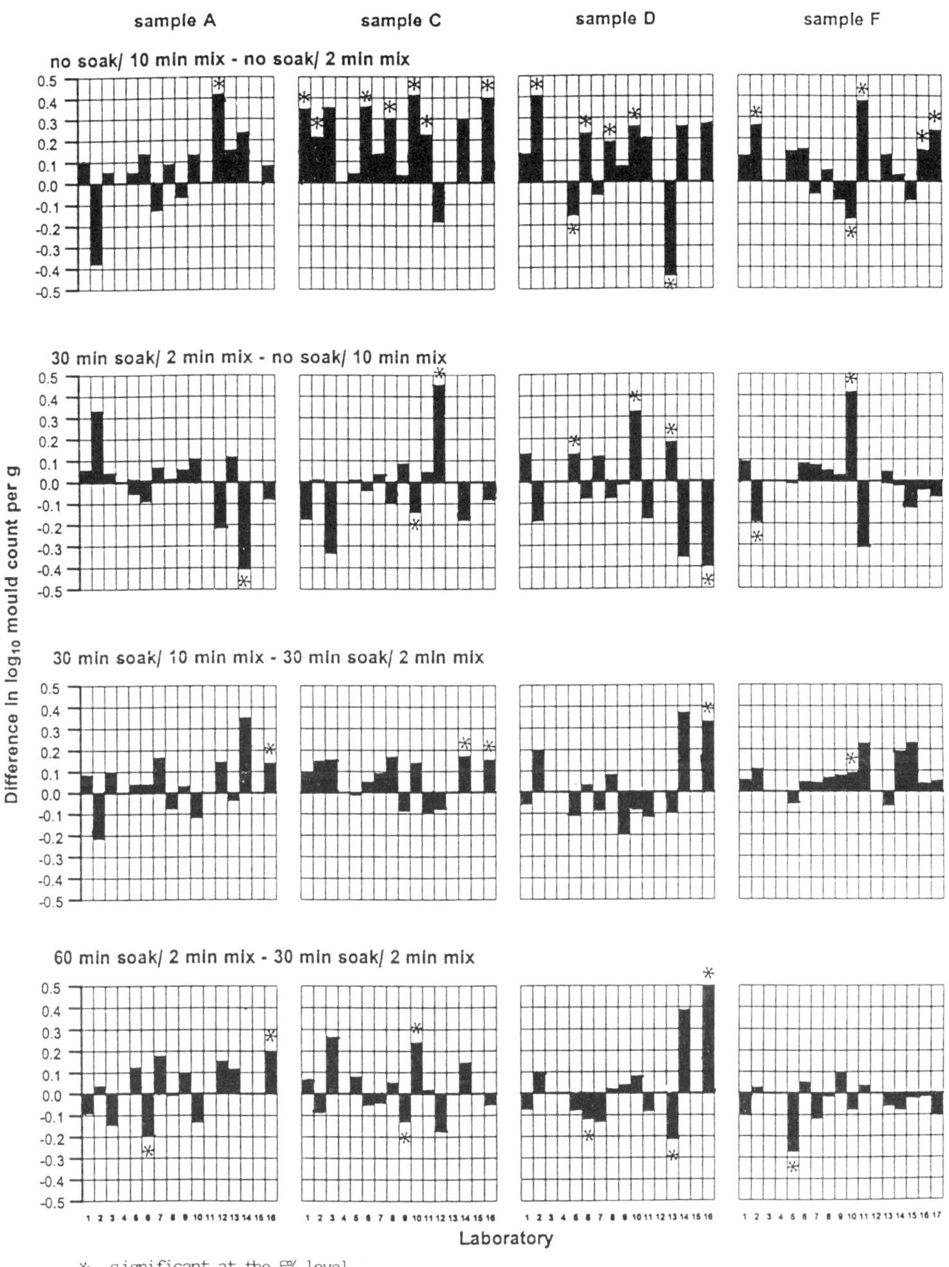

Figure 1. Comparison of various presoaking and mixing treatments

cations were only carried out on the four countable plates resulting from the duplicate subsamples which had been soaked in diluent for 60 min and then mixed with the Stomacher for 10 min. Because of the very low counts obtained, the results with samples B and E have been omitted from the analysis. The mean percentage incidence of the various mould genera and species obtained by the ten laboratories for the other four samples is shown in Table 7.

The mould flora in the four samples varied considerably. With samples A, C and D (wheat, oats and barley) the population was dominated by xerophiles. *Eurotium* species were the most common fungi present in sample C whereas *Wallemia sebi* predominated in Sample D. Both of these genera were found in sample A but a range of others was also present. The flora of sample F (mixed animal feed) was quite different with *Aspergillus candidus* and other Aspergilli making up the majority of the population and only small numbers of *Eurotium* and *Wallemia* being detected.

To a bakery microbiologist these findings are of interest since *Eurotium* and *Wallemia* are the principle causes of spoilage of low a_w products such as cakes. If the high counts of these genera in the cereal are transferred through to the flour, which is likely, the mere handling of such a material in the bakery would increase atmospheric counts to an extent that the shelf life of products is likely to be reduced. Fortunately, experience with counts in flour (Seiler, 1992) suggest that the miller is unlikely to use cereals with such high counts.

Agreement between laboratories on the composition of the mould flora in the different samples was remarkably good. If we take as an example the results from sample F (Table 8) it is apparent that only laboratories 2, 5 and 7 differ substantially from the others. Taken overall, the results are encouraging and indicate that most laboratories are able to separate a range of fungi on DG18.

Table 7. Mean incidence (all laboratories) of fungi in four samples of cereal or cereal product

Mould	% incidence in sample			
	A (winter wheat)	C (spring oats)	D (spring barley)	F (mixed animal feed)
Alternaria spp.	0.2	0.2	0	0
Aureobasidium sp.	0.3	0.7	4.1	7.3
Aspergillus candidus	2.3	0.2	0.4	52.1
A. versicolor	1.4	1.7	1.3	3.4
A. restrictus	2.8	0	2.5	0.1
Aspergillus spp.	8.0	10.4	2.0	6.9
Cladosporium spp.	0.8	1.2	0.6	1.7
Eurotium spp.	39.8	71.0	10.9	5.1
Mucoraceae	0	2.4	0.3	1.8
Penicillium spp.	8.3	7.9	6.6	11.9
Verticillium spp.	1.4	0.2	0.7	3.1
Wallemia sebi	20.7	0.2	57.4	1.7
Others	14.0	3.8	13.3	4.9

Table 8. Fungi identified in sample F (mixed animal feed) by ten collaborating laboratories

	% incidence found by laboratory number									
	1	2	3	5	7	8	9	11	12	16
Aureobasidium	3	18	10	33			3	5		1
Aspergillus candidus	77	22	64			56	91	69	57	85
A. versicolor		7		22				1	2	2
A. restrictus									2	
Aspergillus spp.		4		37	27				1	
Cladosporium spp.		4		1	1	9			2	
Eurotium spp.	7	4	9	3	1	4	3	8	8	4
Mucoraceae		11			2	1	2		1	1
Penicillium spp.	13	30	7	3	33	11		11	6	5
Verticillum spp.			10		14	7				
Wallemia sebi			1	7			6	2	1	
Others					15	12	1		20	1

CONCLUSIONS

Taking the expected variability in subsamples into consideration, most counts obtained for the various samples were acceptable, lying within 0.5 log cycle of the mean.

1. There is good evidence to indicate that, without presoaking, increasing the mixing time from 2 to 10 min causes an increase in recovery of moulds and yeasts in the cereal samples examined.

2. The level of count and variability between laboratories after presoaking for 30 min and mixing for 2 min was similar to that obtained by mixing for 10 min without soaking. The former technique is preferred on gounds of convenience. There appears to be little additional benefit in soaking for 60 min.

3. Serious dilution errors were encountered which are larger with some samples than others and vary with laboratory. The best method of calculating the count needs further consideration. Diluting 1:5 would probably improve the accuracy of mould counts.

5. With few exceptions there was remarkably good agreement between laboratories on the composition of the mould flora in the different samples.

6. DG18 performed well by adequately preventing bacterial contamination and permitting the growth and identification of a wide range of moulds and yeasts.

ACKNOWLEDGEMENTS

The author is grateful for the financial assistance provided by the UK Home Grown Cereals Authority which made it possible to organise and participate in this International collaborative exercise. Thanks are also due to Dr Paula Curtis for the many hours which she spent in making a statistical evaluation of the large amount of data received. The assistance of Mr Frank Petagine and Miss Tania Carter, who carried out the laboratory tests, is also acknowledged.

REFERENCES

FARMILOE, F.J., CORNFORD, S.J., COPPOCK, J.B.M. and INGRAM, M. 1954. The survival of *Bacillus subtilis* spores in the baking of bread. *J. Sci. Food Agric.* **5**, 292.

HOCKING, A.D. and PITT, J.I. 1980. Dichloran-glycerol medium for enumeration of xerophilic fungi from low moisture foods. *Appl. Environ. Microbiol.* **39**: 488-492.

SEILER, D.A.L. 1986. Effect of presoaking on recovery of fungi from cereals and cereal products. *In* Methods for the Mycological Examination of Food, eds. A.D. King, J.I. Pitt, L.R. Beuchat and J.E.L. Corry. p. 26-28. New York: Plenum Press.

SEILER, D.A.L. 1992. Collaborative study to compare fungal counts in flour samples. *In* Modern Methods in Food Mycology, eds R.A. Samson, A.D. Hocking, J.I. Pitt and A.D. King. pp. 67-72. Amsterdam: Elsevier.

COLLABORATIVE STUDY OF MEDIA AND METHODS FOR ENUMERATING HEAT STRESSED FUNGI IN WHEAT FLOUR

L. R. Beuchat and R. E. Brackett

Department of Food Science and Technology
University of Georgia Agricultural Experiment Station
Griffin, GA 30223-1797 USA

In collaboration with: L. S. Carlson, D. E. Conner, K. E. Olson and M. E. Parish

SUMMARY

A collaborative study was performed to evaluate Acidified Potato Dextrose Agar, Chloramphenicol supplemented Plate Count Agar and Dichloran Rose Bengal Chloramphenicol Agar for recovering yeasts and moulds from unheated and heated wheat flour. Spread and pour plating techniques were evaluated. Five commercial flours were analysed by six collaborating laboratories. Overall, surface plating using DRBC was judged to be the most suitable technique for enumeration of fungi in both unheated and heated flour.

INTRODUCTION

The potential presence of stressed fungal cells in foods at the time they are subjected to mycological analysis is well known (Beuchat, 1984; Stevenson and Graumlich, 1978). The effect of stress on cells is largely unknown, but sensitivity to adverse conditions such as inhibitors in media would be expected to result. The collaborative study reported here was designed to determine the suitability of media and plating techniques for recovering unheated and heat stressed yeasts and moulds from wheat flour.

MATERIALS AND METHODS

Product analysed

Five wheat flours commercially marketed by five different companies were analysed. All were purchased from supermarkets in the Griffin, GA area in November, 1989. Each flour was thoroughly mixed, then 25 g subsamples were sealed in polyethylene bags such that a layer approximately 1 cm thick resulted when bags were laid on a flat surface. Subsamples were sent to five collaborators with instructions to store at laboratory temperature (approximately 21 °C) until analysed.

Media evaluated

Three media were evaluated for their suitability to recover yeasts and moulds from unheated (control) and heated flours. Acidified Potato Dextrose Agar (APDA), Difco, Inc., Detroit, MI, was prepared according to directions. Chloramphenicol-supplemented Plate Count Agar (CPCA) was prepared from Plate Count Agar (Difco) according to directions, except that chloramphenicol was added before autoclaving, to give a concentration of 100 μg/ml. Dichloran Rose Bengal Chloramphenicol Agar (DRBC) was the standard formulation (see Appendix) except that pH was adjusted to 5.6 before autoclaving at 121°C for 15 min.

All media were cooled to 47-50°C before pouring. Media used for spread plating were poured and allowed to dry for 24 hr at laboratory temperature before samples were deposited.

Heat treatment
Preliminary studies in the first author's laboratory determined time and temperature conditions resulting in at least 20% reduction in viable counts but not complete lethality. Sample bags of flour (25 g) were heated on a rack in a forced air oven at 55, 60, 65 and 70°C for 1 or 3 hr. Heated samples and unheated controls were analysed immediately after heat treatment by both pour and spread plating on APDA, then incubated upright at 25°C for 5 days.

Procedure for enumeration
Triplicate unheated and heated samples (25 g) were combined with 225 ml of 0.1N potassium phosphate buffer (pH 7.0) and homogenised in a Colworth Stomacher for 1 min, allowed to stand for 15 min and homogenised again for 1 min. Serially diluted samples were both spread and pour plated in duplicate on APDA, CPCA and DRBC within 30 sec of the second homogenisation. Spread plate samples (0.1 mL) were distributed with a sterile bent glass rod. Pour plate samples were 1.0 mL.

Plates were incubated upright for 5 days at 25°C. Plates on which 15-150 colonies had formed were counted. An occasional plate outside this range was included when overgrowth by spreading moulds or incorrect dilution of samples did not result in plates with 15-150 colonies. Collaborators were also asked to make subjective evaluations regarding each system.

Statistical analysis
Data were subjected to the Statistical Analysis System (SAS Institute, Cary, NC) for analysis of variance and Duncan's multiple range test.

RESULTS AND DISCUSSION

Preliminary studies
Treatment of the five wheat flours for 1 or 3 hr at 55°C had little effect on viable fungal populations. Treatment at 60°C caused some reduction, and at 65°C resulted in about 0.2 to 1.5 log reduction in populations, depending upon the flour analysed. Treatment at 70°C for 1 hr generally reduced counts to very low or non detectable levels. One flour was exceptional, containing markedly heat resistant mycoflora, indicating a diversity in the mycoflora of the flours from different sources. Heating at 65°C was selected for the collaborative study.

Collaborative study
Data from collaborating laboratories were composited and analysed for statistically significant differences (Table 1). The mycoflora detected in flours consisted largely of moulds.

Table 1. Comparison of APDA, CPCA and DRBC, and spread and pour plating for enumerating fungi in unheated and heated wheat flour[1,2]

Time (hr) of heating at 65 °C	Population (CFU/g)					
	APDA		CPCA		DRBC	
	Spread	Pour	Spread	Pour	Spread	Pour
0	3.15 ab	3.07 abc	3.18 a	3.02 c	3.14 abc	3.05 bc
1	3.08 a	3.06 ab	3.03 ab	2.87 b	3.06 ab	2.95 ab
3	2.98 a	2.98 a	2.82 ab	2.75 b	2.86 ab	2.84 ab

[1]Composite of data from six laboratories and five flours
[2]Values in the same row which are not followed by the same letter are significantly different ($P < 0.05$)

For unheated flours, significantly lower counts were obtained using pour plates of CPCA and DRBC compared with spread plated CPCA. With flours heated for 1 hr at 65 °C, pour plates of CPCA were inferior to spread plated APDA which was not significantly different from the other test systems. Counts on pour plated CPCA from flours heated for 3 hr were significantly lower than on spread or pour plated APDA, but not significantly different from the other systems. Based on total counts, five of the six test systems performed similarly for recovering heat stressed fungi, the exception being the inferior performance by pour plated CPCA.

Overall, collaborators preferred spread plated DRBC for enumerating fungi in unheated and heated wheat flour. Colony colours and the lack of spreading by filamentous fungi were given as major advantages of DRBC, although identification of isolates on DRBC may be hampered somewhat by the slow formation of reproductive structures.

None of the plating systems examined in this study is suitable for enumerating the xerophilic fungi often present in wheat flour. For example, *Wallemia sebi* and *Eurotium* species are often isolated from grains and flours when appropriate media are used. These and other xerophilic fungi may grow slowly or not at all on the media evaluated in this study. Also, this study involved only one food, and heat stressed mycoflora naturally present in other types of foods may react quite differently from the mycoflora in wheat flour. Additional work is needed to determine the suitability of media and plating techniques for recovering stressed mycoflora from a wide range of food types.

REFERENCES

BEUCHAT, L.R. 1984. Injury and repair of yeasts and moulds. *In* The Revival of Injured Microbes, eds M.H.E. Andrew and A.D. Russell, pp. 293-308. London: Academic Press.

STEVENSON, K.E. and GRAUMLICH, T.R. 1978. Injury and recovery of yeasts and molds. *Adv. Appl. Microbiol.* **23**: 203-217.

INFLUENCE OF METHOD OF STERILISATION AND SOLUTE IN THE RECOVERY MEDIUM ON ENUMERATION OF HEAT-STRESSED *ZYGOSACCHAROMYCES ROUXII*

D. A. Golden and L. R. Beuchat

Department of Food Science and Technology
University of Georgia, Agricultural Experiment Station
Griffin, Georgia 30223-1797, USA

SUMMARY

A study was conducted to determine the influence of solutes and the procedure for sterilising recovery media on colony formation by five strains of sublethally heat stressed *Zygosaccharomyces rouxii*. Media were prepared by autoclaving (110°C, 15 min) and by repeated treatment with steam. An increase in sensitivity was observed when heat stressed cells were plated on Yeast Malt agar (YMA) containing glucose at a water activity (a_w) of 0.880 compared to a_w 0.933 and 0.998. Recovery of *Z. rouxii* was unaffected by medium sterilisation procedures when glucose or sorbitol was used as the solute. Significant differences were sometimes detected in counts on YMA containing sucrose or glycerol, with higher recoveries on steamed media. Significant differences in recovery of unheated cells were caused by solute type in 20% of comparisons, but in 43% of comparisons using heated cells. The enhanced effect of solute on recovery of unheated and heat stressed cells was typically in the order of glucose ≥ sucrose ≥ glycerol ≥ sorbitol and glucose ≥ glycerol ≥ sucrose ≥ sorbitol, respectively. It is concluded that glucose supplemented media sterilised by autoclaving is acceptable for enumerating healthy and heat-stressed *Z. rouxii*.

INTRODUCTION

Detection of xerophilic yeasts in high sugar foods such as dried fruits, syrups, and soft centred confectionery products can be improved by plating on media of reduced water activity (a_w) (Hocking and Pitt, 1980). However, high sugar contents in recovery media can create problems. Maillard reaction products may be formed when reducing sugars and compounds containing amines are heated and some of these products are toxic to animals and microorganisms (Nagao *et al.*, 1983; Shibamoto, 1983). Consequently, if media containing high concentrations of sugar are sterilised by autoclaving (121°C, 15 min), a reduction in recovery of xerophilic yeasts may occur. Heat injured cells may be particularly sensitive to Maillard reaction products.

This investigation was undertaken to determine the suitability of media sterilised by autoclaving (110°C, 15 min) and repeated treatment with steam (100°C) for recovering healthy and heat-injured cells of the xerophilic yeast *Zygosaccharomyces rouxii*.

MATERIALS AND METHODS

Fungi

Five isolates of *Zygosacchromyces rouxii* were used, NRRL Y-2547, NRRL Y-2548, NRRL Y-12691, ATCC 2623, and UGA 36.

Recovery media

Four basal media were used: Yeast Malt extract agar (YMA) made with YM Broth (Difco, Detroit, MI, USA) with 1.5% agar, wort agar (WA), corn meal agar (CMA), and oatmeal agar (OMA). Media were sterilised by autoclaving at 110°C for 15 min or by successive steam treatments (100°C) of 45, 60, and 75 min at 24 h intervals. Each medium was adjusted to 0.936 ± 0.003 a_w by adding 33% glucose, 45% sucrose, or 21% glycerol or sorbitol, and adjusted to pH 7.0 ± 0.2 with 1.0 N HCl before and after sterilising.

Based on these preliminary studies, the main study was carried out using cells heated for 12 min at 47 or 48°C, depending on strain. Unheated and heat-injured cells of the five test strains were appropriately diluted in 0.1 M potassium phosphate buffer and surface plated (0.1 mL) onto autoclaved and steamed media. Plates were incubated for 7 days at 25°C before colonies were enumerated.

Procedure for inducing heat injury

Initial experiments were designed to determine conditions inducing sublethal heat injury in the yeasts studied. Cultures grown in Yeast Malt extract broth (YMB; YM Broth, Difco) plus 33% glucose, pH 5.0, for 48-50 h at 25°C were inoculated (1 mL) into 100 mL of YMB-33 and held at 47, 48, 50, 52, 55, or 58°C for up to 36 min. Samples were withdrawn at 6 min intervals, serially diluted in 0.1 M potassium phosphate buffer (pH 7.0), and surface plated (0.1 mL) on YM agar (YMA; 0.998 a_w) and YMA supplemented with 33% (0.933 a_w), or 49% (0.880 a_w) glucose. Recovery media (pH 7.2 ± 0.2) were prepared in deionised water and autoclaved for 15 min at 110°C.

Sublethal heat injury was detected by assessing the relative sensitivity of unheated and heated cells to reduced a_w in YMA recovery media. In addition, a Biotran III Automatic Count/Area Totaliser was used to determine rates of increase in colony size and number at 12-24 h intervals throughout a 7 d incubation period at 25°C. The mean number of colonies which exceeded 0.2 and 0.5 mm in diameter was recorded at each time of analysis. On the seventh day of incubation, colonies were enumerated manually to determine the total number of colonies without regard to colony diameter.

Measurement of a_w

The a_w of media was measured at 25°C using a Rotronic Hygroskop DT hygrometer.

Statistical analysis

Analysis of variance and Duncan's multiple range tests were used to determine statistical differences between values obtained from experiments using autoclaved and steamed media and differences in the number of colonies formed as influenced by the type of solute used.

RESULTS AND DISCUSSION

Heat injury

Heating at 50°C and above rapidly inactivated the five strains of *Z. rouxii* examined. At 47 and 48°C, inactivation rates were slower, although UGA 36 and NRRL-Y 12691 exhibited substantially greater resistance at 47°C compared to ATCC 2623, NRRL-Y 2547 and NRRL-Y 2548.

Unheated cells of NRRL-Y 2547 and NRRL-Y 2548 were the least tolerant to low a_w (0.880) in YMA + glucose recovery media. Heat injured cells of ATCC 2623, UGA 36, and NRRL-Y 12691 also had increased sensitivity to 0.880 a_w. Typically, recovery of heat injured cells of ATCC 2623, NRRL-Y 2547, and NRRL-Y 2548 was favoured at 0.933 a_w, whereas recovery of heat injured cells of UGA 36 and NRRL-Y 12691 was similar at 0.998 and 0.933 a_w.

Colony development by heat injured cells

Colonies which developed from unheated cells on YMA at 0.998 and 0.933 a_w generally exceeded 0.5 mm in diameter within 3.5 to 4 days incubation, whereas colonies formed from unheated cells on YMA at 0.880 a_w did not exceed 0.5 mm in diameter until after 5.5 to 6.5 days. The number of colonies exceeding 0.5 mm in diameter formed by heat injured cells on YMA at 0.880 a_w was 90 to 95% less than the total number of colonies detected, i.e., using no limits of exclusion based on colony diameter.

Effect of sterilisation procedure

Recovery of any strain of *Z. rouxii* was not significantly affected ($P \leq 0.05$) by medium sterilisation procedure when glucose and sorbitol were used as solutes. In seven out of eighty paired comparisons (four basal media, two solutes, five strains, unheated and heat injured cells), counts were significantly different on autoclaved and steamed media supplemented with sucrose and glycerol. In six of these seven pairs, counts were significantly higher on steamed media.

Effect of solute

Significant differences in populations resulted from the use of different solutes to lower the a_w. When considering unheated cells, significant differences due to solute type were observed in 16 out of 80 comparisons; for heat injured cells, significant differences were noted for 34 of the 80 comparisons based on solute type. When differences did occur, enhanced recovery of unheated cells was typically in the order of glucose ≥ sucrose ≥ glycerol ≥ sorbitol, whereas recovery of heated cells was generally in the order of glucose ≥ glycerol ≥ sucrose ≥ sorbitol. In some instances, data were not obtained when heat injured cells were plated on media containing sorbitol, because colonies on plates on which the undiluted suspension was applied were too numerous to count, but no colonies were detected on plates inoculated with a ten-fold dilution. The additional stress imposed by osmotic shock sustained from exposure of heat injured cells to high a_w in the buffer diluent may have been sufficient to increase their sensitivity to sorbitol in recovery media.

CONCLUSIONS

Observations that heat injured cells of *Z. rouxii* had increased sensitivity and developed more slowly on media at 0.880 a_w emphasise the importance of adequate incubation times and optimum a_w for recovering sublethally heat injured yeasts. It was concluded that autoclaving media containing high concentrations of glucose by heating at 110°C for 15 min did not result in browning products that are toxic to uninjured and heat injured *Z. rouxii*. Therefore, tedious steaming processes used to sterilise media at a_w as low as 0.880

can be replaced by the simpler autoclaving procedure used in this investigation without adversely affecting the total number of colonies formed by unheated and heat-injured *Z. rouxii*. It is recommended that glucose be used to reduce the a_w since this solute is less inhibitory than glycerol, sucrose, and sorbitol. Furthermore, it is recommended that consideration also be given to reducing the a_w of diluents used in procedures for enumerating *Z. rouxii*, especially if injured cells are likely to be present in the low a_w foods being analysed.

REFERENCES

HOCKING, A.D. and PITT, J.I. 1980. Dichloran-glycerol medium for enumeration of xerophilic fungi from low moisture foods. *Appl. Environ. Microbiol.* **39**: 488-492.

NAGAO, M., SATA, S. and SUGIMURA, T. 1983. Mutagens produced by heating foods. *ACS Soc. Symp. Ser.* **215**: 521-526.

SHIBAMOTO, T. 1983. Heterocyclic compounds in browning and browning/nitrite model systems: occurrence, formation mechanisms, flavour characteristics, and mutagenic activity. *In* Instrumental Analysis of Foods, Vol. I, ed G. Charalambous. pp. 229-278. New York: Academic Press.

SPECIFICATIONS FOR FUNGI IN AUSTRALIAN FOODS

S. Andrews

School of Chemical Technology
South Australian Institute of Technology
Ingle Farm, SA 5098, Australia

SUMMARY

A survey form was circulated to 59 food processors throughout Australia to determine their specifications for fungi in a variety of raw materials. These included herbs and spices, flavour mixes including cheese powders, cereal fractions including wheat germ and oat bran, nuts, dried fruits and vegetables, coconut, cocoa and egg powders. Thiry two responses were received and analysed. For each of the food categories surveyed, there was wide variation in the acceptable levels for fungal contamination. However, a contamination level of less than 1000 fungi/gram would be acceptable or marginally acceptable to most food manufacturers.

INTRODUCTION

To assist in development of a database for unacceptable levels of fungal contamination of foods, a survey form (available from the author on request) was circulated to various Australian food manufacturers. The major objective was to determine if a 2 class (accept/reject) or a 3 class (accept/reject/marginally acceptable) evaluation plan was used and at what concentrations of fungal contamination these categories applied. Information on procedures and media used were also requested, along with whether the presence of a specific fungus influenced the acceptable guidelines. Additional information on whether mycotoxin testing was carried out, and for which specific toxins was also requested.

This survey set out to collect data on the specifications of a selection of raw materials that are incorporated into food products, rather than to evaluate an enormous range of consumer items. Some of the "raw materials", for example spices, are also consumer items, but they are used extensively in the preparation of a number of food commodities. Therefore survey forms were sent to producers of raw materials who performed their own microbiological analyses to meet processors specifications. Surveys were also sent to processors who would buy in consignments and analyse them to check that they met their own in-house specifications.

A total of 59 surveys was circulated within Australia to food processors who were considered to have appropriate quality control procedures. Most of the larger food manufacturers were included. Thirty two of these organisations responded. Many of them provided specifications for five or more raw material categories, generating a great deal of data. Although the survey was distributed to overseas collaborators, the response rate was very disappointing. The limited data received has been incorporated into the Tables, but essentially the acceptance plans are based on those of Australian food manufacturers.

The survey requested information on the following raw materials:
(i) herbs and spices, including extracts thereof;
(ii) flavour mixes, including cheese powders, used to prepare dried soups, gravies, sauces, coatings for snack foods etc;
(iii) nuts, especially peanuts and pastes or extracts thereof;
(iv) coconut;
(v) flour, all types;
(vi) cereal fractions such as rolled oats, wheat germ, oat bran, etc.;
(vii) dried fruits, fruit powders; and
(viii) cocoa and chocolate.

Recipients of the survey were also requested to include any other raw materials for which they had specifications for fungal contamination. As a result, data were also received on various gums, milk and egg powders, lecithin, malt extract and similar food ingredients.

Each of the products is considered individually below.

SPICES

Spices have been defined to be those "products of plant origin which are used primarily for the purpose of seasoning food". Technically this includes the use of dehydrated vegetables such as onion, garlic, celery, green pepper etc., however these were surveyed separately and will be discussed under the heading of flavour mixes.

Spices are produced from roots, rhizomes, leaves, bark, seeds, etc. by a combination of cleaning, drying, grinding and storage, much of which is carried out under primitive conditions. Spices are resistant to microbial spoilage if properly dried and stored, however moisture control is not always effective in the tropics where most spices originate. Contamination may also occur from soil, dust, faecal matter of birds, rodents and insects, and occasionally from the use of non-potable water. Consequently, raw spices may contain excessive numbers of fungi.

Microbiological specifications for spices have been set previously (ICMSF, 1974) but these have been found to be inappropriate and no specifications for yeasts and moulds on spices are known to exist at present. Spices with high levels of fungal contamination would normally be rejected by food processors, because incorporation into a high moisture product might cause premature spoilage. Although mycotoxins have been detected in spices, in particular, aflatoxins in nutmeg, the incorporation of a spice into a food product is usually at less than 0.1% w/w, and hence the final toxin concentration is unlikely to be significant. However contamination of a spice with mycotoxigenic fungi is of concern, because of their potential to grow and produce toxin in the food product. Fortunately, the most common fungi found on spices are nontoxigenic.

Some of the companies surveyed were processors of spices, responsible for mixing and blending, or antimicrobial treatments such as ethylene or propylene oxide fumigation, or irradiation, to meet customer specifications. Other organisations requested that suppliers meet in-house specifications. Some companies tested batches on receipt and accepted or rejected on that basis, while others relied on suppliers checking that each batch met specifications.

Twenty three surveys provided information on spices, including three from overseas. All but two had specifications for less than 1000 yeasts and moulds per gram of spice (Table 1). Just over 50% had stricter specifications of less than 100 fungi per gram.

About half included a marginally acceptable category, which allowed up to 2000 fungi per gram provided there was no organoleptic deterioration of the raw material. The variation in acceptance levels by the different organisations is a reflection on the different end uses of the spices.

These specifications would appear to be achievable based on fungal enumerations on commercial spices available in Australia during 1977-1978 (King *et al.*, 1981). In that study most samples had counts of 100/gram or less except pepper, sage and Chinese five-spice which were 10^4-10^5/gram. Samples with low counts may have been treated with antimicrobials or fumigated.

In determining fungal counts, the respondents were equally split on the use of pour plates or spread plates. The choice of medium was strongly influenced by the Australian Standards for the Microbiological Examination of Foods (AS 1766.2.2) which recommends the use of Oxytetracycline Glucose Yeast extract agar (OGY). Of the manufacturers surveyed just over 50% used OGY, about 40% Dichloran Rose Bengal Chloramphenicol agar (DRBC), and another 25% Dichloran 18% Glycerol agar (DG18). A few laboratories used Potato Dextrose Agar (PDA), Malt Extract Agar (MEA), 50-60% Glucose Yeast Extract, or Mycophil agar. Many laboratories used more than one medium was used for each product. Others used specific media for different products, depending on the water activity of the final product.

Table 1. Specifications for fungal contamination in spices

Manufacturer	Acceptable	Unacceptable	Manufacturer	Acceptable	Unacceptable
1	$< 10^4$/g	$> 10^4$/g	13	<1000/g	>1000/g
2	< 500/g	>2000/g	14	< 100/g	>1000/g
3	< 10/g	> 100/g	15	$< 10^4$/g	$> 10^5$/g
4	<1000/g	$> 10^5$/g	16	< 100/g	> 500/g
5	< 50/g	> 500/g	17	< 100/g	> 100/g
6	< 10/g	> 10/g	18	< 200/g	> 200/g
7	<1000/g	>1000/g	19	< 100/g	$> 10^4$/g
8	< 100/g	> 100/g	20	< 100/g	> 100/g
9	< 100/g	> 500/g	21	< 50/g	> 50/g
10	<1000/g	>1000/g	22	< 300/g	> 300/g
11	< 200/g	> 500/g	23	< 10/g	> 100/g
12	< 100/g	> 100/g			

The identity of the fungi on enumeration plates was rarely considered important. Two companies used Aspergillus flavus and parasiticus agar (AFPA) to detect *Aspergillus flavus* and *A. parasiticus*. Most respondents did not test batches of spices for mycotoxins, however about 30% carried out aflatoxin tests on an irregular basis.

All respondents indicated that they based consignment rejection on the belief that premature spoilage would result from contamination levels above those specified. In about 30% of the cases, fungal counts above the specifications were thought to provide an unacceptable risk of inducing a mycotoxicosis if the product was mishandled. Only one respondent was concerned that high levels of fungal contamination in the food may cause an allergenic reaction after consumption.

Some of the companies surveyed did not use dried spices, preferring oleoresins (extracted with solvents) or essential oil extracts (from steam distillation), because the oils or resins can be mixed and blended to produce a more uniform product. Furthermore, the resins and oils are essentially sterile. Many companies use the extracts adsorbed onto carriers such as salt, sugar or cereals, in which case fungal specifications have been set and are usually very low, *ca.* 10-50/g.

FLAVOUR MIXES

Flavour mixes are comprised of many dry ingredients that are mixed and packaged in moisture proof containers or sachets. Ingredients may be extracts of meat, chicken, or fish; dried vegetables or vegetable powders; milk, cheese or egg powders; spices or spice extracts; flours, starches and thickeners, as well as fats and sugars. Most of the flavour mixes considered in this survey were used for the production of dried soups, dips, flavour sachets for casseroles, sauces, gravies, seasonings, flavoured noodles, and coatings for snack foods.

Table 2. Specifications for fungal contamination in flavour mixes including cheese powders

Manufacturer	Acceptable	Unacceptable	Manufacturer	Acceptable	Unacceptable
1	< 200/g	> 500/g	8	< 100/g	> 10^4/g
2	< 10/g	> 100/g	9	< 100/g	> 100/g
3	< 10/g	>1000/g	10	< 10^4/g	> 10^4/g
4	< 100/g	> 100/g	11	< 100/g	> 100/g
5	< 100/g	> 100/g	12	< 50/g	> 500/g
6	<1000/g	>1000/g	13	< 100/g	>1000/g
7	< 100/g	> 100/g	14	< 50/g	> 500/g

The mycoflora of the flavour mixes is essentially that of the dry ingredients plus any contamination that occurs during mixing and packing. Therefore by the choice of selected ingredients with low fungal counts, and mixing and packing under good hygienic conditions, the production of flavour mixes with low fungal counts should be possible. Mixes normally contain less than 7% moisture, preventing fungal growth.

Most respondents considered acceptable flavour mixes must contain less than 100 fungi/g (Table 2). Three of the 14 companies accepted mixes with higher counts, the highest being 10^4 fungi/g by a processor able to use an ethylene oxide facility to reduce the microbial counts when necessary. Most respondents were unwilling to use mixes with $>10^3$ fungi/g.

Despite the temperature of air used to dry vegetables, the product rarely exceeds 35-40°C, killing very few of the contaminants. Reduction in contamination is usually achieved by trimming to remove the outer layers, blanching or dipping in a brine if appropriate. Some respondents had separate specifications for dehydrated vegetables so they have been considered separately.

Some specifications for powdered or dried vegetables were for less than 100 fungi per gram (Table 3) but were more commonly 1-2,000/g. Obviously base levels of fungal contamination of vegetables due to associated soil make lower acceptance levels very difficult to achieve.

Specifications levels for flavour mixes and vegetables are similar to those found in a survey of dried soups by Fanelli *et al.* (1965).

Procedures used for the analysis of flavour mixes and vegetables were similar to those for spices, with most respondents using OGY to enumerate fungi. Very little differentiation of the fungi was made, and only a few organizations tested for mycotoxins.

Table 3. Specifications for fungal contamination in dried or powdered vegetables

Product and manufacturer	Acceptable	Unacceptable
Powdered onions		
1	< 100/g	> 100/g
2	< 100/g	> 100/g
3	<1000/g	>1000/g
4	< 100/g	> 500/g
Onion flakes	< 500/g	> 500/g
French onion dip seasoning	<2000/g	>5000/g
Garlic powder		
1	< 100/g	> 100/g
2	< 100/g	> 500/g
Celery powder (1)	<1000/g	>1000/g
Pumpkin powder (1)		
Beetroot powder (2)	< 100/g	> 500/g
Tomato powder	< 100/g	> 10^4/g
Dried mushrooms	<1000/g	>1000/g
Dehydrated vegetables including carrots, bean, celery		
1	<1000/g	>1000/g
2	<1000/g	>1000/g
3	< 100/g	> 500/g

PEANUTS

The mycoflora of peanuts is essentially derived from the soil in which the nuts have grown. Peanuts are harvested by lifting bushes from the soil and inverting them on the ground to dry. Invasion of nuts by specific fungi may occur before harvest, or during drying if weather conditions are favourable for fungal growth. When dried, no further fungal growth will occur if nuts are correctly stored, however the initial mycoflora will survive. During shelling, kernels are further contaminated with spores from the shells.

Of specific concern with peanuts is the presence of *Aspergillus flavus* and *A. parasiticus* and their potential to produce aflatoxins. Blanching and roasting of nuts reduces fungal loads, but has little effect on aflatoxins present. Electronic colour sorting is used to remove nuts contaminated with aflatoxin, and fungal levels are reduced at the same time.

Most of the organisations surveyed indicated a preference for nuts with a fungal count of less than 200/g but would consider accepting batches up to 10^4/g (Table 4). Some companies only purchase roasted nuts, and had lower specifications. Unlike most of the commodities surveyed, peanuts were examined for specific fungi, *A. flavus* and *A. parasiticus*. Aflatoxin assays were used by all organisations as a criterion for accepting a consignment of nuts. *A. flavus* and *A. parasiticus* are usually detected on OGY or DRBC. Only two organisations used AFPA.

In Australia, shellers test all batches of peanuts for aflatoxins, consignments being acceptable if they contain less than 15 μg/kg total aflatoxin. Imported peanuts are tested for aflatoxins by customs authorities. Some manufacturers test in-house, others use consulting laboratories. Because each lot of peanuts is checked for aflatoxins, little justification exists for setting microbiological criteria for peanuts, except to protect products which incorporate peanuts against premature spoilage. Such specifications are sometimes desirable for food processors (Table 4).

Table 4. Specifications for fungal contamination of nuts and nut extracts

Manufacturer	Acceptable	Unacceptable	Manufacturer	Acceptable	Unacceptable
1	< 100/g	>1000/g	9	< 100/g	> 100/g
2	< 100/g	>1000/g	10	< 10^4/g	> 10^4/g
3	< 10/g	> 100/g	11	< 50/g	> 500/g
4	< 100/g	> 500/g	12	< 50/g	> 500/g
5	< 200/g	> 200/g	13	<2000/g	>5000/g
6	< 20/g	> 100/g	14	<1000/g	> 10^4/g
7	< 100/g	> 10^4/g	15	-[a]	-
8	< 100/g	> 100/g			

[a] Only aflatoxin assays used in plan

COCONUT

Intact coconuts on the tree contain few if any viable microorganisms. However, after harvest, nuts are stored on the ground in contact with soil and perhaps manure. The husking of coconuts causes damage to some shells and leakage of milk may occur. Pierced nuts may become contaminated with soil, and subsequent handling may spread contamination. The critical processing steps are the grinding of coconut meat which allows fungal contamination and dispersal, and the rate of drying of the product, which if too slow may permit fungal growth. Properly dried, coconut has a sufficiently low a_w to prevent fungal growth, but many contaminants will survive.

Table 5. Specifications for fungal contamination of coconut

Manufacturer	Acceptable	Unacceptable
1	< 50/g	> 60/g
2	< 100/g	>1000/g
3	< 20/g	> 100/g
4	< 100/g	> 10^4/g
5	< 10/g	> 100/g
6	< 100/g	> 100/g
7	<1000/g	>1000/g
8	< 100/g	> 500/g
9	<1000/g	>1000/g

Most specifications indicated acceptance up to 100 fungi/g (Table 5), with two up to 1000 fungi/g. Contamination levels higher than the acceptance level are accepted by some companies after organoleptic evaluation: unacceptability of high levels of fungal contamination relates to the action of fungal lipases on coconut lipids.

Procedures used for enumerating fungi in coconut are essentially the same as those detailed above.

COCOA AND CHOCOLATE

The a_w of cocoa and chocolate is sufficiently low to prevent growth of microorganisms. However contamination or spoilage may occur during production. During the fermentation of cocoa beans in heaps, fungal growth may occur on and within the top layers. Visibly mouldy beans should be discarded, but infected beans without obvious mould damage are accepted. After fermentation, beans are dried to a safe water content, below 8%, but poor drying or storage may permit further fungal growth. Depending on time and temperature, many of these fungi will be destroyed during roasting. Subsequently, contamination may occur from soil and poorly sanitized equipment.

Cocoa powder packaged for the retail market may contain as many as 10^5 fungi/g. The presence of such high numbers, if they produce lipases, has a detrimental effect on the product. Due to extraction and processing, cocoa butter and cocoa liquor are expected to have lower levels of fungal contamination than cocoa powder.

With only one exception, specifications for cocoa powder were less than 100/g (Table 6), but most organisations did not reject the product under 1000 fungi/g. Limited data suggested specifications for cocoa butter and cocoa liquor were lower than for cocoa powder, as expected. Specifications for chocolate were similar to those for cocoa powder, as previously summaried by Pivnick (1980).

Xerophilic yeasts concerned manufacturers of chocolate based confectionery. These organisations also enumerated their products on media containing 20-50% sucrose and had separate sets of specifications for yeasts. Most did not test for mycotoxins, although cocoa beans frequently contain *A. flavus* (Llewellyn *et al.*, 1978).

Table 6. Specifications for fungal contamination in cocoa and related products

Product and manufacturer	Acceptable	Unacceptable
Cocoa powder		
1	< 100/g	> 100/g
2	< 100/g	>1000/g
3	< 50/g	> 250/g
4	< 100/g	>1000/g
5	<1000/g	>1000/g
6	< 20/g	> 100/g
Cocoa liquor		
1	< 20/g	> 40/g
2	< 100/g	>1000/g
Cocoa butter	< 20/g	> 20/g
Chocolate		
1	< 100/g	> 10^4/g
2	< 100/g	> 100/g
3	< 10/g	> 100/g

FLOUR AND CEREAL FRACTIONS

During growth and drying in the field, cereal grains are exposed to a wide variety of microorganisms from dust, water, diseased plant material, insects, soil, fertiliser and animal droppings. During storage, the grain must be correctly ventilated to prevent moisture build up in pockets, to prevent fungal growth. During transport and distribution the grains may become further contaminated.

Growth of fungi on grain produces a musty odour due to free fatty acids, affecting gluten quality, which in turn influences the quality of flour and bread. Furthermore contamination of grain with mycotoxigenic fungi is of concern because of the high proportion of cereal based products in the human diet.

Grains for milling are washed, often in chlorinated water at high temperature, tempered, screened and aspirated. Washing reduces the microbial load, but milling introduces contamination related to the cleanliness of mill equipment. Accumulated flour residues in a mill have been reported to contain up to 3.4×10^6 fungi/g, which contributes significantly to the quality of flour generated (Christensen and Cohen, 1950). Subsequent bleaching or heating of flour will reduce mycoflora, but not improve quality.

Nearly half of the companies surveyed had strict acceptance specifications of less than 200 yeasts and moulds/g of flour (Table 7). A similar number specified less than 2000/g but three companies accepted 5×10^3-5×10^4 yeasts and moulds/g (Table 7). Despite set acceptance specifications, some companies would only reject a consignment if yeast and mould counts were greater than 1000-5000/g.

Specifications for cereal fractions such as germ or bran, were similar to those for flour (Table 8), varying from less than 100/g to less than 5×10^4/g, with less than 1000/g

Table 7. Specifications for fungal contamination of flour[1]

Manufacturer	Acceptable	Unacceptable
1	<1000/g	>1000/g
2	<1000/g	>1500/g
3	< 100/g	> 100/g
4	<1000/g	> 10^4/g
5	< 100/g	> 500/g
6	< 100/g	> 500/g
7	< 200/g	> 500/g
8	< 10^4/g	> 10^4/g
9	<5000/g	>5000/g
10	<1000/g	> 10^4/g
11	<5x10^4/g	>5x10^4/g
12	<1000/g	>1000/g
13	< 100/g	> 10^4/g
14	< 100/g	>5000/g
15	<2000/g	>2000/g

[1] Includes baker's flour, oat flour, rye flour, soy flour, white flour, wholemeal flour, and corn flour. The same specifications usually applied to each type of flour within one organisation.

being a reasonable average. Such products were rejected at contamination levels exceeding 10^3 to 5x10^4/g and again were similar to the standard applied for flour.

The variation seen in acceptance specifications for cereal products and flour is a reflection on different grain pretreatments, different types of flours made, and different end uses. A flour made from grain treated with hot chlorinated water, derived from the endosperm and bleached may justifiably be expected to contain less than 100 fungi/g. However, flour of such quality is unnecessary and probably unsuitable for certain bakery products and confectionery. Higher fungal counts are acceptable in other products with a higher proportion of the outer components of the grain and less stringent pretreatments. Because of this complexity, evaluation of the merits of specifications given is not feasible.

As a result of similar considerations, ICMSF (1986) set fungal tolerances for flour and cereal products in the range 10^2-10^4/g but rejected products with >10^5 fungi/g. The current survey supports this recommendation.

Flour and cereal fractions were analysed for fungi in a similar way to spices with an emphasis on OGY, DRBC and to a lesser extent DG18. Differentiating fungi or analysing for mycotoxins were considered justifiable only in exceptional circumstances.

Many of the milling companies surveyed did not carry out any microbiological tests and had no microbiological specifications for flour or cereal fractions. These organisations purchased grain through the Australian Wheat Board or similar Grain Authorities who were responsible for the visual examination of grain samples and rejection or downgrading on the basis of visible mould. High grade grain was assumed to have minimal fungal contamination and fungal enumeration was considered unnecessary. Most of these organisations also commented that they relied on Australia's dry conditions to minimise mould contamination. Hence many producers and users of flour do not consider microbiological specifications necessary.

Table 8. Specifications for fungal contamination on various cereal fractions

Product and manufacturer	Acceptable	Unacceptable
Wheat germ		
1	<1000/g	>1000/g
2	< 100/g	>1000/g
3	< 10^5/g	> 10^5/g
4	< 100/g	> 500/g
Oat bran/rolled oats		
1	< 100/g	> 500/g
2	<$5x10^4$/g	>$5x10^4$/g
3	<1000/g	>1000/g
Wheat bran		
1	< 100/g	> 500/g
2	<1000/g	>1000/g
Starch		
1	< 100/g	> 100/g
2	<$2x10^4$/g	>$2x10^4$/g
3	< 100/g	> 500/g
4	<1000/g	>1000/g
Gluten	< 100/g	> 500/g

DRIED FRUITS

Fruits are dried to a_w levels at which fungi cannot grow. Some dried fruits are later reprocessed, packaged and sold at higher moisture levels. These products rely on heat processing, or treatment with sulphur dioxide or sorbic acid, and effective packaging for microbial stability.

In this survey, the majority of respondents specified an acceptance level of less than 100 yeasts and moulds/g (Table 9), but two companies specified less than 10 fungi/g. A few organisations specified levels as high as 1000/g, only marginally acceptable to most companies with stricter specifications.

These specifications appear to be unrealistic if the results reported by King *et al.* (1981) still reflect the mycoflora of Australian dried vine fruits which consistently cultured 10^3-10^4 fungi/g. More recently, 58% of sultana samples were reported to be contaminated with >1000 *A. niger*/g and 11% with >1000 yeasts/g (MacDonald, 1988). It is not possible to comment on how readily organisations are able to secure samples of dried fruits that meet their specifications.

Table 9. Specifications for fungal contamination on dried fruits and related products

Manufacturer	Acceptable	Unacceptable	Manufacturer	Acceptable	Unacceptable
1	< 100/g	>1000/g	9	< 100/g	> 100/g
2	< 100/g	> 500/g	10	< 100/g	>1000/g
3	< 100/g	>1000/g	11	<1000/g	>1000/g
4	< 10/g	> 100/g	12	< 100/g	> 500/g
5	< 100/g	> 100/g	13	<1000/g	> 10^4/g
6	< 100/g	> 500/g	14	< 10/g	> 100/g
7	<1000/g	>1000/g	15	< 500/g	> 500/g
8	< 100/g	> 100/g			

For dried fruits used in chocolate and confectionery fillings, some companies specified the absence of xerophilic yeasts in dried fruits, in which case yeasts were enumerated on media containing 20-30% sucrose, in addition to the usual enumeration procedures. Companies producing or packing dried fruits containing sulphur dioxide usually had no microbiological specifications. Any fruit with visible mould damage was rejected, otherwise all fruit was considered acceptable, a satisfactory specification in view of the known sterility of sulphured fruit. Specifications for such types of dried fruit came from users, producing higher moisture products. These specifications all appear to have been derived empirically.

MISCELLANEOUS FOODS

In addition to the above food categories, some organisations provided specifications for products such as milk and egg powders, malt and yeast extracts, lecithin and gums. The number of responses in each category was small, but are presented in Table 10.

When used in infant formulae, milk and whey powders have stringent specifications (e.g. <10/g), but for use in bakery products the specifications are not as strict (e.g. <100-<1000/g). Similar specifications existed for malt extract, egg powders, yeast extract and lecithin for incorporation into bakery and confectionery products. One confectionery maker had very low specifications for malt extract, lecithin, egg albumin and milk powder but the justification for this is not apparent.

Some food processors have set specifications on gums for food use, but these varied widely. It is difficult to judge the justification for the specifications presented, as details on the different types of gums and their applications was not sought.

DISCUSSION

For each of the food categories surveyed, wide variations in the acceptance levels for fungal contamination were observed. Variations result from different product formulations as well as the different types of product processing. Further, cost-benefit perceptions

Table 10. Specifications for fungal contamination of miscellaneous foods

Product	Acceptable	Unacceptable
Milk and whey powders	< 20/g	> 40/g
	< 10/g	> 10/g
	< 10/g	> 100/g
	< 100/g	> 100/g
	< 1000/g	> 1000/g
Malt extract	< 20/g	> 20/g
	< 100/g	> 100/g
	< 100/g	> 100/g
Yeast extract	< 1000/g	> 1000/g
Lecithin	< 20/g	> 20/g
	< 1000/g	> 1000/g
Gums	< 1500/g	> 1500/g
	< 500/g	> 500/g
	< 100/g	> 100/g
Egg powder	< 1000/g	> 1000/g
Egg albumin	< 20/g	> 20/g
	< 1000/g	> 1000/g

differ, so that one company will set low specifications to minimise spoilage whereas another may accept a higher level of contamination and risk a higher spoilage rate.

As an overall guideline, a contamination level of less than 1000 fungi/g would be acceptable or marginally acceptable to most food manufacturers. Some routinely set a lower acceptance level usually less than 100 fungi/g but frequently have a marginally acceptable category that will tolerate up to 500-1000/g. Manufacturers that reject raw materials with greater than 100 fungi/g, are usually making products which receive no further processing, or products which will support the growth of fungi. At the opposite extreme one company accepts nearly all raw materials provided they contain less than 10^4 fungi/g because all their finished products are baked, and they rely on thermal inactivation of the fungi.

Companies surveyed were all considered to have acceptable manufacturing practices with good quality control objectives, otherwise the data collected would have been meaningless. Each company was requested to record the specifications by which they operated, and what levels were considered attainable. Some of the specifications appeared very low and were obtainable only through very strict hygienic practices and frequently some form of antimicrobial treatment.

Because of the general nature of the survey it was not possible to learn how each organisation set unacceptable levels of contamination. Often specifications arose from specific stability trials on manufactured commodities. However some of the completed surveys would suggest that some specifications were more or less arbitrary: for example some companies had the same specifications for all of their raw materials, despite the fact that they would be used in a diversity of products receiving different processing steps and with different formulations. Therefore there is some reservation that all of the acceptance levels indicated have been arrived at from careful consideration of product stability trials. It is acknowledged that in products containing a number of raw materials, this becomes a complex issue, nonetheless it is considered that some specification levels for some commodities may be unnecessarily low. Justification for such specifications will be requested in future extension work.

Of interest were the number of organisations which had no microbiological specifications or rarely had their raw material analysed for levels of microbial contamination. Many manufacturers of flour, cereal fractions, dried fruits and vegetables as well as spices were not included in the survey because they carried out no microbial analyses. They all relied on careful selection of their raw materials, based mainly on visual inspection and organoleptic perception, and, in some cases, supplementary moisture determinations and chemical analyses. While it is acknowledged that in most cases the trained purchaser of raw materials does this job very well given reliable sources, microbiological analysis of each consignment of raw material is recommended to enhance the Quality Assurance program of any company.

Although the emphasis in the survey was on levels of fungal contamination that were acceptable to a food manufacturer, this is not the sole criterion or the dominant parameter recommended for a Quality Assurance program. The setting of specifications and adherence to them is only part of good manufacturing practice whereby each of the critical control steps is correctly monitored. In most cases the specifications related to raw materials, so by accepting only raw materials of acceptable quality, and adopting a GMP, a good quality product could be assured. It is anticipated that in future work correlations between the specifications for raw materials and processing along with product formulation may be obtainable to justify the variations recorded for the acceptance levels within individual product categories.

Are we justified in setting specifications for the commodity categories covered in this survey? The answer is an emphatic yes, because although the raw materials surveyed were low moisture foods, essentially microbiologically stable, each was to be incorporated into higher moisture foods that were vulnerable to premature spoilage if incorrectly processed or handled. Furthermore, all of these raw materials may be contaminated by mycotoxigenic fungi. By setting specifications for raw materials, the manufacturer has better control over product spoilage or health hazards.

In this survey, a database has been established from which microbiological criteria on a number of raw materials may be established. In most cases discrimination of the types of fungi present was not considered as important as the total contamination level, although the presence of *Aspergillus flavus*, *A. parasiticus* and xerophilic yeasts were of concern for specific products. Specifications for peanuts included aflatoxin levels, but mycotoxin specifications were not justifiable for the other food categories.

The issue of sampling plans was not addressed in this survey, which already requested a substantial input, and it was considered that requesting any additional information would have lead to a lower response rate. Future work will address this issue.

Of specific interest from the surveys was the influence the Australian Standards for the Microbiological Examination of Foods for Yeasts and Moulds (AS 1766.2.2) had on the analytical procedures the laboratories adopted. In the long term it is essential that the recommended media and procedures for the examination of foods for yeasts and moulds be incorporated in the standard methods prepared in every country participating in international food trade. This will facilitate future attempts to formulate the microbiological criteria for specific food commodities.

REFERENCES

CHRISTENSEN, C.M. and COHEN, M. 1950. Numbers, kinds and sources of molds in flour. *Cereal Chem.* **27**: 178-185.

FANELLI, M.J., PETERSON, A.C. and GUNDERSON, M.F. 1965. Microbiology of dehydrated soups. I. A survey. *Food Technol.* **19**: 83-86.

ICMSF. 1974. Microorganisms in Foods. 2. Sampling for Microbiological Analysis: Principles and Specific Applications. 1st edn. Oxford: Blackwell Scientific.

ICMSF. 1986. Microorganisms in Foods. 2. Sampling for Microbiological Analyses. Principles and Specific Applications. 2nd edn. Oxford: Blackwell Scientific.

KING, A.D., HOCKING, A.D. and PITT, J.I. 1981. The mycoflora of some Australian foods. *Food Technol. Aust.* **33**: 55-60.

LLEWELLYN, G.C., BENEVIDES, J. and EADIE, T. 1978. Differential production of aflatoxin on natural and heat treated cocoa beans. *J. Food. Prot.* **41**: 785-787.

MACDONALD, S. 1988. The Microbiological Quality of Sultanas. Thesis, Royal Melbourne Institute of Technology, Melbourne, Australia.

PIVNICK, H. 1980. Sugar, cocoa, chocolate and confectionery. *In* Microbial Ecology of Foods. Vol. 2. Food Commodities. International Commission on Microbiological Specifications for Foods. New York: Academic Press.

SPECIFICATIONS FOR YEASTS IN AUSTRALIAN BEER, WINE AND FRUIT JUICE PRODUCTS

S. Andrews

School of Chemical Technology
South Australian Institute of Technology
Ingle Farm, S.A. 5098, Australia

SUMMARY

A survey of beverage manufacturers in Australia was carried out to ascertain specifications for yeasts in beer, wine and fruit juice products. Data were received and analysed from processors responsible for more than 80% of Australian beer, wine and fruit juice production. Most processors had specifications for commercial sterility for beer, wine and pasteurised fruit juice products. Greater variation was observed in specifications for fruit juice concentrates and chilled pasteurised juice, reflecting the differences in preservative systems and storage temperatures for these products.

INTRODUCTION

To solicit information on levels of yeast contamination considered to be acceptable or unaccceptable in specific foods, a survey was developed to obtain information from manufacturers on the following:

1. Whether the company considered it important to differentiate between total yeast counts and the presence or number of specific spoilage yeasts in a particular product.
2. If so, what procedures and media were used.
3. Whether specifications for all products were similar, or whether account was taken of variations in processing, packaging and presence of preservatives.
4. Whether the manufacturer uses a two class or three class sampling plan.
5. What the likely course of action would be if a product batch fails to meet company specifications.

The survey forms (available from the author on request) were distributed to manufacturers of a range of fruit juice products, beer and wine, particularly to those with known control procedures. Completed surveys were obtained from most manufacturers - in each product category, it is estimated that the data refers to at least 80% of the beer, wine and fruit juice processed within Australia.

It is recognised that the yeast flora on these products and the quality of the raw materials processed may differ significantly throughout the world. The same survey was issued to seven international collaborators to circulate to representative manufacturers. The response was insufficient to incorporate into this paper, so only specifications for yeasts in products manufactured in Australia are addressed.

The results of the survey are described below.

FRUIT JUICE PRODUCTS

The mycological quality of products based on fruit juice depends on formulation, especially water activity, pH and chemical preservatives; processing; sanitation and manufacturing practice; and storage temperature throughout the distribution system.

The surveys were targeted at the following commodities:

(i) fruit juice concentrates, which usually contain 60-70% sugar, are of pH 2.0 - 2.5 and have been pasteurised;

(ii) fruit juices or fruit based drinks that are commercially sterile, i.e. products pasteurised at 90°C for at least 30 sec, then aseptically packaged;

(iii) fruit juices or fruit based drinks which contain preservatives and have been pasteurised, but are distributed chilled.

In Australia, fruit based drinks are fruit juices diluted to a final concentration of 20-30% fruit juice. The survey did not distinguish between these products, as their mycology is similar. Canned fruit juices were not surveyed.

Fruit juice concentrates

Eight responses were obtained, of which 75% indicated use of a three class sampling plan (Table 1). Most manufacturers specified an acceptance level of less than 100 yeasts/mL but some less than 1-5/mL. The variation in acceptance levels was greatly influenced by the use of preservatives and storage temperature. Specifications were often set by the purchaser, who was usually a juice or drink processor.

Unacceptable levels for concentrates varied from greater than 1/mL up to greater than 1000/mL. Higher levels are specified for products destined for high temperature pasteurisation; prior to this they will be stored frozen, often with a preservative added. Very low unacceptable levels are for concentrates without preservatives to be transported overseas without refrigeration.

A variety of media was used for yeast enumeration, including acidified Potato Dextrose Agar (PDA) and Malt Extract Agar (MEA), WL Nutrient Agar (WL), Orange Serum Agar (OSA) and Mycological Agar. Some manufacturers used more than one medium simultaneously. Only three of the respondents tested for xerophilic yeasts using media with 30-45% glucose or sucrose added. Specifications for xerophilic yeasts varied from not greater than 1/100 mL to a limit of 10/mL, depending on storage temperature and preservative use. Two processors tested for preservative resistant yeasts, one by

Table 1. Acceptable levels (cfu/mL) of yeast contamination in fruit juice concentrates.

Manufacturer	Acceptable	Unacceptable	Marginally acceptable
1	< 100	> 200	100 - 200
2	< 250	> 1000	250 - 1000
3	< 30	> 50	30 - 50
4	< 100	> 1000	100 - 1000
5	< 50	> 50	-
6	< 20	> 20	-
7	< 5	> 100	5 - 100
8	< 40	> 100	40 - 100

adding the appropriate preservative to their enumeration medium, the other by using MEA with 0.5% acetic acid. These processors distribute a preservatised concentrate which is not frozen, and both set a limit of 10/mL of preservative resistant yeasts.

Acceptance levels were based on achievable levels using good manufacturing practice and the expected storage life of the products. Failure to meet these levels was usually attributed to poor factory sanitation, especially after the pasteuriser, and less commonly to underpasteurisation or poor quality raw materials. Unacceptable levels of yeast contamination invariably lead to organoleptic deterioration of the product, which may limit its future processing. Even with acceptable levels of yeast contamination, deterioration of an unfrozen concentrate may occur due to temperature fluctuation allowing surface condensation and yeast growth in the diluted zone.

Commercially sterile fruit juices and drinks

Manufacturers producing commercially sterile fruit juices and fruit based drinks use a two class sampling plan. Three manufacturers responded, two specifying less than 1/mL, the third less than 100/mL because a preservative was incorporated. Such a practice seems to defeat the purpose of high temperature pasteurisation and aseptic filling.

Each manufacturer used pour plates of PDA, OSA, especially for citrus products, or WL. The latter two media detect lactic acid bacteria as well as yeasts. None of the manufacturers indicated that they filtered the product prior to plating.

Three respondents from the U.S.A. indicated that they routinely carried out product incubation by holding at 25-30°C for 7-10 days, and used this test for acceptance/rejection. If a product fails the incubation test it is usually cultured to identify the causative microorganism and its source, but this does not influence the acceptance/rejection plan. Australian manufacturers did not report such testing.

Failure to meet specifications was reportedly due to inadequate pasteurisation, poor sanitation of the filling line, or incorrect filling temperature. One of the respondents highlighted a problem in their acceptance plan for the pasteurised products, suggesting that underpasteurisation may permit survival of heat stressed yeasts, which may then resuscitate over periods up to 10-12 weeks at ambient temperatures. Such products pass the specifications, but spoilage may occur after more than 3 months storage.

Pasteurised, preserved, chilled fruit juices and drinks

Most manufacturers of chilled juices buy concentrates according to microbiological and chemical specifications, dilute, add preservative, pasteurise and package it. These processors have limited technical facilities and rely on the concentrate supplied meeting specifications. In consequence, only five producers of chilled, preservatised juices responded to the survey.

A three class sampling plan is most frequently used, commonly with an acceptance level of less than 500/mL but sometimes as low as 10/mL (Table 2). Acceptance levels often include bacteria as they can spoil this product as readily as yeasts. Pour plates are usually used, of WL, OSA, PDA or MEA.

A differential count for preservative resistance yeasts may be included by the addition of preservative to the medium or by the use of MEA with 0.5% acetic acid. Counts of greater than 10 preservative resistant yeasts/mL are considered unacceptable.

One producer reported incubating samples at 25°C for 2 days, and releasing it after visual and organoleptic evaluation. However, products of this type are usually released immediately before such testing could be carried out.

Table 2. Acceptable, unacceptable and marginally acceptable levels (cfu/mL) of yeast contamination in pasteurised, preservatised, chilled fruit juices and fruit based drinks.

Manufacturer	Acceptable	Unacceptable	Marginally acceptable
1	< 250	> 10,000	250 - 10,000
2	< 500	> 500 or > 10 preservative resistant yeast/mL	
3	< 5	> 20	5 - 20
4	< 250	> 1000	250 - 1000
5	< 200	> 1000	200 - 1000

Unacceptable levels were reportedly set according to preservative type and concentration, the pasteurisation process and expected shelf life. All manufacturers indicated that their unacceptable levels were conservative to allow for distribution and consumer abuse with respect to storage temperature. A count of less than 500 microorganisms/mL is readily achievable with good manufacturing practice.

Two responses from the U.S.A. indicated production of chilled juices without added preservatives: the acceptable levels were less than 1/mL and 10/mL respectively. Both rejected this product at greater than 100/mL. Both manufacturers held product at ambient temperatures for 48 hours before yeast enumeration. Pasteurisation processes were not stated, but must be substantial to meet these standards.

BEER

Contamination of beer with wild yeasts seriously affects quality. This is usually manifest as turbidity or ropiness, sometimes pellicle formation on the surface and usually off flavours. The critical level is considered to be around 4-10 yeast cells per bottle (Brumstead and Glenister, 1962).

Surveys were circulated to all main Australian breweries, requesting information on specifications for lager beers, low alcohol beers and stouts. Responses indicated that standards were the same for all products. A conventional ale produced by one manufacturer is considered separately. Two class sampling plans are used in each brewery. All breweries specify less than 5 yeasts/100 mL, except one with 10 yeasts/100 mL (Table 3). Most breweries determine the level of yeast contamination by filtering 100 mL or more of beer and culturing the filter on WL, although one brewery uses Wort Agar and another Universal Beer Medium (Oxoid CM651).

Differentiation of wild yeasts from brewery yeasts may be made on WL but more commonly, Lysine Medium (Oxoid CM191) or a CuSO4 Yeast Malt Agar (Difco YM) is used. The detection of *Saccharomyces diastaticus* causes rejection of beer because of its ability to ferment maltotetroses and dextrins, and the detection of *Brettanomyces* is also unacceptable because of off-flavour development.

Standards are readily achievable if pasteurisation is correct (60°C for 10-15 min.), and sanitation is good. Some breweries do not use yeast counts on end product as a

Table 3. Acceptable and unacceptable levels of viable yeasts in beer.

Manufacturer	Acceptable	Unacceptable
1	< 0/100 mL	> 0/100 mL
2	< 10/375 mL	> 10/375 mL
3	< 0/100 mL	> 0/100 mL
4	< 5/100 mL	> 5/100 mL
5	< 10/100 mL	> 10/100 mL
6	< 1/100 mL	> 1/100 mL
7	< 4/100 mL	> 4/100 mL
8	Accelerated stability trials only	

clearance criterion but prefer accelerated stability trials. In these cases the beer is held at 25-30°C for 5-7 days and then checked visually and organoleptically. One brewery does stability trials only, but most breweries do both, and use yeast enumeration to monitor plant sanitation and the pasteurisation procedures as well as the finished product.

An ale produced by one brewery undergoes secondary fermentation in the bottle. After primary fermentation in open vats, the ale is primed, bottled and fermented. After 4 weeks of secondary fermentation, the beer is tested. If less than 20 wild yeasts/mL are detected the batch is accepted. The beer is considered marginally acceptable if it contains 20-400 yeasts/mL, provided there are no detectable off flavours. However if *Saccharomyces diastaticus* or *Brettanomyces* spp. are detected, the beer is rejected.

WINE

Growth of residual yeasts in bottled wines may produce turbidity, sediments or off flavours. In the case of sweet wines or "coolers" (fruit juices added to wine bases), sugar metabolism by contaminants may spoil the product or destroy bottles. Deak and Reichart (1986) showed that at a contamination level of 50 ± 23 yeasts per litre there was a 95% chance that wine would be stable for 12 months under conditions in Hungary. However, these limits would not be applicable to Australian conditions, where ambient temperatures are much higher.

Separate surveys were circulated for dry table wines (less than 0.2% sugar) and sweet table wines (>0.2% sugar). However, in all but two responses, the same specifications were used, so all table wines were considered together.

In general the Australian wine industry uses a two class sampling plan. Wine is commonly accepted if it contains no detectable yeasts in a 750 mL sample (Table 4) but some wineries tolerate up to 30 yeasts/750 mL. One winery accepts up to 75 yeasts/750 mL, determined by direct microscopy, but all others enumerate yeasts. In all cases, the wine was filtered and the filter cultured on WL medium. This medium, used because it allows the simultaneous detection of both yeasts and bacteria, may also be supplemented with tomato juice to enhance detection of lactic acid bacteria.

Most wineries filter a 750 mL volume, one bottle, after the cork is aseptically removed. Some wineries filter smaller volumes (e.g. 300 mL), and this is reflected in the specifications. If the count exceeds the acceptance level, another sample is analysed 4-5 days later. If a similar yeast count is detected then, the sample is usually rejected temporarily, because this implies that the yeast can survive in the wine. If the second test meets the acceptance level, the yeast contaminants are assumed to be of non-wine origin and to have died. If the bottled wine is not distributed immediately, rejected product is analysed again 3-4 weeks later, to determine if the contaminants have now died. If living yeasts are still detected, accelerated stability trials would be carried out to determine if the wine is likely to spoil.

Wild yeasts, usually detected in wine on WL or Lysine Medium (Oxoid CM191), are usually carefully examined. Film yeasts or *Rhodotorula* species are not a concern as they will not grow in wine. However, unfamiliar isolates, possibly *Saccharomyces* species or *Zygosaccharomyces bailii* are usually forwarded to a mycologist for identification before an action plan is implemented.

Table 4: Acceptable, unacceptable and marginally acceptable levels of yeast contamination in wine

Manufacturer	Acceptable	Unacceptable	Marginally acceptable
1	< 15/750 mL	> 15/750 mL	0-20/750 mL
2	< 1/2000 mL	> 1/2000 mL	
3	< 0/750 mL	> 20/750 mL	
4	< 0/750 mL	> 0/750 mL	
5	< 30/750 mL	> 30/750 mL	
6	< 10/100 mL	> 10/100 mL	
7	< 0/300 mL	> 0/300 mL	
8	< 10/750 mL	> 10/750 mL	
9	< 0/750 mL	> 0/750 mL	
10	< 5/250 mL	> 5/250 mL	

Yeasts of concern are sometimes inoculated into a "significance broth," which mimics the wine in pH and alcohol content but is slightly more nutritious. If growth occurs within 72 hours, it is considered that the yeast is capable of spoiling the wine (Thomas and Ackerman, 1988). This test has been introduced into several wine bottling plants in the U.K., but it is yet to be adopted in Australia. In time, Significance Broth may replace accelerated stability trials and form part of the acceptance specifications for wine.

As wines may contain up to 300 mg/L of sulphur dioxide or 200 mg/L of sorbic acid, some wineries tolerate a low level of yeast contamination because preservatives may inhibit wild yeasts present in the wine.

Failure to meet these specifications in wineries is usually due to a fault in the operation of the membrane filters or poor sanitation between the filter and the filler. On rare occasions defective packaging may lead to unacceptable yeast contamination.

Direct microscopy is used to examine wine for viable yeasts in most wineries, but is used in place of viable counts in only one of the wineries surveyed. The majority of wineries use acridine orange to stain the cells (viable - green, non viable - orange) but others use Genetian violet (viable - pink, non viable - violet). None of the wineries have automated the process because this method is not part of specifications.

Wine coolers, wine based beverages to which fruit juice, flavour and carbonation may be added, are a microbiological hazard because of their substantial sugar content. This product was also surveyed: yeast specification varied from 0/750 mL to 5/mL. Sulphur dioxide and sorbic acid are added as preservatives, often assisted by carbonation. The presence of *Z. bailii*, however, is very dangerous, so that yeast contaminants in wine coolers need to be carefully identified before the product is released.

DISCUSSION

The specifications described in this paper relate to microbiological testing of end products, which is only part of a satisfactory quality control program. These specifications rely on correct monitoring of critical processing steps and plant sanitiation.

In setting specifications for beer, wine and pasteurised juices, most processors aim for commercial sterility. Sample sizes for commercial sterility tests varied widely from 750 mL for many wineries, to 100 mL for breweries to 1 mL for fruit juices. Product stability trials were routinely carried out on beer and wine and were incorporated into specifications. U.S. processors of pasteurised fruit juices relied upon accelerated stability trials for product acceptance and rarely carried out yeast enumeration on the final product. Such a procedure is not used in Australia: it is recommended that accelerated stability trials be emphasised in developing specification guidelines, because larger sample sizes can be analysed.

Great variation was found in specifications for fruit juice concentrates and chilled preservatised fruit juices. This reflects the different processing conditions, different preservatives and preservative levels, and varied storage temperatures. Therefore, to recommend specifications for these products is difficult unless the types of processing and end uses are differentiated. This is not a problem with wine, beer, and pasteurised juices because manufacturers produce these products to meet specifications for commercial sterility.

In formulating specifications for beer, wine and fruit juice products, it is important to include specifications for preservative resistant yeasts, xerophilic yeasts and heat resistant fungi. These fungi present a high spoilage risk, but many processors did not enumerate them, or include them in specifications. Some processors were aware of the spoilage risks associated with *Zygosaccharomyces bailii*, *Schizosaccharomyces pombe*, *Saccharomyces diastaticus* and *Brettanomyces* spp., but did not routinely monitor them.

Specifications quoted in this paper were all achieved on a regular basis by processors. Obviously each organisation had slightly different cost-benefit ratios which is reflected in the variations observed in the acceptance levels for similar products.

Data provided allow tentative microbiological specifications for acidic beverages to be formulated. Such criteria are for the processors' benefit rather than the consumer, because these products do not usually pose a health hazard unless product explodes. The number of media used to enumerate yeasts, especially for fruit juice products, is of concern. In a separate study WL Nutrient Agar has been evaluated against DRBC and OGY for the enumeration of yeasts from these products. The recovery of yeasts on WL is similar to that on OGY (Andrews, 1992), but other comparative studies are needed. Ultimately, uniform microbiological criteria will require uniform media and methods. In addition, data from overseas producers of wine, beer and fruit juices are required to ascertain if Australian specifications are similar to international specifications.

REFERENCES

ANDREWS, S. 1992. Comparative study of WL Nutrient Agar with DRBC and OGY for yeast enumeration in foods. *In* Modern Methods in Food Mycology, eds R.A. Samson, A.D. Hocking, J.I. Pitt and A.D. King. pp. 61-65. Amsterdam: Elsevier.

BRUMSTEAD, D.D. and GLENISTER, P.R. 1962. The viability of minimal populations of a wild yeast in beer: possible implications for bulk pasteurization. *Proc. Am. Soc. Brew. Chem.* **1962**: 72-76.

DEAK, T. and REICHART, O. 1986. Unacceptable levels of yeasts in bottled wine. *In* Methods for the Mycological Examination of Food, eds A.D. King, J.I. Pitt, L.R. Beuchat and J.E.L. Corry. pp. 215-219. New York: Plenum Press.

THOMAS, D.S. and ACKERMAN, J.C. 1988 . A selective medium for detecting yeasts capable of spoiling wine. *J. Appl. Bacteriol.* **65**: 299-308.

3

METHODS FOR XEROPHILIC FUNGI

COLLABORATIVE STUDY ON MEDIA FOR ENUMERATION OF XEROPHILIC FUNGI

Ailsa D. Hocking

CSIRO Division of Food Processing
P.O. Box 52
North Ryde, NSW 2113
Australia

With contributions from C. Anderson, S. Andrews, N. Aran, L.R. Beuchat, N. Braendlin, N.J. Charley, D.E. Conner, T. Deák, J.C. Frisvad and A.D. King.

SUMMARY

Dichloran 18% glycerol agar (DG18) has been gaining acceptance for enumeration of xerophilic fungi from reduced moisture foods. In an attempt to more selectively restrict growth of *Eurotium* colonies in relation to other more slowly growing xerophiles, two other media were compared with DG18. These media were DG18 + 2% NaCl, and DRBC + 7.5% NaCl.

A total of 92 counts was carried out on 62 different samples. Foods and commodities enumerated included cereals and cereal fractions, spices, nuts, confectionery, bakery products, dried vegetables and animal feeds. Counts were carried out using the dilution plating technique with surface plating. Plates were incubated upright at 25°C for 7 days.

The highest counts were obtained on DG18 (58.8%) followed by DG18+2% NaCl (33.3%), with DRBC+7.5% NaCl having the highest counts in only 9.6% of samples. Many participants commented that this latter medium was highly inhibitory, particularly to *Wallemia sebi* and *Aspergillus candidus*. However, it did induce compact colony formation in some *Eurotium* species, and where *Eurotium* was the dominant flora, DRBC+7.5% NaCl often gave the highest counts. DG18 gave best development of hyphal colours in *Eurotium* species.

The consensus was that DG18 was still the most satisfactory of the media tested for enumeration of fungal flora in reduced a_w foods.

INTRODUCTION

The use of media of reduced water activity (a_w) for enumeration of fungi in low moisture foods began with Christensen (1946), who used Malt Salt Agar (MSA) for the detection of fungi in flour. Since then, numerous formulations for reduced a_w media have been published, using either salt, sucrose or glucose to control the a_w of the medium. Salt based media have the disadvantage that they can be inhibitory to some species of fungi, and they often gel poorly. Sugar based media give better recoveries of fungi, but rapid, spreading growth by *Eurotium* species often masks slower growing fungi like *Aspergillus* series *Restricta*. Sorbose (the oxidation product of sorbitol) has been used in media to restrict radial growth rates of fungi (Moore and Stewart, 1972; Trinci and Collinge, 1973), but sorbitol was observed to have little effect on the spreading growth habit of *Eurotium* species when used in experimental media formulations in preparation for this collaborative study (Hocking, unpublished).

Dichloran 18% Glycerol agar (DG18; Hocking and Pitt, 1980) was designed to overcome some of these problems, and has been gaining acceptance for enumeration of xerophilic fungi from reduced moisture foods. It was originally designed to maximise enumeration of common xerophilic fungi such as *Eurotium* species, members of *Aspergillus* ser. *Restricta* and *Wallemia sebi*, fungi which are not readily enumerated on high a_w media such as Dichlroan Rose Bengal Chloramphenicol agar (DRBC). DG18 gives excellent recovery of these fungi, but *Eurotium* species still grow much more quickly than some of the other xerophiles. Moreover, the colonies have irregular margins which makes accurate enumeration difficult if the colonies are crowded on the plate.

In an attempt to more selectively restrict growth of *Eurotium* colonies in relation to the other more slowly growing xerophiles, a number of modifications of DG18 and DRBC were investigated. Mixtures of 5% sorbitol plus 10% glycerol, and 5% sorbitol plus 15% glycerol both with and without rose bengal were tested with pure cultures of *Eurotium chevalieri* and mixtures of *E. chevalieri* and *Mucor racemosus*. While these media showed promise, they did not appear to be significantly better than DG18 at restricting growth of *Eurotium* species, and they were considerably more trouble to prepare.

Two other media were then compared with DG18. These were DG18 + 2% NaCl, and DRBC + 7.5% NaCl. Media with added salt have been used in the past, and are still in common but decreasing use, as salt is considered to be inhibitory to some species, and to sublethally damaged cells. However, these media had the advantage that they could be prepared from commercially available basal media, simply by the addition of the appropriate amounts of NaCl. The salt concentrations were chosen because 7.5% NaCl is the amount often added to PDA to prevent seeds sprouting (Mislivec and Bruce, 1977), but only 2% NaCl was added to DG18 because it was not considered desirable to significantly reduce the a_w. The final a_w values of the two media were DG18 + 2% NaCl, 0.945 a_w, and DRBC + 7.5% NaCl, 0.95 a_w.

These two media exhibited improved control of *Eurotium* species, so it was decided to subject them to collaborative testing.

MATERIALS AND METHODS

Enumeration of fungi

Ten laboratories participated in the collaborative study. They were requested to compare DG18, DG18 + 2% NaCl, and DRBC + 7.5% NaCl, using the spread plating technique. Plates were to be incubated at 25°C for 7 days. Each laboratory performed counts on 6 to 10 samples in which xerophilic fungi could be expected to be present. Each laboratory was asked to carry out two counts on samples of domestic wheat flour, as a reference point. Some laboratories performed duplicate counts on all samples.

Laboratories were asked to identify fungi were possible, and to comment on the ease of use of the three media, their efficiency in controlling spreading of *Eurotium* colonies, and their general effectiveness as media for enumeration of fungi from low moisture foods.

Statistical analysis

The results were subjected to statistical analysis. Counts based on the three media were compared by the usual two way analysis of variance for all 101 samples combined, and also for the following sub-groups: cereals and cereal fractions (47 counts); animal feeds (15); herbs and spices (14); nuts and oil seeds (9); fruit and vegetable products (10) and miscellaneous (6). Using the residual mean square from the analysis of variance, least significant differences were calculated to determine whether media gave statistically different counts when averaged over all relevant samples.

RESULTS

Fungal counts

A total of 101 dilution counts was performed, for 81 samples on the three test media. Results are presented in summary form only. The largest group of samples examined was wheat flour, whole grains and grain fractions. For wheat flour, fungal counts were generally in the range $5x10^2$ to $5x10^3$ (Figure 1). In whole grains, (which included wheat, barley, oats, and rice), and grain fractions (oat meal, rye meal, rice flour, rice meal), fungal counts were generally between 10^4 and $5x10^5$ (Figure 1).

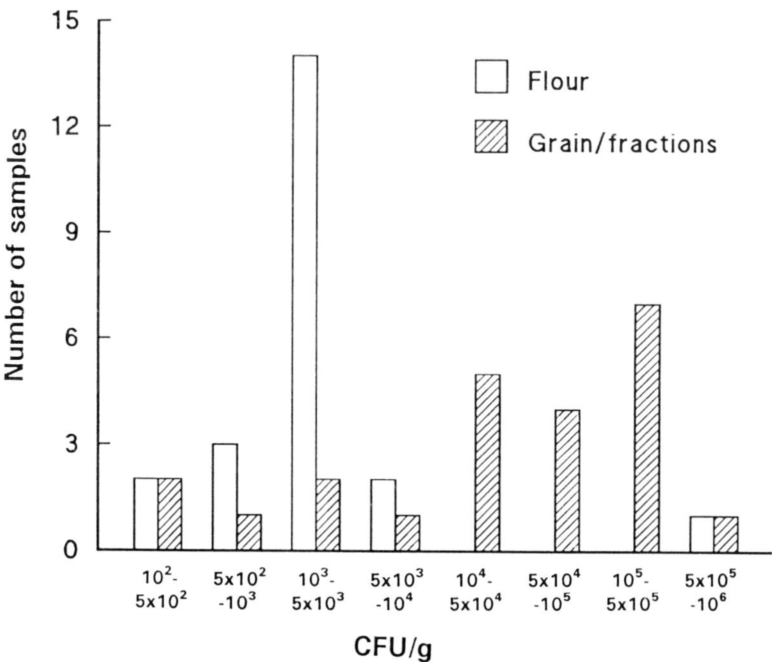

Figure 1. Fungal counts obtained on DG18 agar from wheat flour samples (22 counts) and whole grains and grain fractions (23 samples). Whole grains included wheat, barley oats and rice. Grain fractions included wheat bran, pollard, wheat meal, oat meal, rye meal, rice meal and rice flour.

Table 1. Summary of counts obtained on DG18 for various food groups

Food type	Number of samples	Average count	Range
Animal feeds	7	$2.1x10^5$	$2.1x10^4$-$3.3x10^5$
Pepper (black and white)	7	$5.1x10^5$	$1.3x10^3$-$1.8x10^6$
Spices	10	$7.8x10^4$	$2.0x10^2$-$6.3x10^5$
Oilseeds, nuts	10	$1.5x10^5$	$3.5x10^2$-$9.0x10^5$
Fruits, fruit products	6	$4.9x10^3$	$2.7x10^2$-$1.8x10^4$

A summary of counts obtained on DG18 for the other samples is shown in Table 1. Some samples of black pepper, animal feeds and mixed spices had quite high fungal populations ($>10^5$). Fruit based products had the lowest counts.

Comparison of media

A summary of the geometric means of counts for all samples, and for groups of samples, on the three test media, is shown in Table 2. Overall, counts were higher on DG18 than the other two media, but the differences between DG18 and DG18+2% NaCl were often not significantly different.

Table 2. Summary of geometric means of fungal counts obtained on DG18, DG18+2% NaCl and DRBC+7.5% NaCl for all samples, and for samples by subgroup.[1]

	DG18	DG18+2% NaCl	DRBC+7.5% NaCl
Overall	4.40[a]	4.28[b]	3.83[c]
Cereals	3.94[a]	3.90[a]	3.41[b]
Animal feeds	6.32[a]	6.31[a]	6.18[a]
Herbs and spices	4.28[a]	4.14[a]	3.88[b]
Nuts and oil seeds	4.25[a]	4.18[a]	3.63[b]
Fruits and vegetables	3.41[a]	2.98[a]	1.98[b]
Miscellaneous	5.36[a]	4.80[a]	4.12[a]

[1] Means with the same superscript within the same row were not significantly different ($P<0.05$). Occasionally the significance level was taken as 0.06.

DRBC +7.5% NaCl

Overall, DRBC+7.5% NaCl consistently gave the lowest counts. *Eurotium* diameters were restricted on this medium (20 mm compared with 30-40 mm for DG18 and DG18+2% NaCl), making them easier to count, but counts were often lower than on the other two media. DRBC+7.5% NaCl gave very good control of *Mucor*, *Rhizopus* and

Syncephalastrum, but most participants commented that other fungi such as *Wallemia sebi*, *Penicillium* species and yeasts were strongly inhibited on this medium, and *Fusarium* was not detected.

Many participants noted that DRBC+7.5% NaCl required longer incubation times, up to two weeks, because of very slow colony development, and that colonies were sparse on this medium. Even after prolonged incubation, counts on DRBC+7.5% NaCl were usually lower than on the other two media.

DG18 + 2% NaCl

The differences observed between DG18 and DG18+2% NaCl were small. *Eurotium* counts were sometimes higher on DG18+2% NaCl, but species were easier to differentiate on DG18 because of better colour development in the encrusting hyphae. DG18+2% NaCl sometimes needed more than 7 days incubation because of slower colony develoment. However, one participant noted that it restricted the colony diameter of *S. racemosum* better than DG18.

General comments

For ease of counting, DRBC+7.5% NaCl was best because it restricted growth of *Eurotium* colonies most effectively, but the counts were consistently lower than other 2 media. DG18+2% NaCl was often slightly easier to count than DG18, but the counts on these two media were not significantly different for the commoditiy groups tested (Table 2), and any benefit was outwieghed by the inconvenience of having to add an extra ingredient to a preformulated medium. *Aspergillus* and *Penicillium* grew on all three media, but recoveries on DRBC+7.5% NaCl were often lower than the other two media. All media controlled the growth of *Rhizopus* adequately.

ACKNOWLEDGEMENT

The author is grateful to Mr John Best, CSIRO Institute of Animal Production and Processing, Biometrics Unit, for statistical analysis of the data.

REFERENCES

CHRISTENSEN, C.M. 1946. The quantitative determination of molds in flour. *Cereal Chem.* **23**: 322-329.

HOCKING, A.D. and PITT, J.I. 1980. Dichloran-glycerol medium for enumeration of xerophilic fungi from low moisture foods. *Appl. Environ. Microbiol.* **39**: 488-492.

MISLIVEC, P.B. and BRUCE, V.R. 1977. Direct plating versus dilution plating in qualitatively determining the mold flora of dried beans and soybeans. *J. Assoc. Off. Anal. Chem.* **60**: 741-743.

MOORE, D. and STEWART, G.R. 1972. Effects of 2-deoxy-D-glucose, D-glucosamine, and L-sorbose on the growth of *Coprinus lagopus*. *J. Gen. Microbiol.* **71**: 333-342.

TRINCI, A.P.J. and COLLINGE, A. 1973. Effect of L-sorbose on growth and morphology of *Neurospora crassa*. *J. Gen. Microbiol.* **78**: 179-192.

OBSERVATIONS ON THE EFFECT OF RECOVERY CONDITIONS ON COLONY FORMATION BY *WALLEMIA SEBI* AND *CHRYSOSPORIUM FARINICOLA*

L. R. Beuchat

Department of Food Science and Technology
University of Georgia, Agricultural Experiment Station
Griffin, GA 30223-1797, USA

SUMMARY

Recovery of heat stressed *Wallemia sebi* conidia at 0.82-0.92 a_w was retarded in the presence of NaCl, while glucose and sorbitol at 0.82-0.97 a_w had similar, less detrimental effects. Colonies were largest on yeast extract glucose agar (YGA) containing sorbitol and smallest on YGA containing NaCl. Maximum recovery of heat-stressed *W. sebi* conidia was observed at 0.92 a_w. Colony formation by uninjured conidia was reduced at 0.92 a_w when incubated at 20°C compared with 25°C, while at 30°C, numbers recovered at 0.97 a_w were reduced. The effect of incubation temperature was more marked on heat stressed conidia. The sensitivity of heat-stressed *W. sebi* conidia increased as the pH of the recovery medium was decreased from 6.55 to 3.71. Aleuriospores of *Chrysosporium farinicola* from 20 day old cultures grown at 25°C were more heat resistant than those from 14 day cultures. Recovery of heat stressed aleuriospores was greater as the a_w of YGA was decreased from 0.95 (glucose, glycerol) and 0.94 (sorbitol) to 0.89 and 0.88, respectively. Populations recovered on NaCl-supplemented YGA were much lower than those detected at 0.92 a_w; no colonies were formed at 0.88 a_w. Tolerance to a_w above 0.88-0.89 as influenced by solute type was in the order of glucose > sorbitol > glycerol > NaCl. A general increase in recoverable aleuriospores was observed when plates were incubated at 20°C compared with 25 or 30°C. This study indicates that sorbitol may be the solute of choice for formulating xerophile enumeration media.

INTRODUCTION

Xerophilic fungi are present on a wide variety of foods, although spoilage is most commonly associated with intermediate moisture fruits, dried fish and meats, confectionery products, jams and jellies, syrups, nuts and cereal grains (Pitt, 1975). Media with reduced water activity (a_w) are necessary for enumerating xerophilic fungi in these foods.

The effect of sublethal injury from physical stress on nonxerophilic fungal spores and vegetative cells is well known (Beuchat, 1984), but the behaviour of xerophile spores exposed to such conditions is unknown. Improved media for enumerating xerophilic fungi is important; however, such media must be formulated to resuscitate injured as well as healthy cells. The work reported here is a distillation of studies designed to determine the effects of enumeration conditions, i.e. media formulation, temperature and incubation time, on heat stressed conidia of *Wallemia sebi* (Fries) von Arx (Beuchat and Pitt, 1990a) and aleuriospores of *Chrysosporium farinicola* Burnside (Skou) (Beuchat and Pitt, 1990b).

MATERIALS AND METHODS

Preparation of conidia used as test inoculum

Wallemia sebi FRR 1471 was grown on a yeast extract glucose medium (YGA) containing yeast extract (10.0 g), glucose (50.0 g), K_2HPO_4 (4.0 g), agar (15.0 g) and 1000 mL distilled water (final pH 6.2). Four day old cultures grown at 25°C were harvested by adding 3 mL of sterile 0.1% peptone water and dislodging conidia by gently rubbing with a sterile glass rod. Suspensions containing 1-3 x 10^7 conidia per mL served as inocula for all studies involving heat treatment.

Inocula of *Chrysosporium farinicola* FRR 377 aleurioconidia were prepared by techniques similar to those above, except that the medium contained 32.6% (w/v) glucose and was sterilised by steaming for 30 min on 3 successive days, and the aleurioconidia were harvested from 14 and 20 day old cultures.

Determination of heat sensitivity of conidia

The ability of *W. sebi* conidia to survive heat treatment when suspended in peptone water at 48, 50, 52 and 54°C was determined. The conidial inoculum (0.5 mL) was mixed with peptone water (4.5 mL) adjusted to the desired heating temperature in a water bath. The suspension, contained in 13 x 100 mm test tubes, was positioned such that its surface was at least 2 cm below the level of the constantly circulating water in the bath. Samples were withdrawn after 10, 20 and 30 min of heat treatment, immediately serially diluted in peptone water and surface plated (0.1 mL) in duplicate on YGA. Colonies were counted after 5 days incubation at 25°C.

Aleurioconidia of *C. farinicola* were similarly treated except that growth was on YGA plus glucose (0.92 a_w) and incubation for 12 days at 25°C.

Determination of susceptibility of conidia to heat injury

Conidia of *W. sebi* were heated in peptone water at 48°C for 10, 20, 30, 40 and 50 min as described above. Recovery media consisted of YGA (a_w 0.97) and YGA plus NaCl, glucose or sorbitol to give 0.96, 0.92, 0.87 and 0.82 a_w. Colonies were counted and colony diameters recorded after 6 days incubation at 25°C. Measurements of colony diameters of unheated *W. sebi* on similar media were also made after 9 days incubation at 25°C.

Aleuriospores of *C. farinicola* were treated similarly except that heating was at 52°C for 0, 10, 20, 30 and 40 min. Dilutions (0.1 mL) were plated in duplicate on YGA supplemented with glucose, sorbitol, glycerol or NaCl at a_w values ranging from 0.88 to 0.95 (pH 6.2 to 6.9). Colonies were counted after 12 to 14 days incubation at 25°C. The diameters of colonies formed on YGA containing various concentrations of solutes were measured after 12 days incubation at 25°C.

Influence of incubation temperature and pH of recovery medium on resuscitation of heat-stressed conidia

Conidia of *W. sebi* heated in peptone water at 48°C for 0 and 30 min were spread plated on YGA (a_w 0.97) and YGA plus NaCl of 0.96, 0.92, 0.87 and 0.82 a_w. Plates were incubated at 20, 25 and 30°C, and colonies were counted after 6 days. Heated conidia were also spread plated on YGA (0.97 a_w) adjusted to pH 6.55, 5.85, 4.49 and 3.71 with 5M phosphoric acid; pH adjustment was made after YGA was heat sterilised. Colonies were counted after 6 days incubation at 25°C.

Suspensions of aleuriospores of *C. farinicola* heated at 52°C for 30 min were spread plated on CYA plus 0, 20 and 40% sucrose (0.997, 0.964 and 0.936 a_w respectively). Suspensions were then either diluted immediately in 0.1% peptone or held for 20 min at 23°C before diluting and surface plating on various YGA media containing glucose (a_w 0.89, 0.92 and 0.95) and NaCl (a_w 0.88, 0.92 and 0.94). Colonies were counted after 14 days incubation at 25°C.

RESULTS AND DISCUSSION

Wallemia sebi

Data from studies on heat-stressed *W. sebi* conidia were fully presented by Beuchat and Pitt (1990a). Heating *W. sebi* conidia at 48°C for 30 min resulted in *ca.* 96% reduction in viable populations on YGA, an effective recovery medium. Glucose and sorbitol had very similar effects on recovery of heat stressed *W. sebi* conidia, although the presence of sorbitol in YGA appeared to enhance resuscitation. Heated conidia were more sensitive to NaCl, with less colonies formed on all NaCl supplemented YGA compared with glucose or sorbitol supplemented YGA at the same a_w. The adverse effect of NaCl was more pronounced as the time of heating increased.

Colonies of *W. sebi* were generally largest on sorbitol supplemented YGA and smallest on YGA containing NaCl. Colony size was largest on YGA at about a_w 0.92, regardless of solute. Colonies formed on glucose and sorbitol supplemented YGA at this a_w were nearly equal in diameter, but those formed on glucose supplemented YGA were easier to count because of their distinct margins. Colonies on YGA containing sorbitol had a feathered floccose appearance, making them more difficult to count, especially on plates containing more than about 100 colonies.

Although the number of colonies formed on YGA supplemented with glucose, sorbitol or NaCl at a_w 0.806-0.974 was essentially the same after 9 days at 25°C, the rate of development was much slower on NaCl supplemented YGA. This is in agreement with observations that germination of *W. sebi* conidia is little affected by solute type but that growth is considerably slower on NaCl supplemented media than on sugar supplemented media (Wheeler *et al.*, 1988).

Recovery of unheated *W. sebi* conidia at 25°C was essentially the same on YGA plus NaCl (0.82-0.97 a_w). At 20°C, the number of unheated conidia recovered at a_w 0.82 was reduced about 10 fold, while at 30°C, the number recovered at a_w 0.97 was reduced to less than 10^2 cfu/mL. This indicates that *W. sebi* conidia may be more tolerant of a_w higher than the optimum (0.92) at lower incubation temperatures but less tolerant of a_w greater than 0.92 when the incubation temperature is higher than 25°C. This effect was magnified for heat stressed conidia. Incubation at 30°C was clearly unfavourable to resuscitation of these conidia, even at the lower end (a_w 0.82) of the a_w range investigated. The suitable a_w range for recovery of heat stressed conidia narrowed as the incubation temperature was raised from 25°C to 30°C.

On YGA at 0.97 a_w, adjustment of pH values from 3.71 to 6.55 did not affect colony formation by unheated *W. sebi* conidia at 20 and 25°C. However, incubation at 30°C resulted in populations less than 10^2 cfu/mL, irrespective of pH, confirming that colony formation on YGA (pH 6.2, a_w 0.97) does not occur at 30°C.

Heat stressed conidia were more sensitive as the pH of YGA decreased from 6.55 to 3.71. This effect was more pronounced at 20°C than at 25°C, indicating that 25°C is nearer to the optimum temperature for recovery of heat stressed conidia. The adverse effect of low pH on recovery and colony formation by heat stressed conidia of non xerophilic moulds has been documented (Beuchat, 1984). The repair process is either inhibited or prevented when sublethally injured conidia are exposed to hydrogen ion concentrations which may otherwise have no adverse effect on germination and growth of uninjured conidia.

Chrysosporium farinicola

Data on studies on heat stressed *C. farinicola* aleuriospores were presented fully by Beuchat and Pitt (1990b). *C. farinicola* aleuriospores from 20 day old cultures were more resistant to heat inactivation compared to those harvested from 14 day old cultures. This difference was magnified as the treatment temperature was raised from 48 to 56°C.

Recovery of heat treated *C. farinicola* aleuriospores increased as the a_w of YGA was decreased from 0.95 (glucose, glycerol) and 0.94 (sorbitol) to 0.89 and 0.88, respectively. In YGA plus NaCl, recovery was greatest at 0.92 a_w, with much reduced numbers at 0.94 a_w, and no colonies at 0.88 a_w. Above 0.88 to 0.89 a_w, solutes were tolerated in the order glucose > sorbitol > glycerol > NaCl. Younger aleuriospores were also more sensitive to the lethal effects of glycerol at high a_w (0.95) and NaCl at 0.88 and 0.94 a_w, i.e. at a_w values further from growth optima on each solute.

The influence of a_w on colony size correlated with an increase in colony numbers formed by heat stressed aleuriospores on the same media. Maximum colony size on YGA with NaCl or sorbitol was observed between 0.88 to 0.93 and 0.88 to 0.94 a_w, respectively; on YGA with glucose and glycerol, colony size was greatest at or below 0.89 a_w. Observations on *C. farinicola* colony development on media containing glucose, sorbitol, glycerol and NaCl tended to confirm the general sensitivity of *Chrysosporium* species to low molecular weight polyols and ionic solutes.

The interacting effects of the type of solute in recovery medium and incubation temperature on colony formation by unheated and heat stressed (30 min, 52°C) *C. farinicola* aleuriospores was determined. Within the range of a_w tested, increased numbers of colonies were correlated with decreased a_w, regardless of heat treatment or type of solute. Incubation at 20°C generally resulted in an increase in recoverable aleuriospores compared to incubation at 25°C or 30°C for 14 days followed by 20°C for 10 days, particularly in media containing glycerol and NaCl. The combined stress induced by NaCl and high temperature appeared to have a lethal effect on *C. farinicola* aleuriospores. The optimum temperature for recovering *C. farinicola* was 20-25°C.

The effects of holding unheated and heated *C. farinicola* aleuriospores for 20 min in diluent containing sucrose before plating on YGA containing various solutes were determined. There appeared to be little if any effect of holding time (0 or 20 min) and concentration of sucrose in diluent on viability of aleuriospores, but as observed previously, recovery media containing glucose and sorbitol were superior to glycerol and NaCl at a given a_w.

CONCLUSIONS

Conidia of *W. sebi* and aleuriospores of *C. farinicola* are susceptible to sublethal injury upon exposure to elevated temperatures. Both the type of solute used and the a_w influences the ability of these spores to revive and form colonies. In addition, the incubation temperature during recovery can greatly affect colony development. Results indicate that media and incubation times routinely used to enumerate fungi in foods are inadequate for enumerating *W. sebi* and *C. farinicola*, particularly when spores may be in a state of sublethal heat injury.

ACKNOWLEDGEMENTS

The author is grateful to the Sir Frederick McMaster Fellowship program, CSIRO, which provided financial support to enable this investigation to be conducted at CSIRO Division of Food Processing, North Ryde, NSW 2113, Australia, and to Ms Sonya Dyer for her technical assistance.

REFERENCES

BEUCHAT, L.R. 1984. Injury and repair of yeasts and moulds. *In* The Revival of Injured Microbes, eds M.H.E. Andrew and A.D. Russell. pp. 293-308. *Soc. Appl. Bacteriol. Symp. Ser.* No. 12. London: Academic Press.

BEUCHAT, L.R. and PITT, J.I. 1990a. Influence of solute, pH and incubation temperature on recovery of heat-stressed *Wallemia sebi* conidia. *Appl. Environ. Microbiol.* **56**: 2545-2550.

BEUCHAT, L.R. and PITT, J.I. 1990b. Influence of water activity and temperature on survival of and colony formation by heat-stressed *Chyrsosporium farinicola* aleuriospores. *Appl. Environ. Microbiol.* **56**: 2951-2956.

PITT, J.I. 1975. Xerophilic fungi and the spoilage of foods of plant origin. *In* Water Relations of Foods, ed. R.B. Duckworth. pp. 273-307. London: Academic Press.

WHEELER, K.A., HOCKING, A.D. and PITT, J.I. 1988. Effects of temperature and water activity on germination and growth of *Wallemia sebi*. *Trans. Br. Mycol. Soc.* **90**: 365-368.

MYCOLOGY AND SPOILAGE OF HAZELNUTS

Judith L. Kinderlerer[1] and Mary Phillips-Jones[1, 2]

[1]*Food Research Centre and Department of Biomedical Sciences*
Sheffield City Polytechnic
Pond Street
Sheffield, S1 1WB, U.K.

[2] *Present address:*
Department of Molecular Biology and Biotechnology
University of Sheffield
Sheffield, S10 2TN, U.K.

SUMMARY

Twenty-eight xerophilic fungal species were isolated and identified from whole hazelnut kernels which were obtained from Turkey. Most isolates belonged to the genus *Aspergillus* including *Eurotium* spp. and *Penicillium*. A high incidence of *Wallemia sebi* was found. *Aspergillus flavus* was detected in both the 1985 and 1988 crops. Two species, *Eurotium repens* and *Penicillium crustosum*, were capable of causing considerable spoilage of hazelnut oil and would be expected to cause spoilage in the nuts. Roasting hazelnuts is an effective control step.

INTRODUCTION

Hazelnuts (the nut of *Corylus avellana* L.) are grown in many countries including Turkey, Italy, Spain, Korea and the USSR. The bulk of UK imports are obtained from Turkey. The nuts may be imported in shell, or they may be shelled and the imported kernels left whole or chopped. The testa can be left on the kernels or removed by blanching. The chopped nuts are frequently roasted before they are incorporated into a variety of food products.

We have been investigating chemical changes caused by fungal growth in oilseeds at low water activity (0.7 - 0.9 a_w). This contribution considers spoilage of hazelnuts due to mould growth. Three approaches have been used: (1) the isolation of the mycoflora by serial dilution and surface plating onto various selective media; (2) direct isolation of fungi after the commodity was allowed to go mouldy under defined conditions; and (3) model spoilage experiments using *Eurotium* and *Penicillium* species which were inoculated into liquid shake cultures with hazelnut oil as the sole carbon source.

MATERIALS AND METHODS

Origin of hazelnuts

The hazelnuts used in this project were all imported from Turkey. Whole shelled nuts from the 1985 and 1987 harvests and chopped, roasted nuts from the 1985 harvest were obtained from wholesale sources. Normal samples from the 1986 harvest were unobtainable, as the trees were in blossom at the time of the nuclear accident at Chernobyl in the USSR. However, a sample was provided by the Atomic Energy Authority at Harwell, the consignment having been rejected due to radioactive contamination.

Enumeration of xerophilic fungi

Hazelnut samples (10 g) were added to 90 mL of either 40% (w/v) sucrose or 0.1% peptone water in sterile Stomacher bags. Samples were homogenised in the Stomacher for 2 min. After futher dilutions, samples (0.1 mL) were surface plated at least in duplicate on the media described below. After incubation at 25°C for 10 days, colonies were enumerated and subcultured for identification.

Three media of low water activity (a_w) were used. Malt extract agar with 40% (w/v) sucrose (MA40), Malt extract Yeast extract 50% Glucose agar (MY50G, Pitt and Hocking, 1985) and Dichloran 18% Glycerol agar (DG18). MA40 was prepared from Malt extract agar (Oxoid) according to the manufacturer's instructions with the addition of 40% (w/v) sucrose (final a_w, 0.96).

Enumeration of *Aspergillus flavus* in intact hazelnuts

To determine the incidence of infection of hazelnuts by *A. flavus*, 20 batches of 5 g of kernels were incubated in Petri dishes lined with filter paper to which 10 mL of 20% glycerol was added. The plates were incubated in a desiccator for 5 d at 30°C. Yellow green colonies that developed were counted and then picked off and subcultured onto AFPA. Colonies which gave an orange reverse after incubation for 2 d at 30°C were counted as *A. flavus*.

Determination of a_w, fat and solids content of hazelnuts

Whole kernels (including the testa) were ground to a powder before measuring the water activity. For determination of fat, 3 g of the powdered nut was extracted for 1.5 h with 40/60° petroleum ether in a Soctec extractor. Solids were obtained by difference.

Storage of hazelnuts at increased a_w values

Water was added to raise the a_w of chopped hazelnuts to give 0.69, 0.72 and 0.83 a_w respectively. Samples were stored in sealed containers for six months at 25°C.

Effect of a_w on germination of spoilage fungi

A thin layer of agar (2%) to which Czapek broth containing sucrose (3%) was added, was spread over sterile microscope slides which were then equilibrated to the desired a_w levels by prolonged incubation over suitably adjusted glycerol solutions in vacuum desiccators at 25°C. For *Eurotium* and *Aspergillus* species, and *Wallemia sebi*, slides were equilibrated to 0.73, 0.75, 0.77, 0.80 and 0.90 a_w. For *Penicillium* species, slides were equilibrated to 0.95 or 0.90 a_w. After equilibration, slides were inoculated with spores and returned to the desiccators. The slides were examined microscopically on a daily basis for germination, which was considered to have occurred when germ tubes were observed in >20% of spores.

Growth of *Penicillium* and *Eurotium* species on hazelnut and other oils

Fungi from hazelnuts were grown in shake flask culture according to the method described by Hatton and Kinderlerer (1990). *Penicillium* species were grown in 25 mL Czapek broth containing 1 g hazelnut oil in flasks rotated at 200 rev/min for 72 hr at 25°C. *Eurotium* species were grown under similar conditions for 16d in 25 mL Czapek broth with 2.5% NaCl and 1 g hazelnut oil. Inoculum levels used were $2.5x10^7$ spores, and all experiments were carried out in triplicate.

RESULTS AND DISCUSSION

Isolation and identification of xerophilic mycoflora

Table 1 shows xerophilic fungal counts obtained for the 1985 hazelnut crop on the three media used, MA40, DG18 and MY50G. Similar counts were found on each medium. Chopped nuts had the highest counts on all media. Fewer fungi were detected on whole kernels, and very low counts were obtained from roasted chopped nuts. Chopping increases the surface area available for contamination, and also spreads fungal propagules present. Sucrose (40%) was a better diluent than peptone (0.1%) for the isolation of xerophilic fungi by dilution methods (Table 2).

Table 1 shows that roasting hazelnuts is an effective method for reducing viable fungal populations. The Hazard Analysis Critical Control Point (HACCP) system would define roasting as a critical control point (CCP_1) (ICMSF, 1988). In formulated and multi-ingredient foods containing hazelnuts it is clearly preferable to use roasted rather than unroasted kernels.

Table 1. Xerophilic mycoflora from the 1985 crop of Turkish hazelnuts[1]

	Total Count	*Eurotium repens*	*Penicillium* spp.
Whole kernels	3.41 ± 0.42	2.76 ± 0.13	3.00 ± 0.24
Chopped kernels	4.61 ± 0.15	4.34 ± 0.20	4.25 ± 0.15
Chopped roasted kernels	<2.69	ND[2]	<2.69

[1] Counts are means obtained after direct plating on Malt Agar with 40% Sucrose, DG18 and MY50G. . Results are expressed as $\log_{10}$ ± SD/g.
[2] ND: not detected.

Table 2. Comparison of 40% sucrose and 0.1% peptone as diluents for the isolation of xerophilic and xerotolerant fungi present on whole hazelnut kernels (1985 crop)[1]

Diluent	MA40	DG18	MY50G	Mean±SD
40% sucrose	3.85	3.48	3.80	3.71±0.20
0.1% peptone	3.30	3.30	2.70	3.10±0.35

Results are expressed as $\log_{10}$ ± SD/g.

Table 3. Fungi detected in whole hazelnuts from the 1985 Turkish crop

Absidia corymbifera (Cohn) Sacc. & Trotter	*Mucor circinelloides* van Tieghem
Aspergillus caesiellus Saito	*Paecilomyces variotii* Bainier
Aspergillus flavus Link[1]	*Penicillium bilaii* Chalabuda
Aspergillus niger van Tieghem	*Penicillium brevicompactum* Dierckx
Aspergillus penicillioides Spegazzini	*Penicillium citrinum* Thom
Aspergillus restrictus G. Smith	*Penicillium crustosum* Thom
Aspergillus sydowii (Bain. & Sart.) Thom & Church	*Penicillium aurantiogriseum* Dierckx
Aspergillus versicolor (Vuill.) Tiraboschi	*Penicillium fellutanum* Biourge
Chrysosporium xerophilum Pitt[2]	*Penicillium glabrum* (Wehmer) Westling
Eurotium echinulatum Delacr.	*Penicillium paxilli* Bainier
Eurotium herbariorum Mangin	*Penicillium rugulosum* Thom
Eurotium repens de Bary	*Penicillium spinulosum* Thom
Eurotium rubrum König *et al.*	*Trichoderma* spp.
Eurotium umbrosum (Bain. & Sart.) Malloch & Cain	*Wallemia sebi* (Fr.) von Arx

[1] Only isolated after the hazelnuts had become mouldy.
[2] Isolated after the nuts were stored in a sealed container for 6 months at an initial water activity of 0.90.

The fungal species isolated from the 1985 hazelnut crop are listed in Table 3. Twenty-eight species from 75 isolations were identified. *Eurotium repens*, *Penicillium* species and *Wallemia sebi* were the most common species on dilution plates. In our experience the number of species isolated was increased by using direct plating as well as allowing fungi to grow on the commodity under controlled conditions.

Enumeration of *Aspergillus flavus*

When hazelnuts were incubated for 5 days at 30°C in the presence of 20% glycerol both *A. flavus* and *A. niger* were isolated from all samples. The incidence of *A. flavus* on the surface of whole hazelnut kernels is given in Table 4. The results are also expressed as the reciprocal of the count per gram. This value indicates the number of potential sites of infection per unit weight of nuts, giving an index of the possibility of formation of aflatoxins. The mycoflora of the 1986 harvest was similar to that found for other years in spite of exposure to irradiation resulting from the nuclear accident at Chernobyl.

Table 4. Incidence of *Aspergillus flavus* in 100 g batches of hazelnuts[1]

Year of harvest	Colonies/100 g *A. flavus*	Weight (g) per *A. flavus* colony ± SD
1985	7.2 ± 1.9	7.62 ± 3.91
1986	8.0 ± 2.0	4.34 ± 1.37
1987	8.4 ± 1.8	5.97 ± 1.58

[1] Incidence was determined according to Kinderlerer and Clark (1986).

Shelf-life of hazelnuts: the effect of water activity on fungal growth
The water activities (Table 5) and compositions (Table 6) of hazelnuts stored for 1-3 years at ambient temperatures are shown below. The a_w for all samples was below 0.5 except for the sample listed as 1987 (good quality). No visible signs of fungal growth were seen on any of the samples and a_w was too low for spores to germinate.

Fungi developed when chopped hazelnuts were stored for 6 months at controlled a_w levels. At the lowest water activity (0.69 a_w) *E. repens* and *A. restrictus* grew. At 0.72 a_w, *E. repens*, *A. penicillioides*, *A. restrictus* and *Chrysosporium xerophilium* grew, while nuts held at 0.83 a_w were covered by *Eurotium* species.

Table 5. Water activity of whole Turkish hazelnuts

Year of harvest	Water activity[1]
1985	0.426 ± 0.007
1986	0.436 ± 0.010
1987	0.485 ± 0.042
1987 (good quality)	0.694 ± 0.005

[1]Values are the arithmetic mean of 3 samples ± SD. Values were determined in 1988.

Table 6. Composition of whole hazelnut kernels analysed in 1988[1]

Year of harvest	Fat % (m/m)	Solids (m/m)	Water % (m/m)
1985	44.1 ± 5.6	51.6 ± 1.9	4.3 ± 3.8
1986	45.2 ± 6.0	50.1 ± 9.2	4.7 ± 3.4
1987	46.2 ± 6.2	50.1 ± 8.7	4.0 ± 1.8
1987	49.4 ± 4.3	46.8 ± 1.4	3.8 ± 1.0

Results are the arithmetic mean of 3 samples ± SD

Effect of a_w on germination of spoilage fungi
Tables 7 and 8 give the time for germination of some of the fungi isolated from hazelnuts at reduced a_w. *Eurotium rubrum* germinated at lower a_w than other species.

Growth of *Penicillium* and *Eurotium* species on hazelnut and other oils
Penicillium and *Eurotium* species isolated from hazelnuts were grown on hazelnut oil as the sole carbon source (Table 9). As indicated by biomass production, both *Penicillium crustosum* and *Eurotium repens* grew very rapidly, indicating that these species have the potential to use hazelnut oil as a sole source of carbon.

Table 7. Time (d) for conidial germination of xerophilic hazelnut mycoflora on a slide culture at 25°C

Species	Water activity (a_w)				
	0.73	0.75	0.77	0.80	0.90
Eurotium rubrum	10	<6	<6	<6	1
Eurotium repens	>21	<6	<5	<6	1
Eurotium umbrosum	>21	<6	<5	<6	1
Eurotium echinulatum	>21	8	5	10	1
Eurotium herbariorum	>21	7	<5	3	1
Wallemia sebi	>21	10	ND	<6	2
Aspergillus versicolor	>21	>26	>20	7	
Aspergillus niger	>21	>26	>20	>26	

Table 8. Time (d) for conidial germination of *Penicillium* species isolated from hazelnuts on a slide culture at 25°C

Species	Water activity (a_w)	
	0.85	0.90
Penicillium spinulosum	17	3
P. crustosum	24	2-3
P. fellutanum	24	2
P. paxillii	24	2
P. brevicompactum	42	2
P. bilaii	42	2
P. rugulosum	>42	4
P. citrinum	>24	-
P. glabrum	>42	2

Table 9. Growth (mg/dry weight) of *Eurotium* and *Penicillium* species from hazelnuts on hazelnut oil

Species	No of isolates	Biomass (mg dry wt) ± SD
Eurotium repens	5	306.4 ± 47.3
Eurotium rubrum	2	57.4 ± 21.0
Penicillium bilaii	1	8.0 ± 0.8
Penicillium brevicompactum	2	17.6 ± 5.3
Penicillium citrinum	2	16.3 ± 12.5
Penicillium crustosum	5	508.0 ± 93.8
Penicillium fellutanum	3	8.4 ± 4.3
Penicillium paxillii	1	6.0 ± 3.2
Penicillium rugulosum	1	6.0 ± 2.4
Penicillium spinulosum	4	5.4 ± 2.5

ACKNOWLEDGEMENTS

We are grateful to Dr A.H.S. Onions for identification of some of the *Penicillium* species, and to Stewart Johnson for assistance with the fat and water activity analysis. This work was supported by the Ministry of Agriculture, Fisheries and Food and the results are the property of the Ministry and Crown copyright.

REFERENCES

HATTON, P.V. and KINDERLERER, J.L. 1991. Toxicity of medium chain fatty acids to *Penicillium crustosum* Thom and their detoxification to methyl ketones. *J. Appl. Bacteriol.* **70**: 401-407.

ICMSF. 1988. Microorganisms in Foods. 4. Application of the hazard analysis critical control point (HACCP) system to ensure microbiological safety and quality. p. 29. Oxford: Blackwell Scientific Publications.

KINDERLERER, J.L. and CLARK, R.A. 1986. Microbiological quality of desiccated coconut. *J. Hygiene*, **96**: 19-26.

PITT, J.I. and HOCKING, A.D. 1985. Fungi and Food Spoilage. Sydney: Academic Press.

MOULD COUNTS AND MYCOFLORA IN SAMPLES OF SPICES AS INFLUENCED BY MEDIUM AND PLATING TECHNIQUE

Kerstin Åkerstrand

Biology Section
National Food Administration
S-751 26 Uppsala, Sweden

SUMMARY

Twelve samples of spices were examined on two media; antibiotic supplemented Czapek Dox Agar and Dichloran 18% Glycerol agar (DG18). Counts from six of the samples were $>10^3$, while for three samples counts were not detectable. *Aspergillus* and *Eurotium* species dominated the mycoflora. Results obtained on the two media were comparable, but DG18 is the preferable medium for enumerating fungi in spices.

INTRODUCTION

Samples of spices were examined to estimate mould counts, to describe the mycoflora, and to evaluate two media for efficacy in mycological analysis of spices.

METHODS

Samples. Consumer packs of 10 kinds of spices were obtained at markets. Pepper, allspice and cloves were obtained both as ground spices and as whole seeds, and both kinds were analysed.

Plating. For dilution plating, 10 g samples were homogenised for two min in 90 g 0.1% peptone water in a Colworth Stomacher. Ten-fold dilutions were pour plated.

For direct plating, whole spices were surface disinfected with a solution of sodium hypochlorite containing 0.05% active chlorine for two minutes. Disinfected particles were rinsed twice in sterile water before plating 25 particles on each medium.

Media. Czapek Dox Agar (Oxoid) and Dichloran 18% Glycerol agar (DG18) (Oxoid) were used. Before pouring plates, chlortetracycline and chloramphenicol were added, each at a concentration of 50 mg/L of medium.

Incubation. Plates were incubated upright at 25°C for 7 days.

RESULTS AND DISCUSSION

Six samples had total fungal counts of $>10^3$ (Table 1). Thyme was the most heavily contaminated with a count of 10^6 per gram. Three samples contained no detectable viable fungi, probably indicating they had been fumigated at some time.

Table 1. Fungal counts and mycoflora in samples of spices[1]

Spice	Count/g Czapek	Count/g DG18	Species	Dilution plating Czapek	Dilution plating DG18	Direct plating Czapek	Direct plating DG18
Black pepper 1	4.0x10^3	1.6x10^4	*Aspergillus flavus*	+++	+++	+	+
			A. tamarii	+	+		
			A. niger	+	-		
			Eurotium spp.	+	++	++	+++
			Penicillium spp.	++	++	+	+
			Zygomycetes	+	-		
Black pepper 2	<10	<10	-				
White pepper	3.2x10^4	4.5x10^4	*A. flavus*	++	++	+++	+++
			A. tamarii	++	++	+++	+++
			A. niger	++	++	-	++
			A. candidus	+	++		
			A. clavatus	+	+	-	-
			A. nidulans	-	-	++	++
			Eurotium spp.	+	++		
			Penicillium spp.	+++	+++	+	++
			Paecilomyces spp.	++	-	+	-
			Zygomycetes	-	-	+	-
Allspice	5.0x10^3	5.9x10^3	*A. niger*	+++	+++		
			A. fumigatus	++	+		
			A. flavus	+	+		
			Eurotium spp.	-	++	+	+
			Zygomycetes	+	-	-	-
Cloves	<10	<10	-				
Nutmeg 1	2.0x10^1	3.0x10^1	*Eurotium* spp.	+	+		
Nutmeg 2	1.6x10^3	1.1x10^4	*Eurotium* spp.	+++	+++		
			A. flavus	+	-		
			A. niger	+	+		
			Penicillium spp.	+	+		
			Zygomycetes	+	+		
Paprika	<10	<10	-				
Thyme	1.0x10^5	9.3x10^4	*A. fumigatus*	++	++		
			A. niger	++	++		
			A. candidus	+	+		
			Aspergillus spp.	+	+		
			Eurotium spp.	++	++		
			Penicillium spp.	+	+		
			Paecilomyces spp.	+	+		
Rosemary	1.3x10^2	1.3x10^2	*A. candidus*	+	-		
			A. niger	+	-		
			Eurotium spp.	+	++		
			Penicillium spp.	+	-		
			Paecilomyces spp.	++	+		
			Zygomycetes	-	+		

[1] +++ dominant species; ++ relatively frequent; + detected in low numbers

Table 1. (continued)

Spice	Count/g Czapek	Count/g DG18	Species	Dilution plating Czapek	Dilution plating DG18	Direct plating Czapek	Direct plating DG18
Oregano	$5.0x10^4$	$5.1x10^4$	*A. niger*	++	++		
			A. versicolor	+	+		
			A. nidulans	-	+		
			Eurotium spp.	-	+		
			Phoma spp.	++	-		
			Zygomycetes	+	-		
Cinnamon	$6.0x10^1$	$1.0x10^2$	*A, niger*	+	+		
			Eurotium spp.	-	+		
			Zygomycetes	+	+		

[1] +++ dominant species; ++ relatively frequent; + detected in low numbers

Counts were generally higher on DG18 than on Czapek Dox agar with antibiotics, but observed differences were mostly small. In most samples, the mycoflora was very similar and was dominated by *Aspergillus* and *Eurotium* species. However, more species were sometimes detected on antibiotic amended Czapek Dox agar than on DG18. Counts of *Eurotium* species were generally higher on DG18 and sometimes inhibited the growth of other species. Otherwise, DG18 is an excellent medium for examining spices. Colonies are mostly distinct and easily counted, and rapidly growing species are more restricted than on Czapek Dox agar with antibiotics. On the other hand, Czapek Dox agar with antibiotics is a very good medium for differentiation and identification of *Aspergillus* and *Penicillium* species. When enumeration and presumptive identification of fungi is the main aim, however, it is not necessary to use more than one medium. In that case, DG18 is preferable.

EVALUATION OF THE MYCOLOGICAL QUALITY OF DRIED AND INTERMEDIATE MOISTURE FRUITS

Janet E. L. Corry, Kathleen Regan and Josephine B. Head

J Sainsbury PLC
Scientific Services Division
Stamford Street
London, SE1 9LL, U.K.

SUMMARY

The mycological quality of three intermediate moisture fruit products (a_w 0.70-0.76) and seven dried fruit products (a_w <0.66) was examined. The most satisfactory media for examining all the products were Dichloran 18% Glycerol agar (DG18) and Glucose Tryptone Chloramphenicol agar (GTC). Direct plating was best for examining the dried products and dilution plating with 50% glucose as diluent for the intermediate moisture products.

On the basis of viable counts for yeasts and moulds, all products except cut mixed peel appeared to be satisfactory. It was recommended that sorbic acid be added to this product to prevent proliferation of xerophilic yeasts during normal storage life. There was some evidence that hygienic conditions during production of some products, eg. Californian raisins and the mixed dried fruit, were suboptimal. The API yeast identification systems were found to be unsuitable for the yeasts encountered in this survey.

INTRODUCTION

The aim of this study was to examine the mycological quality of various dried and intermediate moisture fruits marketed by a major supermarket chain in the U.K. In the course of the investigation dilution plating and direct plating methods were compared using a variety of plating media.

MATERIALS AND METHODS

Products examined

The products examined are listed in Table 1. All were retail packs except for the mixed fruit complaint sample, which had been received in an unlabelled glass jar with no evidence of origin. Small patches of white mycelium were visible through the glass.

Media and conditions of incubation

The media used were prepared according to King *et al.* (1986) unless otherwise stated. They were: Dichloran Rose Bengal Chloramphenicol agar (DRBC, Oxoid); Glucose Tryptone Chloramphenicol agar (GTC); Dichloran 18% Glycerol agar (DG18, Oxoid); Malt extract Yeast extract 50% Glucose agar (MY50G); Malt extract Yeast extract 70% Glucose Fructose agar (MY70GF) (Pitt and Hocking, 1985) and Aspergillus Flavus and Parasiticus agar (AFPA).

Table 1. Products tested

Product	Preservative	Country of origin	Water activity
Cut mixed peel	Sulphur dioxide	Italy	0.76
Ready-to-eat prunes	Sorbic acid, sulphur dioxide	USA (California)	0.72
Glace cherries	Sorbic acid, sulphur dioxide	France	0.72
Prunes	None	USA (California)	0.66
Mixed fruit	None	Greek sultanas and currants Californian raisins Italian mixed peel	0.62
Mixed fruit (complaint)	-	Not known	0.59
Currants	Sulphur dioxide	Greece	0.55
Sultanas	None	Australia	0.55
Seedless raisins	None	USA (California)	0.52
Seeded raisins	Sulphur dioxide	Australia	0.50

DRBC and GTC are high a_w media; DRBC contains dichloran and rose bengal to inhibit spreading fungi. DG18 also contains dichloran, but has an a_w of about 0.95. MY50G and MY70GF are designed to isolate xerophilic fungi and have a_w levels of about 0.89 and 0.76 respectively (Pitt and Hocking, 1985). Plates were incubated upright at 25°C, sealed in plastic bags to prevent drying, and examined daily for up to 4 weeks.

Spread plates

The initial dilution was prepared by adding 10 g of sample to 90 mL diluent (0.1% peptone, 0.05% Tween with 50% w/v glucose added) and stomaching for 2 min. Decimal dilutions (1+9 mL) were prepared down to 10^{-4}, and spread plated in duplicate, 0.1 mL per plate.

Direct plating

Assessment of surface contamination. Three or four separate fruits (depending on size) were pressed gently onto the surface of the agar plates, then removed and surface sterilised in order to determine internal contamination (see below).

Determination of internal contamination. The fruit was placed in a sterile Stomacher bag and weighed, then covered with 70% (v/v) ethanol at 21-23°C and left for 2 min. The ethanol was decanted and replaced with a measured volume of 10% (v/v) aqueous sodium hypochlorite in the proportion of nine parts hypochlorite to one part fruit. After 2 min, the hypochlorite was decanted and the fruit rinsed with several washes of sterile distilled water.

Three or four of the surface sterilised fruits were placed on one plate of each of the six media.

Yeast and mould identification

Yeasts were identified using the API 20C AUX and ATB 32C systems (API, UK Ltd., Grafton Way, Basingstoke), and the profiles obtained submitted to the API computer service for interpretation. Moulds were identified according to Samson and van Reenen-Hoekstra (1988) and Pitt and Hocking (1985).

Determination of a_w

Water activity (a_w) was measured using a Novasina Thermoconstanter Humidat - TH2 (Humitec Ltd., Crawley, Sussex). Values for a_w obtained are listed in Table 1. The products fell into two groups: those with $a_w > 0.7$ and those with $a_w < 0.66$.

RESULTS

Results for all products except prunes, glace cherries and Australian raisins are summarised in Tables 2 and 3. *A. flavus* was not detected in any sample.

Intermediate moisture products ($a_w > 0.7$).

Cut mixed peel (Tables 2 and 3) contained high numbers (up to 10^6) of yeasts as determined by dilution plating (Table 2). Most contamination was on the fruit surface (Table 3, cut mixed peel A) with up to 75% reduction in count following surface sterilisation. No growth was observed on ready to eat Californian prunes on any agar by either plating method. On glace cherries, a *Penicillium* sp. (100 cfu/g) was identified on DRBC by dilution plating only.

Dry products ($a_w < 0.66$)

As shown in Tables 2 and 3, dilution plating of the mixed fruit sample returned as a complaint showed growth of *Aspergillus niger* on DG18 only. However, using direct plating, a wider variety of microorganisms were observed on three of the media used (GTC, AFPA, DG18) including bacteria, yeasts and moulds.

Dilution plating of sound mixed fruit revealed low numbers of yeasts (100 cfu/g) on AFPA (Tables 2 and 3). No other growth was observed using this method. Surface contamination (Table 3, Mixed fruit, A) consisted of yeasts which were cultured on DRBC and AFPA at rates of 25% (1/4) and on GTC at a rate of 75% (3). Internal contamination (Table 3, Mixed Fruit, B - post surface sterilisation) was with yeasts (25% on GTC) and *Eurotium* species (25% on MY50G). *Aspergillus niger* was detected on currants by dilution plating and direct plating on DRBC (Tables 2 and 3), with a 100% contamination rate in sample B (Table 3). A *Bacillus* sp. was identified as a surface contaminant on GTC at a rate of 50%.

A *Bacillus* sp. was identified from Australian sultanas on GTC (Table 2). An *Aspergillus* sp. was present at low levels (100 cfu/g) on DG18. Surface contamination with *A. niger* was observed on GTC (25%) and DG18 (25%), and yeasts on GTC (75%), and DRBC (25%). *Bacillus* sp. was identified on GTC (75%). After surface sterilisation, go growth was observed on any agar. No growth was observed from Australian seeded raisins and Californian prunes on any agar by either plating method, nor from Californian seedless raisins by dilution plating. However, *A. niger* surface contamination was observed on GTC (100%) and DG18 (100%), and internal contamination on GTC (100%), DRBC (100%) and AFPA (100%) (Table 3).

Identification of yeasts

Zygosaccharomyces and *Candida* spp. were identified using API 20C AUX and API ATB 32C on cut mixed peel (Table 2). *Candida rugosa, C. inconspicua, C. valida, C.*

lipolytica and *C. glabrata* were provided by the API database as identifications for profiles obtained. However, none of these species are normally associated with fruits, and therefore no attempt was made to assign yeasts to species level.

Table 2. Counts (log_{10}/g) on dried fruits and fungi obtained from dilution plates[1]

Medium	Sample				
	Cut mixed peel	Mixed fruit (complaint)	Mixed fruit	Currants	Australian sultanas
DRBC	4.8[a,f,g]	<2.0	<2.0[d]	3.0	<2.0
GTC	6.5[a]	<2.0	<2.0	<2.0	2.0[b]
AFPA	4.0[a]	<2.0	2.0[a]	<2.0	<2.0
DG18	5.0[a,e]	2.0[d]	<2.0	<2.0	2.0[c]
MY50G	5.1	<2.0	<2.0	<2.0	<2.0

[1] Superscript letters indicate genera isolated: a, yeasts; b, *Bacillus* spp.; c, *Aspergillus* spp.; d, *A. niger*; e, *Penicillium* spp.; f, *Zygosaccharomyces* spp.; g, *Candida* spp.

Table 3. Percentage infection of dried fruits by direct plating[1]

Medium	Sample											
	Cut mixed peel		Mixed fruit (complaint)		Mixed fruit		Currants		Australian sultanas		Californian raisins	
	A[2]	B[2]	A	B	A	B	A	B	A	B	A	B
DRBC	50[a]	25[a]	0	0	25[a]	0	0	100[d]	25	0	0	100[d]
GTC	100[a]	25[a]	50[b]	0	50[b]	25[a]	50[b]	0	75[a],25[d]	0	100[d]	100[d]
AFPA	100[a]	25[a,e]	25[a,c]	25[f]	75[a]	0	0	0	0	0	0	100[d]
DG18	100[a]	50[a]	100[c,d]	100[g]	0	0	0	0	75[d]	0	100[d]	0
MY50G	100[a]	0	0	0	0	25[f]	0	0	0	0	0	0
MY70GF	0	0	0	0	0	0	0	0	0	0	0	0

[1] Three or four fruits were examined per product per medium. Superscript letters indicate genera isolated: a, yeasts; b, *Bacillus* spp.; c, *Aspergillus* spp.; d, *A. niger*; e, *Penicillium brevicompactum*; f *Eurotium* spp; g, *Penicillium* spp.
[2] A, surface count; B, internal count (after surface sterilisation)

DISCUSSION

Direct vs dilution plating

Misleading results may be obtained using dilution plating for examining foods contaminated with fungi, due to disruption of sporulating structures during homogenising (Andrews, 1986). Results presented here clearly illustrate that not all contaminants are observed using this method except where contamination is due to yeasts. However, results obtained by direct plating relate only to the proportion of infected fruit tested and give no indication to the true extent of fungal contamination of the product (Andrews, 1986).

Interpretation of dilution plating results can be affected by aerial contaminants in the laboratory and by spores released by colonies within the sealed plate. Mislivec and Bruce (1977) recommended direct plating in preference to dilution plating particularly when handling toxigenic samples, since aerial contamination and handling are reduced. Based on results presented here, direct plating is the preferred method for dried fruits and dilution plating for candied fruits.

Mycological quality of dried and intermediate moisture fruits

The products tested in this survey fall into two categories: intermediate moisture fruits with relatively high a_w (>0.7), and dried fruits with low a_w (<0.66). Glace cherries, mixed cut peel and ready to eat prunes fall into the first group, and the mixed fruit, sultanas, raisins and conventional prunes in the second category.

Intermediate moisture products. All of the intermediate moisture products examined contain preservatives. Cut mixed peel is cooked and preserved with 100 μg/kg sulphite only, glace cherries are cooked and preserved with 100 μg/kg sulphite plus 1000 μg/kg sorbate while ready to eat prunes are dried in the processing plant and preserved using 500-1000 μg/kg sorbate after fumigation with aluminium phosphide.

Only cut mixed peel was found to contain significant levels of fungi. This can be attributed to the relatively low level of preservative used. Glace cherries and ready to eat prunes had higher a_w levels, but insignificant levels of fungal contamination.

Dry Products. In general dried fruits sold in the U.K. are grown and sun dried by small farmers, cleaned by cooperative bulk packing stations and then further cleaned and packed for retail sale either in the U.K. or in the country of origin. The Australian products (sultanas and seeded raisins) are generally sun dried on racks raised off the ground, which protects the fruit to some extent from contamination due to soil, insects and vermin. After 10-14 days the racks are shaken and the fruit, minus stalks falls onto paper on the ground below. Australian fruit is then cleaned, sorted and packed to export specifications. The method of drying Californian and Greek fruit is likely to cause more contamination since they are generally sun dried on paper or PVC placed directly on the ground, exposed to contamination from soil, birds, rodents and insects. Californian prunes are mechanically dehydrated.

The results of this survey indicate that none of these low a_w products contained large numbers of viable fungal propagules. All were within the post harvest limits for yeasts and moulds of 10^3 and 10^4 cfu per gram respectively, recommended by ICMSF (1984). No fungi were detected on the Australian stoned raisins or Californian dried prunes. This could be due to the method of drying, or a result of fumigation with phosphine. Australian raisins are seeded passing between rollers after heating, which may inactivate fungal propagules.

Aspergillus niger, considered by Pitt and Hocking (1985) to be resistant to sun drying, was found in 100% of the Greek currants, one sample of mixed fruit and the Californian raisins. Twenty five to 75% of the Australian sultanas contained *A. niger*. If the incidence of this fungus is taken as an index of quality, the Australian dried fruit examined was superior to the Greek and Californian. However, the presence of *A. niger* does not pose any hazard to health as it does not produce mycotoxins.

Detection of *Bacillus* sp. in currants and mixed fruit when using GTC agar (which is not designed to detect bacteria) may indicate less hygienic conditions during drying and/or processing.

Experience with other dried foods such as flour and grain, shows that counts tend to decrease during storage (Seiler 1986) which in combination with fumigants and a hot water wash may explain the low counts obtained with low a_w dried fruits in this survey.

Comparison of the results of direct plating of the low a_w products (Table 3) showed that the mixed fruit complaint sample was of poorer quality than the sound sample, with 100% of fruits contaminated with moulds both internally and externally when examined using DG18 medium. These were probably the xerophilic moulds responsible for the visible spoilage in the jar. The a_w of the complaint sample was apparently too low for even xerophilic mould growth (Table 1), so it is probable that growth had occurred in areas of relatively high a_w, caused by localised temperature fluctuations.

Future studies may obtain better indications of the mycological quality of this type of product by using tests for total mycelial load, e.g. by determination of ergosterol (Seitz *et al.*, 1979) or use of a total mould immunoassay (Notermans and Heuvelman, 1985; Robertson *et al.*, 1988).

CONCLUSIONS

Tests on the products with relatively high water activity (mixed peel, glace cherries and ready to eat prunes) indicated that the mixed peel contained insufficient preservative. Addition of sorbate would inhibit the development of xerophilic yeasts during normal storage life. Tests on the products with lower water activity indicated that the quality was good. Of the media tested in this study, GTC and DG18 were found most useful for direct or dilution plating (with a diluent containing 50% glucose). Direct plating was most appropriate for the dry products.

Because drying and other processes such as treatment with fumigants, exposure to sunlight and washing in hot and or chlorinated water reduce numbers of viable fungal propagules, quality tests for nonviable fungi may be worth consideration. For similar reasons, absence of viable aflatoxin producing fungi in a product is no guarantee of the absence of aflatoxin.

The API 20C AUX and ATB 32C systems for the identification of yeasts should not be used in conjunction with the API database since this is only appropriate for clinical isolates.

ACKNOWLEDGEMENT

We are indebted to Mr A.P. Williams, Leatherhead Food R.A., for supplying cultures of various species of xerophilic moulds and for general advice.

REFERENCES

ANDREWS, S. 1986. Dilution plating versus direct plating of various cereal samples. *In* Methods for the Mycological Examination of Food, eds A.D. King, J.I. Pitt, L.R. Beuchat and J.E.L. Corry. pp. 40-45. New York: Plenum Press.

ICMSF (International Commission on Microbiological Specifications for Foods). 1974. Microorganisms in Foods 2. Sampling for Microbiological Analysis: Principles and Specific Applications. Toronto: Uni versity of Toronto Press.

KING, A.D., PITT, J.I, BEUCHAT, L.R. and CORRY, J.E.L., eds. 1986. Methods for the Mycological Examination of Food. New York: Plenum Press.

MISLIVEC, P.B. and BRUCE, V.R. 1977. Direct plating versus dilution plating in qualitatively determining the mold flora of dried beans and soybeans. *J. Assoc. Off. Anal. Chemists* **60**: 741-743.

NOTERMANS, S. and HEUVELMAN, C.J. 1985. Immunological detection of moulds in foods by using the enzyme-linked immunosorbent assay (ELISA): preparation of antigens. *Int. J. Food Microbiol.* **2**: 247-258.

PITT, J.I. and HOCKING, A.D. 1985. Fungi and Food Spoilage. Sydney: Academic Press.

PITT, J.I. and HOCKING, A.D. 1989. Modern media and methods in food mycology. *Culture, Oxoid* **10**: (2) 1-2.

ROBERTSON, A. PATEL, N. and SARGEANT, J.G. 1988. Immunofluorescent detection of mould - An aid to the Howard mould counting technique. *Food Microbiol.* **5**: 33-42.

SAMSON, R.A. and VAN REENEN-HOEKSTRA, E.S. 1988. Introduction to Food-borne Fungi. 3rd edn. Baarn, Netherlands: Centraalbureau voor Schimmelcultures.

SEILER, D.A.L. 1986. Baseline counts for wheat, flour and bran. *In* Methods for the Mycological Examination of Food, eds A.D. King, J.I. Pitt, L.R. Beuchat and J.E.L. Corry. pp. 194-197. New York: Plenum Press.

SEITZ, L.M., SAUER, D.B., BURROUGHS, R., MOHR, H.E. and HUBBARD, J.D. 1979. Ergosterol as a measure of fungal growth. *Phytopathology* **69**: 1202-1203.

4

HEAT RESISTANT FUNGI

INFLUENCE OF PRETREATMENT OF RASPBERRY PULP ON THE DETECTION OF HEAT RESISTANT MOULDS

R. A. Samson[1], Ellen S. van Reenen-Hoekstra[1] and B. J. Hartog[2]

[1]*Centraalbureau voor Schimmelcultures*
PO Box 273
3740 AG Baarn, The Netherlands

[2]*CIVO-TNO*
PO Box 360
3700 AJ Zeist, The Netherlands

SUMMARY

Frozen raspberry pulp was defrosted and preheated within 2 minutes using microwaves. Samples were heated in a waterbath for 15 minutes at 50, 60, 70, 75, 80, 85, 90 and 100°C. After rapid cooling in ice the pulp was mixed 1:1 with double concentrated Oxytetracyline Glucose Yeast extract agar (OGY) and incubated at 25°C for up to 14 days. Heat treatment at 50 and 60°C caused a decrease in viable counts; mesophilic moulds, including *Rhizopus* and *Penicillium* spp. were found. After treatment at 60°C, some ascomycetes, *Byssochlamys, Hamigera, Eupenicillium* and *Neosartorya* spp. were detected. At 70°C and higher temperatures, only heat resistant moulds from these genera were isolated. The raspberry pulp contained three species of *Hamigera*, including an undescribed taxon. Detection methods, contamination levels and the composition of the mycoflora are discussed.

INTRODUCTION

Following an outbreak of mould contamination of marmalades, a mycological investigation of the raw material consisting of frozen raspberry pulp was carried out. Because the marmalades were contaminated by *Eupenicillium* species, heat-resistant moulds could be expected. Beuchat and Rice (1979) and Hocking and Pitt (1984, 1986) described methods for screening for heat resistant moulds. For our analysis the procedures were slightly modified. The aim of the study was to examine the effect of heat treatment on the qualitative and quantitative detection of heat resistant moulds in the raw fruit. During the investigation an unexpected number of heat resistant species were observed. The procedure was repeated with more emphasis on the total mycoflora and the results are reported in this paper.

MATERIAL AND METHODS

From each of the cartons containing frozen raspberry pulp, 500 g samples were taken aseptically. Portions of 100 g were defrosted in a refrigerator overnight. Before heat treatment the portions were put on melting ice to achieve the same initial temperature of

0°C. To accurately achieve the desired temperatures of 50, 60, 70, 75, 80, 85, 90 and 100°C, the samples were preheated in a microwave oven for different time intervals. To ensure a uniform temperature, the samples were shaken between application of the microwaves, then placed in a waterbath for 15 minutes at the desired temperature.

After rapid cooling to 50°C in melting ice, the pulp was mixed 1:1 with double concentrated Oxytetracycline Glucose Yeast Extract Agar (OGY) and poured into Petri dishes of 14 cm diameter. Incubation time was up to 14 days at 25°C in darkness. Colonies were counted and moulds isolated and identified to species level, using Malt Extract and Oatmeal agars.

RESULTS AND DISCUSSION

The total colony count (see Table 1) decreased from 350 cfu/100g pulp at 50°C to 12 cfu/100 g after heat treatment of 90°C. After treatment of 100°C no fungal colonies were observed although on one occasion one colony of an *Aspergillus* species belonging to the section *Fumigati* was detected. *Rhizopus* colonies were found after heating at 50°C, and *Penicillium* species such as *P. roqueforti* and *P. glabrum* survived a temperature of 60°C.

Heat resistant species of *Eupenicillium, Neosartorya* and *Hamigera* were detected between 50 and 90°C (Table 2). Two species of *Byssochlamys* were detected: at 60°C *B. nivea* was isolated, while at 70°C and higher, *B. fulva* was more common (Table 2).

Cultivation and identification of *Eupenicillium* isolates proved to be difficult. This was due to poor production of ascospores or even a total absence, despite prolonged incubation for up to 6 weeks. One isolate was identified as *Eupenicillium brefeldianum* and one as close to *E. lapidosum*. Isolates representing another unidentified taxon produced only sterile sclerotia.

All isolates of *Neosartorya* were identified as *N. fischeri* var. *glabra*. This variety of *N. fischeri* has been regularly observed in other surveys of heat resistant moulds (Jesenska *et al.*, 1991; our unpublished observations). Surprisingly three species of *Hamigera* Stolk & Samson (= *Talaromyces* sensu Pitt, 1979) were isolated and identified as *Hamigera avellanea, H. striata* and an undescribed new taxon (Figure 1). At 90°C, one isolate of

Table 1. Total colony counts and counts for genera present after 15 minutes heating at the temperatures indicated

Heating temp. (°C)	Viable counts/100 g						
	Total count	*Penicillium* spp.	*Neosartorya* spp.	*Eupenicillium* spp.	*Hamigera* spp.	*Byssochlamys* spp.	*Talaromyces* spp.
50	350	>200	18	110	4	-	-
60	268	32	40	180	20	2	-
70	218	-	22	166	28	2	-
75	198	-	34	146	16	2	-
80	97	-	26	63	7	1	-
85	58	-	22	22	10	4	-
90	12	-	2	1	5	1	1
100	-	-	-	-	-	-	-

Table 2. The occurrence of heat resistant mould genera in relation to heat treatment

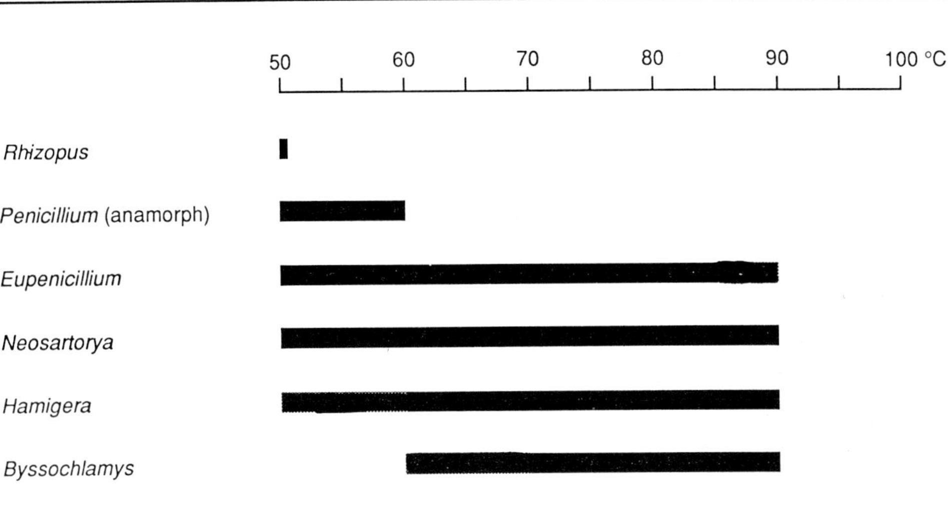

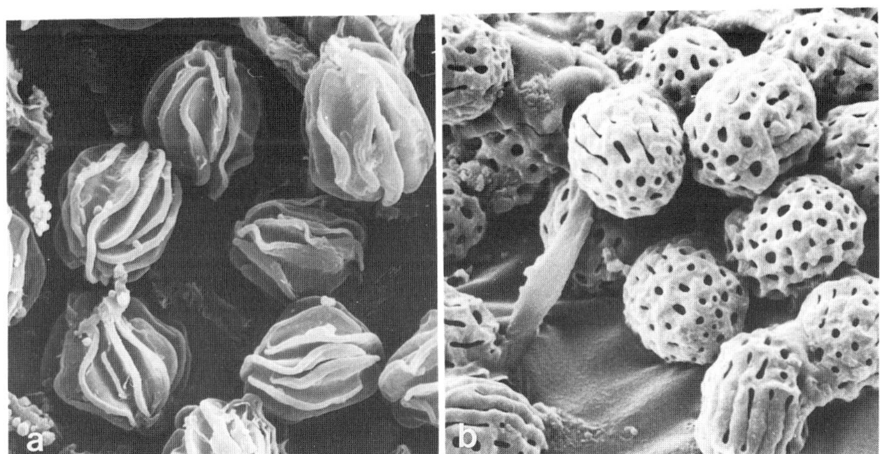

Figure 1. Scanning electron micrographs of the ascospores of (a) *Hamigera striata* and (b) an undescribed taxon (magnification x 2400).

Talaromyces emersonii was found. This species is a true thermophile, but the heat resistance of its ascospores is unknown.

From this experiment it was concluded that the sample of fruit pulp contained several ascomycetous heat resistant species. However, only *Eupenicillium* caused spoilage of the food product. This is probably due to the relatively large number of ascospores present in comparison with the other species. During this experiment species of *Hamigera* were reported as food-borne heat resistant moulds for the first time.

The procedure proposed by Hocking and Pitt (1984, 1986) proved to be very suitable for this study. We confirmed that large samples are important for the enumeration of heat resistant moulds. In addition, an incubation time of longer than 7 days is required for enumeration and identification of heat resistant moulds.

REFERENCES

BEUCHAT, L.R. and RICE, S.L. 1979. *Byssochlamys* spp. and their importance in processed fruits. *Adv. Food. Res.* **25**: 237-288.

HOCKING, A.D. and PITT, J.I. 1984. Food spoilage fungi. II. Heat resistant fungi. *CSIRO Food Res. Q.* 44: 73-82

HOCKING, A.D. and PITT, J.I. 1986. Techniques for the enumeration of heat resistant fungi. *In* Methods for the Mycological Examination of Food, eds A.D. King, J.I. Pitt, L.R. Beuchat and J.E.L. Corry, pp. 138-142. New York: Plenum Press.

JESENSKA, Z., PIEKOVA, E. and SEPITKOVA, J. 1991. Thermoresistant propagules of *Neosartorya fischeri*; some ecologic considerations. *J. Food. Prot.* **54**: 582-584.

PITT, J.I. 1979. The Genus *Penicillium* and its Teleomorphic States *Eupenicillium* and *Talaromyces*. London: Academic Press.

DIFFERENTIATION OF FOOD-BORNE TAXA OF *NEOSARTORYA*

P. V. Nielsen[1] and R. A. Samson[2]

[1] *Department of Biotechnology*
Technical University of Denmark
2800 Lyngby, Denmark

[2]*Centraalbureau voor Schimmelcultures*
PO Box 273
3740 AG Baarn
The Netherlands

SUMMARY

Recent studies have shown that the three varieties of *Neosartorya fischeri*, (*N. fischeri* var. *fischeri*, *N. fischeri* var. *glabra* and *N. fischeri* var. *spinosa*) are of different relevance to the food processing industry. Therefore, *N. fischeri* varieties isolated from food products were examined by High Pressure Liquid Chromatography (HPLC) and Scanning Electron Microscopy (SEM). Heat resistance at 80, 85 and 90°C was determined by heating in apple juice and enumerating on Czapek Yeast extract Agar (CYA) supplemented with sucrose (100 g/l) and rose bengal (20 ppm). Known isolates of the three *N. fischeri* varieties were tested as standards. The three varieties were clearly distinguished from each other and from other *Neosartorya* species by ascospore ornamentation and characteristic secondary metabolite profiles. *N. fischeri* var. *glabra* was further separated into three chemotypes. None of the food isolates were *N. fischeri* var. *fischeri* which has never been reported to spoil fruit products. *N. fischeri* var. *fischeri* was less heat resistant (D_{85}, 6-10 min; Z value, 9-10°C) than *N. fischeri* var. *spinosa* (D_{85}, 10-96 min; Z, 5-7°C) and *N. fischeri* var. *glabra* (D_{85}, 10-21 min; Z, 6-14°C). A food isolate of *Neosartorya aureola* was also highly heat resistant (D_{85}, 10 min; Z, 12°C). Differences in secondary metabolite profiles and heat resistance show the importance of correct identification of the species and varieties causing food spoilage.

INTRODUCTION

The Ascomycete genus *Neosartorya* includes seven species (*N. aurata, N. aureola, N. fennelliae, N. fischeri, N. quadricincta, N. spathulata,* and *N. stramenia*) and three varieties *N. fischeri* var. *fischeri*, *N. fischeri* var. *glabra,* and *N. fischeri* var. *spinosa* (Raper and Fennell, 1965, Samson *et al.*, 1990). Recently Peterson (1992) raised *A. fischeri* var. *thermomutatus* to species level as *N. pseudofischeri* on the basis of differences in ascospore morphology and genetic similarity studies. So far only isolates of *Neosartorya fischeri* have been reported as heat resistant spoilage fungi in foods and beverages (Kavanagh *et al.*, 1963; McEvoy and Stuart, 1970; Splittstoesser and Splittstoesser, 1977; Hocking and Pitt, 1984; Jesenska *et al.*, 1984; Jesenska and Petrikova, 1985; Scott and Bernard, 1987; Splittstoesser and Churey, 1989). As *N. fischeri* is widely distributed in soil it predominantly causes spoilage in heat processed food products containing fruits which have been contaminated with soil. Most species in the genus *Neosartorya* are able to produce toxic secondary metabolites (Samson *et al.*, 1990, Table 1).

Table 1. Reported production of secondary metabolites by *Neosartorya* species[1]

Taxon	Mycotoxin and antibiotics
N. aurata	Helvolic acid
N. aureola	Tryptoquivaline
N. fennelliae	Trypacidin, viridicatumtoxin
N. fischeri var. *fischeri*	Fumitremorgin A, B, C, verruculogen, terrein, tryptoquivalin
N. fischeri var. *glabra*	Avenaciolide, mevinolins, tryptoquivaline, trypacidin
N. fischeri var. *spinosa*	Tryptoquivaline, terrein
N. quadricincta	Cyclopaldic acid
N. stramenia	Canadensolide

[1] Turner (1971); Turner and Aldridge (1983); Samson *et al.* (1990).

Recent studies in our laboratories have indicated that *N. fischeri* var. *glabra* and *N. fischeri* var. *spinosa* are much more frequently isolated from spoiled heat processed fruit based products than *N. fischeri* var. *fischeri* and other *Neosartorya* species. This study reports on the differences in heat resistance of relevant *Neosartorya* taxa, along with their secondary metabolite profiles, ascospore ornamentation and macromorphology.

MATERIAL AND METHODS

Fungi

The isolates used in this study are listed in Table 2. Some isolates e.g. H-37, H-73 and M-1 were re-identified as *N. fischeri* var. *glabra*, *N. fischeri* var. *spinosa* and *N. pseudofischeri*, while one isolate (WR-1) could not be accommodated among the known taxa. They were cultured on Oatmeal Agar (OA), Czapek Yeast extract Agar (CYA), Malt Extract Agar (MEA), Yeast Extract Sucrose agar (YES) and Czapek Yeast extract agar with 20% sucrose (CY20S) at 25 and 30°C. Morphological characters were examined after one and two weeks. Secondary metabolite production was examined on these media by the agar plug method (Filtenborg *et al.*, 1983), and by high performance liquid chromatography (HPLC) with diode array detection (Frisvad and Thrane, 1987). For HPLC analysis isolates were cultured on 3 plates each of CYA, MEA, YES and CY20S for 14 days at 30°C and extracted with 150 ml chloroform methanol (2:1) for 2 minutes in a Colworth Stomacher. After filtration and evaporation the procedure was as described by Frisvad and Thrane (1987).

Heat resistance

For heat resistance determinations, ascospores were harvested from the surface of OA incubated at 30°C for 32 days. Suspensions in peptone water (0.1%) with Tween 80 (0.1%) were treated twice for 5 min in a water bath sonicator with ice water to break up cleistothecia and asci. After settling (2 min), ascospore suspensions were filtered through sterile glass wool, diluted with sterile water to total counts of *ca.* 10^6 spores per millilitre

Table 2: Isolates examined

Species	Isolate and Source
N. aureola	FRR 2269 = CBS 583.90, ex Fijian passionfruit juice.
N. fischeri var. *fischeri*	IBT 3023 = CBS 832.88, ex soil, Denmark IMI 16143 = IBT 4868, ex soil, England
N. fischeri var. *glabra*	IMI 102173 = IBT 4867, ex canned strawberries, Eire IBT 3004, ex soil, Urmston, UK CBS 112.55, ex garden soil, South Australia IBT 11025 = RO 27-3, ex fruit filling, the Netherlands H-73 = CBS 585.90 = IBT 6558, ex surface of a mechanical grape harvester, New York State, USA CBS 582.90, ex raspberry pulp, Poland
N. fischeri var. *spinosa*	CBS 483.65 = IBT 3022, ex soil, Nicaragua IBT 3001, ex annatto seeds, Brazil FRR 2334 = CBS 580.90 , ex Brazilian passionfruit juice H-37 = CBS 581.90 = IBT 6559, ex soil, vineyard, New York State, USA
Neosartorya pseudofischeri	M-1 = CBS 586.90 = IBT 6472, ex cherry filling, USA
Neosartorya species	WR-1 = CBS 584.90, ex grape drink, USA

and stored frozen. Aliquots (0.25 mL) of the ascospore suspensions were deposited into a series of 13 x 100 mm screw cap test tubes containing 2.25 mL apple juice held at 80, 85 and 90°C. Suspensions were maintained at these temperatures under constant agitation in a water bath shaker. After various heating times, tubes were withdrawn and immediately cooled to approximately 10°C in cold water. Viable ascospores were enumerated on CYA, pH 3.5 with increased sucrose (10%) and rose bengal (20 mg/L). Colonies were counted after 4 and 6 days incubation at 30°C.

Scanning electron microscopy (SEM)

For SEM, mature cleistothecia were transferred to aluminium stubs which were covered with double sided adhesive tape. A small drop of water containing some Tween 80 was added and the cleistothecia crushed. The suspension was air dried, coated with gold and examined by a JEOL 840 scanning electron microscope.

RESULTS AND DISCUSSION

Ascospore morphology

Separation of the taxa of *Neosartorya* accepted by Raper and Fennell (1965) is feasible by light microscopy based on colour of ascomata and ascospore morphology. The varieties of *N. fischeri*, however, are difficult to distinguish: differentiation is mainly based on ascospore ornamentation, which can be best elucidated by SEM (Figure 1).

Ascospores of *N. fischeri* var. *fischeri* are described by Raper and Fennell (1965) as having convex surfaces bearing anastomosing ridges. Under SEM, the ascospores were typically reticulate (Figure 1a). Ascospores of *N. fischeri* var. *spinosa* (Figure 1b) showed

Figure 1. Scanning electron micrographs of ascospores. A. *N. fischeri* var. *fischeri* CBS 832.88, B. *N. fischeri* var. *spinosa* (H-37 = CBS 581.90), C. *N. fischeri* var. *glabra* (H-73 = CBS 585.90), D. *N. aureola* (FRR 2269 = CBS 583.90), E. *Neosartorya pseudofischeri* (M-1 = CBS 586.90), F. *Neosartorya species* (WR-1 = CBS 584.90) (magnification 2700 x)

the spines observed by Raper and Fennell (1965), but also showed some gradation from rough to distinctly spinulose. Ascospores of *N. aureola* (Figure 1d) were also spinose but this species can be separated from *N. fischeri* var. *spinosa* by yellow to orange ascomata rather than white to cream. The finely roughened surface of *N. fischeri* var. *glabra* ascospores (Figure 1c) was not recognisable in the light microscope and was described as smooth or nearly so by Raper and Fennell (1965). The different chemotypes of this variety were not distinguishable under SEM.

Isolate M-1 (= CBS 586.90) obtained from cherry filling in the USA has a distinct ascospore morphology consisting of triangular projections identical with that of the ex type

Table 3. Secondary metabolites produced by the examined isolates as determined by HPLC-DAD[1]

	FUM	MEV	TRY	VIR	XAN	FUA	FUB	NE1	G1
N. aureola									
FRR 2269	-	-	+	-	-	+	-	-	-
N. fischeri var. *fischeri*									
CBS 832.88	+	-	+	-	-	-	+	+	-
IMI 16143	+	-	+	-	-	-	+	+	-
N. fischeri var. *glabra* I									
IMI 102173	-	-	-	-	-	-	-	-	+
IBT 3004	-	-	-	-	-	-	-	-	+
CBS 582.90	-	-	-	-	-	-	-	-	+
N. fischeri var. *glabra* II									
CBS 112.55	-	+	-	-	-	-	-	-	-
IBT 11025	-	+	-	-	-	-	-	-	-
N. fischeri var. *glabra* III									
CBS 585.90	-	-	-	-	+	-	-	-	-
N. fischeri var. *spinosa*									
CBS 483.65	-	-	+	-	-	+	-	+	-
IBT 3001	-	-	+	-	-	+	-	+	-
FRR 2334	-	-	+	-	+	+	-	+	-
CBS 581.90	-	-	+	-	-	-	-	+	+
Neosartorya pseudofischeri									
M-1	-	-	+	-	+	-	+	-	-
Neosartorya species									
WR-1	-	-	+	+	-	-	-	+	+

[1] FUM = fumitremorgins A, B and C and verruculogen, MEV = compounds with UV spectra similar to the mevinolins, TRY = tryptoquivalins, VIR = viridicatumtoxin and XAN = xanthocillins. FUA, FUB, NE1 and G1 are undescribed secondary metabolites with characteristic UV-spectra and retention indices (Figure 3).

culture of *N. pseudofischeri* (Peterson, 1992). This ornamentation is also seen in a soil-borne isolate IBT 3002 (Samson *et al.*, 1990). Based on ascospore morphology *N. fischeri* var.*glabra* and *N. fischeri* var. *spinosa* appear to be closer related to each other and *N. aurata* and *N. pseudofischeri* than to *N. fischeri* var. *fischeri*. Ascospores of WR-1 (= CBS 584.90) have a distinct cerebriform surface structure (Fig. 1 E) resembling the ascospores produced by the heterothallic *N. fennelliae* (Samson *et al.*, 1990).

Secondary metabolite profiles

Secondary metabolite data correlated well with the morphological data (Table 3). All isolates examined produced characteristic secondary metabolite profiles as determined by HPLC-DAD. Only *N. fischeri* var. *fischeri* produced the neurotoxic fumitremorgins A, B, C and verruculogen, whereas all taxa except *N. fischeri* var. *glabra* produced the potentially tremorgenic tryptoquivalines. Despite very similar ascospore morphology, isolates included in *N. fischeri* var. *glabra* produced three distinct secondary metabolite profiles (Figure 2). The two isolates in chemotype I produced almost exclusively one compound, termed G1 (Figure 3d). This compound was also identified in the extract of WR-1 (= CBS 584.90) and of H-37 (= CBS 581.90). Isolates of *N. fischeri* var. *glabra* chemotype II produced high amounts of compounds with UV spectra similar to the mevinolins. These compounds have not been detected in other isolates in the genus *Neosartorya*. Xanthocillins were only produced by *N. fischeri* var. *glabra* chemotype III, one isolate of *N. fischeri* var. *spinosa* (FRR 2334) and by *N. pseudofischeri*. WR-1 (= CBS 584.90) produced viridicatumtoxin, a toxin previously isolated from cultures of the heterothallic *N. fennelliae*.

Heat-resistance

Heat resistance studies showed that food isolates of *N. fischeri* and reference isolates from the same varieties or chemotypes as the food isolates were clearly more heat resistant than species not relevant to foods (Table 4). The *N. aureola* isolate from food was exceptional, having a heat resistance comparable to isolates from sources other than foods. Some variation in heat resistance was observed within each taxon, especially in *N. fischeri* var. *spinosa*, which also showed the greatest variability in secondary metabolite profiles. It is notable that the three chemotypes of *N. fischeri* var. *glabra* possessed quite different resistances: these differences are not clearly understood.

Large differences in heat resistance for isolates of *N. fischeri* have been reported previously. McEvoy and Stuart (1970) found that 20 min at 85°C caused virtually no destruction of *N. fischeri* var. *glabra* isolated from canned strawberries. At 90°C, 1.9% survived 20 min in water, indicating a D value of approximately 11 min. Splittstoesser and Splittstoesser (1977) reported that survival of *N. fischeri* CBS 584.90 (= WR-1) ascospores after heating for 60 min at 85°C ranged from 0.14% in water and 3.9% in Concord grape juice to 68% in 40% glucose. Scott and Bernard (1987) reported that ascospores of *N. fischeri* isolated from apple juice had a D value of 1.4 minute at 87.8°C and a Z value of 5.6°C when heated in apple juice. From this the D value can be calculated as 4.4 min at 85°C and 0.6 min at 90°C. Beuchat (1986) found that two isolates of *N. fischeri* had D values of 30.1 to 116 min at 85°C in various food products, while another *N. fischeri* isolate had D values of 15.1 to 19.4 min under the same conditions.

The large variation between heat resistance data presented by different authors can partly be explained by differences in media used for ascospore production, heating menstruum and ascospore age, as these factors have been proven to have a significant

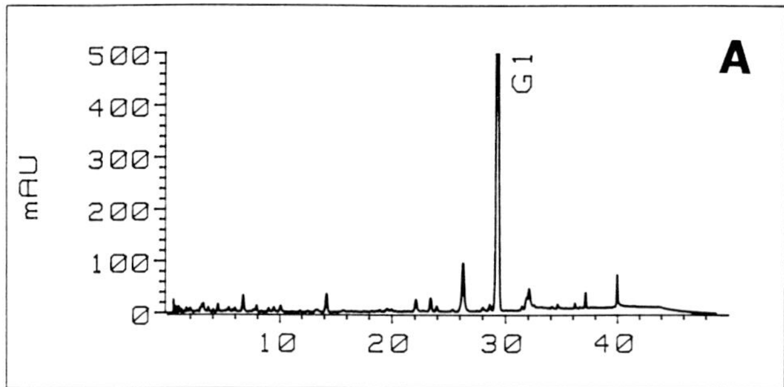

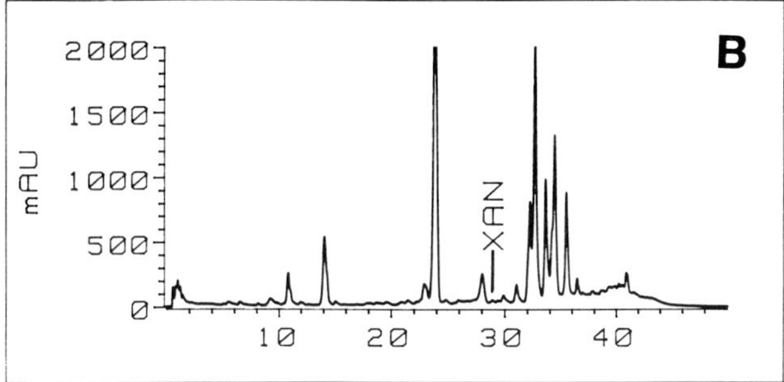

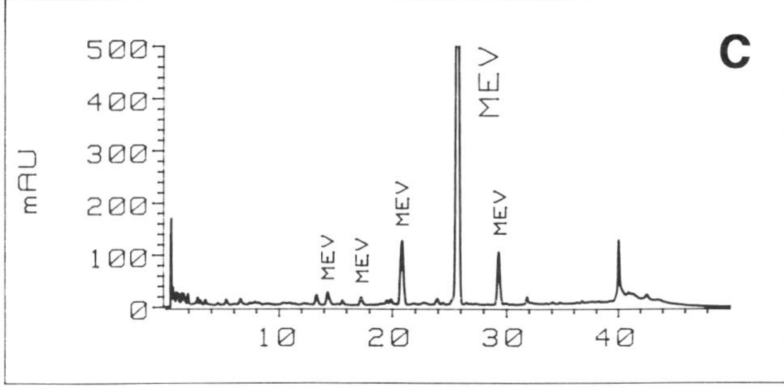

Figure 2. HPLC chromatogram of the three chemotypes of *N. fischeri* var. *glabra* showing the absorbance at 225 nm. A. chemotype I, IMI 102173, producing the unknown metabolite G1; B. chemotype III, H73, producing xanthocillin; C. chemotype II, CBS 112.55, producing metabolites with UV spectra similar to the mevinolins.

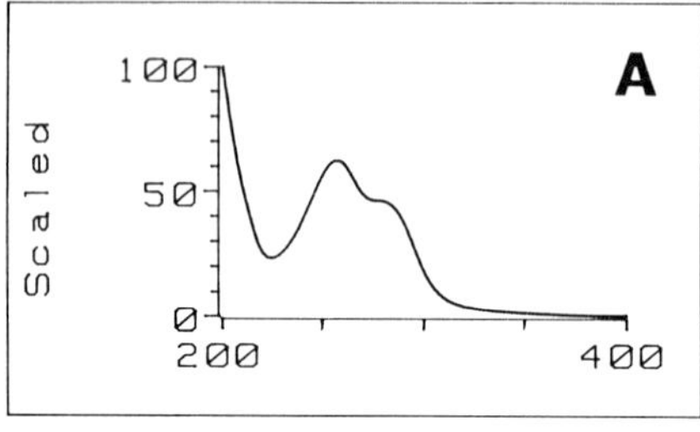

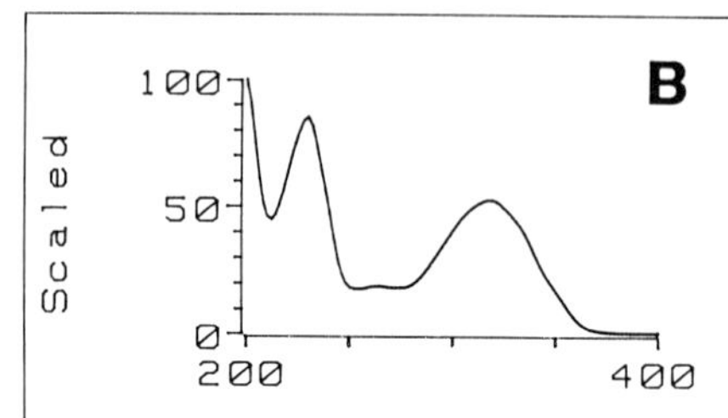

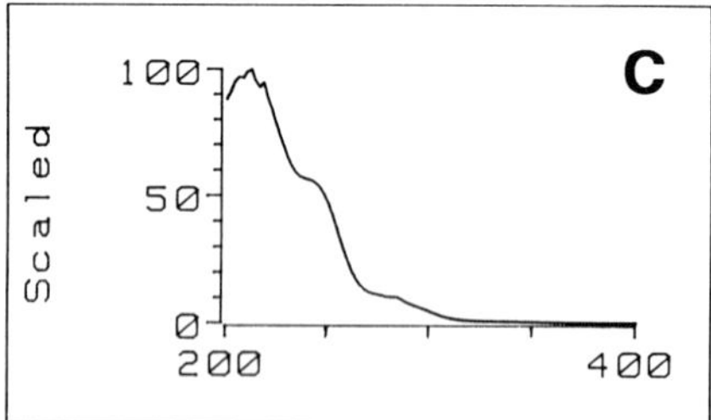

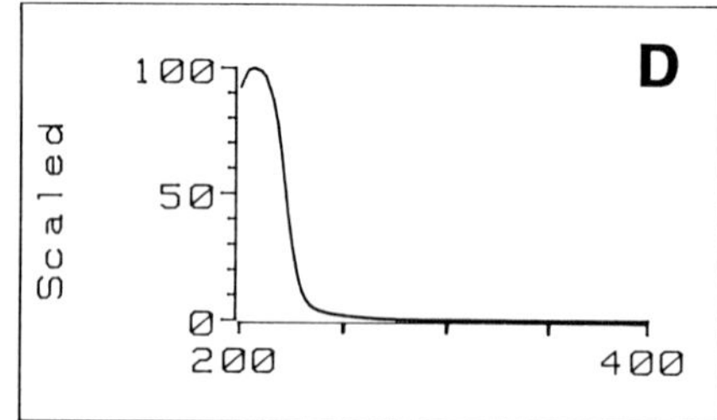

Figure 3: UV-spectra of four characteristic but unknown metabolites. A. FUA (retention index RI:844, retention time RT:9.78, *N. fischeri* var. *spinosa*, IBT 3001); B. FUB (RI:1078, RT:22.05, *N. fischeri* var. *fischeri*, CBS 832.88); C. NE1 (RI:1028, RT:19.57, *N. fischeri* var. *fischeri*, CBS 832.88); D. G1 (RI:1261, RT:29.28, *N. fischeri* var. *glabra* I, IMI 102173).

Table 4. Heat resistance of the examined isolates

		D_{80}	D_{85}	D_{90}	Z
N. aureola	FRR 2269[1]	28	10	4	12
N. fischeri					
v. *fischeri*	IBT 3023	23	6	2	9
v. *fischeri*	IMI 16143	23	10	2	10
v. *glabra* I	IBT 3004	>120	20	4	6
v. *glabra* I	IMI 102173[1]	>120	21	3	7
V. *glabra* I	CBS 582.90	>120	16	5	6
v. *glabra* II	CBS 112.55	27	14	6	14
v. *glabra* II	IBT 11025	32	-	-	-
v. *glabra* III	H73	34	10	6	13
v. *spinosa*	H37	>120	96	9	5
v. *spinosa*	FRR 2334 [1]	>120	21	4	6
v. *spinosa*	CBS 483.65	>120	11	5	7
v. *spinosa*	IBT 3001	>120	10	3	6
N. pseudofischeri	M-1[1]	>120	25	4	7
Neosartorya species	WR-1[1]	>120	15	8	8

[1] Isolated from food products

influence on heat-resistance (McEvoy and Stuart, 1970; Splittstoesser and Splittstoesser, 1977; Beuchat, 1986; Conner and Beuchat, 1987a,b). A difference may even exist between different spore crops (Splittstoesser and Churey, 1989). Also the enumeration medium affects the recovery of heat stressed ascospores (Splittstoesser and Splittstoesser, 1977). The influence of heating technique and method of calculating heat resistance (King, 1986) is not known as a detailed study so far has not been carried out. Data in Table 4 illustrate the importance of correct identification as isolates of *N. fischeri* var. *fischeri* are less resistant that the other varieties. Furthermore isolates of the chemotypes II and III of *N. fischeri* var. *glabra* are also less resistant than those belonging to chemotype I. Consequently it is important to standardise these parameters to improve interlaboratory comparisions.

Despite the large variation in heat resistance observed it is clear that all isolates tested in this study are able to survive the heat treatment routinely applied to fruits and fruit products, often 3 min at 90°C (Beuchat, 1986).

CONCLUSIONS

N. fischeri var. *fischeri* was clearly different from the other *N. fischeri* varieties based on ascospore morphology, secondary metabolites and heat resistance. Similarities in ascospore morphology and heat resistance, the other varieties closely related. Based on secondary metabolite production, the new varieties, CBS 584.90 and 586.90, were more closely related to *N. fischeri* var. *spinosa* than to *N. fischeri* var. *glabra*. In theory all *Neosartorya* isolates examined are able to cause spoilage in heat processed fruit products. In reality only the most heat stable varieties have been encountered as food spoilage fungi, excluding the potentially most toxic variety *N. fischeri* var. *fischeri*. Differences in secondary metabolite production and heat resistance underline the importance of correct identification of fungi isolated from foods during quality control and from spoiled products.

ACKNOWLEDGMENTS

The authors thank Dr S. W. Peterson for providing data of his recent research on *N. pseudofischeri* and Drs A. D. Hocking and D. F. Splittstoesser for providing food-borne *Neosartorya* isolates.

REFERENCES

BEUCHAT, L.R. 1986. Extraordinary heat resistance of *Talaromyces flavus* and *Neosartorya fischeri* ascospores in fruit products. *J. Food Sci.* **51**: 1506-1510.

CONNER, D.E. and BEUCHAT, L.R. 1987a. Heat resistance of ascospores of *Neosartorya fischeri* as affected by sporulation and heating medium. *Int. J. Food Microbiol.* **4**: 303-312.

CONNER, D.E. and BEUCHAT, L.R. 1987b. Efficacy of media for promoting ascospore formation by *Neosartorya fischeri*, and the influence of age and culture temperature on heat resistance of ascospores. *Food Microbiol.* **4**: 229-238.

FILTENBORG, O., FRISVAD, J.C. and SVENDSEN, J.A. 1983. Simple screening method for molds producing intracellular mycotoxins in pure culture. *Appl. Environ. Microbiol.* **45**: 581-585.

FRISVAD, J.C. and THRANE, U. 1987. Standardized high performance liquid chromatography of 182 mycotoxins and other fungal metabolites with alkylphenone retention indexes and UV-VIS spectra (diode array detection). *J. Chromatogr.* **404**: 195-214.

HOCKING, A.D. and PITT, J.I. 1984. Food spoilage fungi. II. Heat-resistant fungi. *CSIRO Food Res. Q.* **44**: 73-82.

KAVANAGH, J., LARCHET, N. and STUART, M. 1963. Occurrence of a heat-resistant species of *Aspergillus* in canned strawberries. *Nature, London* **198**: 1322.

JESENKA, Z., HAVRANEKOVA, D. and SAJBIDOROVA, I. 1984. On some problems of moulds on some products of canning industry. *Cesk. Hyg.* **24**: 102-109.

JESENSKA, Z. and PETRIKOVA, D. 1985. Microscopic fungi as an agent causing spoilage of preserved fruit. *Cesk. Hyg.* **30**: 175-177.

KING, A.D. 1986. Incidence, properties and detection of heat resistant fungi. *In* Methods for the Mycological Examination of Foods, eds A.D. King, J.I. Pitt, L.R. Beuchat and J.E.L. Corry, pp. 153-155. New York: Plenum Press.

McEVOY, I.J. and STUART, M.R. 1970. Temperature tolerance of *Aspergillus fischeri* var. *glabra* in canned strawberries. *Ireland J.Agric. Res.* **9**: 59-67.

PETERSON, S.W. 1992. *Neosartorya pseudofischeri* sp. nov. and its relationship to other species in *Aspergillus* Section *Fumigati*. *Mycological Research*, in press.

RAPER, K.B. and FENNELL, D.I. 1965. The Genus *Aspergillus*. Baltimore, Maryland: Williams and Wilkins.

SAMSON, R.A., NIELSEN, P.V. and FRISVAD, J.C. 1990. The genus *Neosartorya*: differentiation by scanning electron microscopy and mycotoxin profiles. *In* Modern Concepts in *Penicillium* and *Aspergillus* classification, eds R.A. Samson and J.I. Pitt, pp. 455-697. New York: Plenum Press.

SCOTT, V.N. and BERNARD, D.T. 1987. Heat resistance of *Talaromyces flavus* and *Neosartorya fischeri* isolated from fruit juice. *J. Food Prot.* **50**: 18-20

SPLITTSTOESSER, D.F., and SPLITTSTOESSER, C.M. 1977. Ascospores of *Byssochlamys fulva* compared with those of a heat resistant *Aspergillus*. *J. Food Sci.* **42**: 685-688.

SPLITTSTOESSER, D.F., and CHUREY, J.J. 1989. Effect of low concentrations of sorbic acid on the heat-resistance and viable recovery of *Neosartorya fischeri* ascospores. *J. Food Prot.* **52**: 821-822.

TURNER, W.B. and ALDRIDGE, D.C. 1983. Fungal metabolites II. London: Academic Press.

ACTIVATION AND GERMINATION OF *NEOSARTORYA* ASCOSPORES

D. F. Splittstoesser and J. J. Churey

NYS Agricultural Experiment Station
Cornell University
Geneva, NY 14456 USA

SUMMARY

Thoroughly washed ascospores of a heat resistant *Neosartorya* isolate exhibited little activation when heated at 70-85°C in distilled water or phosphate buffer. In dilute grape juice a significant percentage could be activated in 10 min at 55°C. This stimulation was not due to glucose, fructose, tartaric acid or malic acid, all major constituents of grape juice, nor to its low pH. Other fruit and vegetable juices enhanced activation by varying degrees.

Germination of ascospores as evidenced by a loss in heat resistance was initiated after an incubation of 2 h at 30°C and was largely complete by 3 h. It occurred as readily in distilled water as in grape juice indicating that exogenous nutrients were not required. Germination was inhibited by 6.55% NaCl but other solutes such as sucrose had less effect at the same water activity (0.960).

INTRODUCTION

The successful detection of heat resistant ascospores requires that their dormancy first be broken (activation) and then that the spores germinate and produce new vegetative growth.

It has long been recognised that heat stimulates ascospore activation (Hull, 1933) and that the heating medium can have a major effect. Thus activation of *Byssochlamys fulva* ascospores was enhanced in potato sucrose agar (Gillespy, 1938), grape juice (Splittstoesser and Splittstoesser, 1977), acetaldehyde (Palmer and McEvoy, 1974), and 0.1N HCl (Splittstoesser *et al.*, 1972). Activation requirements may vary with strains and perhaps even spore crops. Beuchat (1986) observed that some strains of *Talaromyces flavus* and *Neosartorya fischeri* were activated at 70°C while others required a temperature of 85°C.

In this study, some factors affecting activation and germination have been investigated. By using loss of heat resistance as the criterion for germination (Lingappa and Sussman, 1959; Gould, 1969), we have separated the factors that affect this early event from those that might affect subsequent outgrowth.

MATERIALS AND METHODS

Ascospore preparation

A heat resistant fungus isolated by us from a spoilage outbreak in a thermally processed cherry pastry filling was used in this study. It was designated M-1 (= CBS 586.90) and identified as a new species of *Neosartorya*, *N. pseudofischeri* W.S. Peterson (Nielsen and Samson, 1992). Ascospores were harvested from the surface of malt extract agar plates incubated for 28 d at 30°C. The suspensions, in sterile aqueous 0.1% Tween 80 solution, were treated for 20 min in a Bransonic water bath sonicator (Smith Kline, Philadelphia, PA) to break up clumps and asci. Crude ascospore suspensions were stored frozen. Although repeated freezing and thawing did not seem to affect activation or germination, stock suspensions were thawed and refrozen no more than twice.

Activation

In a typical trial, the stock spore suspension (0.1 mL) was added to sterile distilled water (1 mL) at 1°C, then centrifuged for 5 min at 14,000 rpm. Washing was repeated four times with the final suspension being made in the test diluent. When heating was at 70°C or higher for 30 to 60 min, the suspension (0.3 to 0.5 mL) was heated in 16 x 100 mm screw cap tubes. The capillary tube procedure (Splittstoesser and Churey, 1989) was used for low temperature, short time treatments such as 55°C for 10 min.

The influence of NaCl on heat activation was studied by washing the spores four times in different concentrations of NaCl before suspending them in the desired saline solution.

After cooling, appropriate decimal dilutions were plated on potato dextrose agar, pH 3.5, containing 8.4 mg/L of rose bengal. Colonies were counted after 3 and 5 d incubation at 30°C. Plates containing under 40 colonies sometimes required the longer incubation.

Germination

Ascospores were first activated by heating in 5° Brix Concord grape juice at 70°C for 60 min, then washed four times in distilled water and suspended in the germination medium. In most experiments, 1 mL volumes in 100 x 16 mm screw cap tubes were incubated at 30°C. At various time intervals, tubes were heated in a 60°C water bath for 30 min, a treatment found to destroy all germinated ascospores. The suspensions were then plated as described above. The percentage of spores that had germinated was calculated on the basis of viable counts before and after heating at 60°C.

Grape juice analysis

The sugar and acid content of Concord grape juice was determined with a Hewlett Packard 1090 Series M HPLC equipped with diode array and refractive index detectors. Two columns (Hewlett Packard Hypersil ODS and BioRad 87H) used in sequence were preceded by a cation H^+ guard column. The solvent was 0.001 N sulphuric acid with 20% (v/v) acetonitrile; the flow rate was 0.6 mL per min. Separations were carried out at 65°C.

RESULTS AND DISCUSSION

Activation

The suspending medium had a marked effect on the activation of ascospores of *N. pseudofischeri*: considerably higher recoveries were observed in grape juice at 70, 80, and 85°C than in distilled water or phosphate buffer (Table 1). The lower recoveries observed at 85°C were probably the result of both activation and spore destruction occurring at that temperature. More resistant ascospores may have given higher relative counts at this temperature (Beuchat 1986).

When ascospores were not washed thoroughly before heating, a higher percentage were usually activated in water. This indicates that some stimulatory substance was carried over from the propagation medium, or that some metabolic product of the fungus stimulated activation.

To determine whether the sugar and acid constituents of grape juice affect activation, a solution of glucose, fructose, tartaric and malic acids was prepared that contained the same concentrations, as measured by HPLC, that were present in 5° Brix Concord juice. Ascospores were then heat activated in different dilutions of grape juice and the sugar acid solution. The results (Table 2) showed that as little as 0.1° Brix grape juice stimulated activation while none of the sugar acid solutions had an effect. These results suggest that other components of grape juice, perhaps tannins or other polyphenolic compounds, may be the active principals.

Table 1. Effect of heating medium and temperature on per cent activation of *N. pseudofischeri* ascospores heated for 30 min at various temperatures

Medium	°C	Per cent colony formation[1]
Na_3PO_4, 0.1M, pH 7	70	0.62
	80	10
	85	3.8
Grape Juice, 5° Brix	70	100
	80	79
	85	33
Distilled Water	70	1.4
	80	34
	85	12

[1] Relative to 30 min at 70°C in 5° Brix grape juice

Table 2. Influence of various concentrations of grape juice or sugar acid solutions on germination of *N. pseudofischeri* ascospores heated at 70°C for 30 min

Concentration	Percent colony formation[1]	
°Brix	Grape juice[2]	Sugar/acid medium
Water	4.3	4.5
0.1	52	--
0.5	--	5.2
1.0	45	--
2.5	--	3.8
5.0[2]	83	4.9

[1] Relative to colony formation after heating for 60 min at 70°C in 5° Brix grape juice
[2] 5° Brix grape juice contains per litre: glucose, 17.15 g; fructose, 19.00 g; tartaric acid, 1.69 g; malic acid, 0.96 g

The pH was not a factor because both the juice and sugar acid solutions had comparable pH values, ranging from 3.6 to 3.7. Furthermore, grape juice neutralised to pH 7 with NaOH was as stimulatory as juice of pH 3.5 (data not shown).

A number of different fruit and vegetable juices were tested to determine their effect on ascospore activation (Table 3). All gave recoveries that were 10 to 20 times greater than that observed in distilled water. All must have contained substances that enhanced activation but did not possess inhibitors that interfered with ascospore detection. The accuracy of plate count procedures is such that it would be difficult to conclude that many of the differences in recovery shown in Table 3 were significant.

Studies on ascospores heated in grape juice have shown that over 80% were activated during the first 10 min heating at 70°C and that almost all were activated by 30 min at 70°C (data not shown). It seemed likely, therefore, that juices may differ in their effects on the kinetics of activation, but that these differences might not be apparent when ascospores were subjected to 30 min treatment at 70°C. Heating ascospores in grape juice for 10 min at 55°C resulted in only 16 to 20% being activated (data not shown) and thus it appeared that this treatment might reveal differences in activation rates between juices.

Milder heat activation (55°C for 10 min) revealed major differences between juices at commercial concentrations and when adjusted to 5° Brix (Table 4). The fact that cranberry juice was most stimulatory was unexpected because cranberry cocktail is a beverage in which cranberry extract is diluted with water, high fructose corn syrup, and ascorbic acid. In future studies the effect of ascorbic acid on activation will be investigated.

The influence of sodium chloride, of interest because a reduced water activity might suppress ascospore activation, varied with the heating medium. Although 26% (w/w) NaCl in water reduced the percentage of activated spores to about one third compared to distilled water, all concentrations permitted only low levels of activation (Table 5). In the

Table 3. Influence of different fruit and vegetable juices on colony formation by ascospores of *N. pseudofischeri* after heating at 70°C for 30 min

Juice	pH	°Brix	Count, per cent[1]
Water	5.1	0	4.7
Apple	3.4	12	38
Apricot	3.8	14	74
Banana	4.0	23	52
Carrot	5.8	9	83
Cherry	3.6	14	28
Cranberry	2.6	14	29
Grape, white	3.7	16	67
Grapefruit	3.3	10	54
Orange	3.7	12	52
Pear	3.6	13	33
Pineapple	3.4	13	82
Prune	3.8	19	42
Tomato	4.2	7	100

[1] Percentage relative to tomato juice

presence of dilute grape juice, the highest concentration of salt (equivalent to 0.759 a_w) only slightly reduced the percentage of spores that were activated. These results indicate that a reduction in water activity suppresses activation only slightly and, therefore, the activation of spores in undiluted juice concentrates may not present a problem.

Germination

Ascospores germinated as readily in distilled water as in nutrient media such a fruit juice, indicating clearly that exogenous nutrients were not required for germination (Table 6). In these experiments, ascospores had been washed four or more times after heat activation, and thus the carryover of nutrients from the grape juice was minimal.

The data in Table 6 are somewhat atypical in that little or no germination was usually detected during the first hour of incubation at 30°C. This is illustrated in Table 7, which shows the effect of incubation temperature on the rate of germination. These results show that germination took place over a temperature range from 20 to 50°C at least, with 40°C optimal.

Table 4. Influence of single strength and diluted fruit juices on activation of *N. pseudofischeri* ascospores after heating at 55°C for 10 min

Medium	Count, per cent[1]	
	Single strength juice	5° Brix dilution
Water	2.7	
Grape, white	9.7	2.8
Grape, Concord	57	30
Cranberry	100	100
Pear	14	3.9
Cherry	90	16

[1] Percentage relative to cranberry juice

Table 5. Effect of NaCl on colony formation by *N. pseudofischeri* ascospores heated at 70°C for 60 min in water or dilute grape juice (0.45° Brix)

NaCl	Count, per cent[1]	
%	Water	Grape juice
0	5.5	--
6	4.5	108
13	3.5	92
26	1.8	83

[1] Percentage relative to 5° Brix grape juice

Table 6. Germination of activated (70°C for 60 min) *N. pseudofischeri* ascospores in water and 5° Brix Concord grape juice

		Viable count x 10^4		
Medium	Time at 30°C (hr)	Before heating	After heating (60°C, 30 min)	Per cent germination[1]
Water	0	120	170	0
	1	100	23	77
	2	340	8	99
	3	300	16	95
Grape juice	0	130	170	0
	1	180	53	70
	2	81	9	88
	3	300	14	95

[1] Percentage germination calculated as (viable count before heating - viable count after heating)/viable count before heating

Table 7. Influence of holding temperatures and time on germination of *N. pseudofischeri* ascospores previously activated at 70°C for 30 min

	Per cent germination[1]		
	Holding time, hr		
Temp. °C	1	2	3
20	--	2.2	69
30	0	24	97
40	0	72	98
50	0	48	85

[1] See footnote, Table 6.

Concentrations of sodium chloride above 5% were found to inhibit germination when added to the incubation medium (Table 8). In the presence of 13% NaCl, for example, no germination was detected at 3 h but at 48 h 63% of the spores had germinated. It is not known whether 16% NaCl, the next concentration used, completely blocked germination or if it slowly occurred even at this concentration.

Solutions of glycerol and sucrose inhibited germination to a lesser extent than sodium and lithium chloride at the same water activities (Table 9). Hence, the effect of sodium chloride appears to be due to other factors in addition to a lowering of water activity.

Table 8. Influence of NaCl concentration on germination of *N. pseudofischeri* ascospores held at 30°C for 3 and 48 h after heating at 70°C for 60 min

% NaCl (w/w)	a_w	Time at 30°C (hr)	Per cent germination[1]
0	1.0	3	97.6
		48	>99.9
5	0.97	3	99.6
6	0.96	3	36
13	0.91	3	0
		48	63
16	0.83	3	0
		48	0

[1] See footnote, Table 6

Table 9. Influence of salt and sugar solutions of 0.960 a_w on the germination of *N. pseudofischeri* ascospores held for 3 hr at 30°C after heating at 70°C for 60 min

Solute	Per cent w/w	Germination per cent[1]
NaCl	6.55	15
LiCl	4.5	24
Sucrose	39.66	65
Glycerol	16.91	82
Water		93
Grape juice	5° Brix	94

[1] See footnote, Table 6

REFERENCES

BEUCHAT, L.R. 1986. Extraordinary heat resistance of *Talaromyces flavus* and *Neosartorya fischeri* ascospores in fruit products. *J. Food Sci.* **51**: 1506-1510.

GILLESPY, T.G. 1938. Studies on the mould *Byssochlamys fulva*. II. *Ann. Rep. Fruit Veg. Pres. Res. Sta.*, Campden, pp. 67-78.

GOULD, G.W. 1969. Germination. *In* The Bacterial Spore, eds G. W. Gould and A. Hurst, pp. 397-444. London: Academic Press.

HULL, R. 1933. Investigation of the control of spoilage of processed fruit by *Byssochlamys fulva*. *Ann. Rep. Fruit Veg. Pres. Res. Sta.*, Campden.

LINGAPPA, Y. and SUSSMAN, A.S. 1959. Changes in the heat resistance of ascospores of *Neurospora* upon germination. *Am. J. Bot.* **46**: 671-678.

NIELSEN, P.V. and SAMSON, R.A. 1992. Differentiation of food-borne taxa of *Neosartorya*. *In*

Modern Methods in Food Mycology, eds R.A. Samson, A.D. Hocking, J.I. Pitt and A.D. King. pp. 159-168. Amsterdam: Elsevier.

NIELSEN, P.V., BEUCHAT, L.R. and FRISVAD, J.C. 1989. Growth and fumitremorgen production by *Neosartorya fischeri* as affected by food preservatives and organic acids. *J. Appl. Bacteriol.* **66**: 197-207.

PALMER, V.M. and McEVOY, I.J. 1974. Chemical and heat activation of ascospores of *Aspergillus fischeri* var. *glaber*, *Byssochlamys nivea* and *Byssochlamys fulva*. *Proc. Soc. Gen. Microbiol.* **2**: 6.

PITT, J.I. and CHRISTIAN, J.H.B. 1970. Heat resistance of xerophilic fungi based on microscopical assessment of spore survival. *Appl. Microbiol.* **20**: 682-686.

SPLITTSTOESSER, D.F. and SPLITTSTOESSER, C.M. 1977. Ascospores of *BysByssochlamys fulva* compared with those of a heat resistant *Aspergillus*. *J. Food Sci.* **42**: 685-688.

SPLITTSTOESSER, D.F. and CHUREY, J.J. 1989. Effect of low concentrations of sorbic acid on the heat resistance and viable recovery of *Neosartorya fischeri* ascospores. *J. Food Prot.* **52**: 821-822.

SPLITTSTOESSER, D.F., WILKISON, M. and HARRISON, W. 1972. Heat activation of *Byssochlamys* ascospores. *J. Milk Food Technol.* **35**: 399-401.

COMPARISON OF *NEOSARTORYA FISCHERI* VARIETIES BASED ON PROTEIN PROFILES AND IMMUNOACTIVITIES

Hélène Girardin[1,2] and J. P. Latgé[2]

[1]*Laboratoire du génie de l'hygiène et des procédés alimentaires*
INRA
91300 Massy, France

[2]*Unité de mycologie*
Institut Pasteur
75724 Paris Cedex 15, France

SUMMARY

Three varieties of *Neosartorya fischeri* have been described on the basis of ascospore ornamentation. This study describes work carried out to determine if protein and antigenic patterns could discriminate between the three varieties, and separate this group from *Aspergillus fumigatus*. SDS-PAGE (sodium dodecyl sulphate-polyacrylamide gel electrophoresis) was performed on water soluble extracts (WS) from mycelium of *N. fischeri* and *A. fumigatus*. Silver nitrate and Coomassie blue staining showed a very large range of fractions varying in size from 90 kDa to 10 kDa. A band of 30 kDa and a high molceular weight fraction at approximately 90 kDa were prominant in all *Neosartorya* extracts. WS extracts of ascospores demonstrated protein patterns very different from mycelial extracts. Ascospore profiles were more suitable for differentiating between the three varieties, with bands between 35 and 60 kDa. Antigenic properties of *N. fischeri* were studied by raising antisera in rabbits and mice immunised with mycelium and ascospores respectively. High ELISA titres were obtained with both extracts, but cross reaction between the different varieties of *N. fischeri* and *A. fumigatus* were observed in direct ELISA tests. Immunoblotting experiments demonstrated that all extracts reacted with the different rabbit antisera. The antigenic profiles varied with the mycelial extracts and antisera.

INTRODUCTION

Heat resistant fungi are often found in spoilage of fruit juices and other heat processed fruit products. The contamination of such products is generally caused by *Byssochlamys fulva*, *B. nivea*, *Neosartorya fischeri* or *Talaromyces flavus*. *N. fischeri* has become an industrial problem by spoiling processed fruit products (Splittstoesser and Splittstoesser, 1977; Hocking and Pitt, 1984; Jesenka *et al.*, 1984; Jesenka and Petrikova, 1985; Scott and Bernard, 1987) and producing mycotoxins (Patterson *et al.*, 1981; Horie and Yamazaki, 1981).

Until now, *N. fischeri* has only been detected by conventional methods. The development of immunoassays for this species would allow a quicker and more sensitive detection than cultural techniques. To achieve this, the antigenic map of *N. fischeri* must be established so that antigen(s) specific to this species can be located.

Neosartorya is an Ascomycete genus related to *Aspergillus fumigatus*. *N. fischeri* has several varieties distinguished by ornamentation of the ascospores (Raper and Fennell, 1965). Polysaccharide antigens excreted by *A. fumigatus* and related species do not permit discrimination between *Aspergillus* species, and even between *Aspergillus* and *Penicillium* (Notermans *et al.*, 1988, Debeaupuis *et al.*, 1990). Therefore, it is of interest to determine if protein and antigenic profiles can distinguish the varieties of *N. fischeri* from each other and from *A. fumigatus*. This study is centred on the comparison of water soluble protein and glycoprotein components extracted from mycelia of *A. fumigatus* (Wilson and Hearn, 1982; Hearn *et al.*, 1989, 1990) and *N. fischeri* and ascospores from *N. fischeri* varieties.

MATERIALS AND METHODS

Fungi and culture conditions

Twenty nine isolates of *N. fischeri* and five isolates of *A. fumigatus* were examined (Table 1). Mycelia were obtained from cultures on glucose (2%) peptone (1%) broth incubated for 6 d at 25°C in shake flasks (100 rpm). Ascospores of *N. fischeri* (six isolates representative of *N. fischeri* var. *fischeri*, *N. fischeri* var. *glabra* and *N. fischeri* var. *spinosa*) were obtained from 3 to 5 week old cultures on 2% malt extract agar at 25°C. Mycelia were recovered by paper filtration and extensive washing with distilled water. Ascospores were harvested from agar by aspiration (suspensions of ascospores were checked under the microscope for absence of significant contaminating mycelium and conidia), and freed from cleistothecia by disruption in a ground glass tissue homogeniser prior to gauze filtration.

Antigen preparation

Water soluble extracts (WS) were prepared by disruption of mycelia or ascospores in distilled water at 5°C in a cell homogeniser (MSK Braun, Melsungen, Germany) with 1 mm diameter glass beads for approximately 3-4 min (homogenisation in phosphate buffered saline, PBS, gave the same protein recovery: 20-80 μg/ml of growth medium). Soluble constituents were separated from insoluble residues by ultracentrifugation at 100,000 g for 1 h. When necessary, supernatants were concentrated under vacuum in a centrifugation concentrator (Speed Vac, Savant Instruments, Farmingdale, NY, USA) before application on SDS-PAGE gels.

Rabbit antisera

Antisera were produced in female New Zealand White rabbits using totally disrupted freeze dried mycelium (cytoplasm and wall) from individual isolates of *N. fischeri* var. *fischeri* F9, *N. fischeri* var. *glabra* G6, and *N. fischeri* var. *spinosa* S7. Freeze dried mycelium (100 mg) in complete Freund's adjuvant (1:1 v/v) was injected subcutaneously, intraperitoneally and in the footpads. Subsequent subcutaneous injections in incomplete Freund's adjuvant were performed 3 wks later. Rabbits were boosted intraveneously (500 μg of mycelium) every 10 days, and were bled after high antibody (Ab) titres were obtained as assayed by ELISA (O.D. >1 for antiserum dilution 1:1000). Anti-*A. fumigatus* A2 mycelium antiserum was a gift from M. Moutaouakil (see Table 2).

Mouse antisera

Disrupted ascospores from *N. fischeri* var. *fischeri* F9, *N. fischeri* var. *glabra* G6, *N.*

fischeri var. *spinosa* S7 were injected subcutaneously (50 μg of proteins) into Balb/C mice. In the first injection, the ascospore extracts were mixed with complete Freund's adjuvant (1:1 v/v) whereas in the following ones they were mixed with incomplete adjuvant. Injections were performed every 15 days. A mouse and a rabbit injected with extracts from *N. fischeri* var. *spinosa* S7 died during the immunistion.

Table 1. Isolates of *Aspergillus fumigatus* and *Neosartorya fischeri* studied

Isolate	Species	No. in text and Figures
IBT 3003	*Neosartorya fischeri* var. *fischeri*	F1
IBT 3005	*N. fischeri* var. *fischeri*	F2
IBT 3007	*N. fischeri* var. *fischeri*	F3
IBT 3008	*N. fischeri* var. *fischeri*	F4
IBT 3009	*N. fischeri* var. *fischeri*	F5
IBT 3012	*N. fischeri* var. *fischeri*	F6
IBT 3013	*N. fischeri* var. *fischeri*	F7
CBS 544.65 (= WB 181)	*N. fischeri* var. *fischeri*	F8
CBS 681.77	*N. fischeri* var. *fischeri*	F9
CBS 525.65 (= WB 4161)	*N. fischeri* var. *fischeri*	F10
WB 4075 (= IMI 16143)	*N. fischeri* var. *fischeri*	F11
CBS 404.67	*N. fischeri* var. *fischeri*	F12
CBS 832.88	*N. fischeri* var. *fischeri*	F13
IBT 3004	*N. fischeri* var. *glabra*	G1
IBT 3006	*N. fischeri* var. *glabra*	G2
IBT 3015	*N. fischeri* var. *glabra*	G3
CBS 111.55 (= WB 2163)	*N. fischeri* var. *glabra*	G4
CBS 112.55 (= NRRL 2392)	*N. fischeri* var. *glabra*	G5
CBS 165.63	*N. fischeri* var. *glabra*	G6
LCP 68.2007	*N. fischeri* var. *glabra*	G7
LCP 87.3513	*N. fischeri* var. *glabra*	G8
LCP 88.3577	*N. fischeri* var. *glabra*	G9
IBT 3001	*N. fischeri* var. *spinosa*	S1
CBS 483.65 (= WB 5034)	*N. fischeri* var. *spinosa*	S2
CBS 297.67	*N. fischeri* var. *spinosa*	S3
WB 4076 (= IMI 16061)	*N. fischeri* var. *spinosa*	S4
CBS 865.70	*N. fischeri* var. *spinosa*	S5
CBS 448.75 (= NHL 5083)	*N. fischeri* var. *spinosa*	S6
CBS 161.88 (= TRTC 50994)	*N. fischeri* var. *spinosa*	S7
LCP 76.3116	*N. fischeri* var. *spinosa*	S8
LCP 88.3574	*N. fischeri* var. *spinosa*	S9
CBS 113.26	*Aspergillus fumigatus*	A1
CBS 143.89	*A. fumigatus*	A2
CBS 144.89	*A. fumigatus*	A3
CBS 192.65	*A. fumigatus*	A4
CBS 331.90	*A. fumigatus*	A5

Table 2. List of antiserum abbreviations

Antisera	Abbreviation
Rabbit anti-mycelium of *N. fischeri* var. *fischeri* F9	Anti-F9m
Rabbit anti-mycelium of *N. fischeri* var. *glabra* G6	Anti-G6m
Rabbit anti-mycelium of *N. fischeri* var. *spinosa* S7	Anti-S7m
Rabbit anti-mycelium of *A. fumigatus* A2	Anti-A2m
Mouse anti-ascospore of *N. fischeri* var. *fischeri* F9	Anti-F9a
Mouse anti-ascospore of *N. fischeri* var. *glabra* G6	Anti-G6a
Mouse anti-ascospore of *N. fischeri* var. *spinosa* S7	Anti-S7a

Enzyme-linked Immunosorbent Assay (ELISA)
WS were diluted in 0.05 M carbonate buffer at pH 9.6 (1 μg/ml) and coated onto microtitre plates (Greiner, Bischwiller, France) which were incubated for 2 h at 37°C and then overnight at 4°C. Antisera from immunised or control rabbits and mice, diluted in PBS Tween 20 (PBS-T) BSA (1%) were applied to washed microtitre wells. The starting dilution was 1:100. After incubating for 1 h at 37°C and washing, the peroxidase conjuguate [anti-rabbit IgG (H+L)-peroxidase and anti-mouse IgG (H+L)-peroxidase; Biosys, Compiègne, France] was added (0.1% in PBS-T BSA). Peroxidase was revealed by H_2O_2 and orthophenylenediamine hydrochloride (OPD) in 0.05 M citrate buffer at pH 5.2. The reaction was stopped by adding 50 μl of 3.6 N sulphuric acid.

ELISA inhibition
WS of mycelia and ascospores from either *A. fumigatus* or *N. fischeri* isolates, diluted in PBS-T containing 1% BSA, were used in inhibition assays. WS were incubated (1:1), from a starting concentration of 5 mg/ml, overnight at 37°C with rabbit antisera (1:1000) and added to washed WS coated wells prepared as described above. After 1 h incubation at 37°C and washing, the peroxidase conjuguate was added, followed by incubation and staining by OPD as described above.

Protein analysis
Total protein was determined using the Bio-Rad kit (Munich, Germany) based on a Coomassie blue binding protein assay with bovine serum albumin as a standard.

Polyacrylamide gel electrophoresis (SDS-PAGE)
Extracts of WS of mycelia and ascospores were subjected to electrophoresis in a vertical slab unit (HSI; San Francisco, CA, USA) (18x16x0.01 cm) and a Bio-Rad Mini Protean TM II unit (8.2x10x0.01 cm) respectively, using a separating gel of 12% or 7.5% polyacrylamide, adapted from Laemmli (1970). Denaturating buffer (2% SDS, 20% glycerol, 5% mercaptoethanol and a trace of bromophenol blue in 62.5 mM Tris/HCl buffer, pH 6.8) was added (1:4) to the samples and boiled for 4 min. Approximately 50 μg of protein in maxi-gels and 10 μg of protein in mini-gels were loaded per lane. A Tris/glycine running buffer (pH 8.3) containing 0.1% SDS was used. Electrophoresis was carried out at a constant current of 30 mA in maxi-gels (Hearn *et al.*, 1989). Mini PAGE

were run at a voltage of 100 V during stacking migration and 200 V during separation. Low molecular weight protein standards (Electrophoresis Calibration kit; Pharmacia, Upsala, Sweden) were run in parallel.

Protein and glycoprotein detection

Separated components were stained for protein in 0.1% Coomassie Brilliant Blue R-250 solution of 40% methanol, 10% acetic acid and 3% glycerol for 1 h at room temperature and destained in the same solvent without dye. Silver nitrate staining was also performed on mini-gels (Garfin, 1990).

Electroblotting

The electrophoretically separated proteins and glycoproteins were electrotransferred to nitrocellulose membranes (0.2 μm; Schleicher and Schuell, Dassel, Germany) overnight at 30 V in a transblotting chamber (HSI; San Francisco, CA, USA), using the method of Towbin *et al.* (1979).

Antigenic reactivity of WS fractions

The free binding sites on the membrane were saturated with either Tris saline buffer (TSB, 50 mM Tris solution containing 0.9% NaCl, 0.2% EDTA, 0.05% Tween 20) containing 10% nonfat dry milk or PBS Tween (0.05%) containing 5% nonfat dry milk for 1 h at 37°C. Nitrocellulose was probed with rabbit antiserum diluted 1:1000 in the same buffer for 4 h. at 4°C. Blots were then exposed to goat anti-rabbit IgG conjugated to peroxidase (Biosys, Compiègne, France) at a 1:1000 dilution in TBS or PBS-T. Antigen/antibody complexes were localised by staining for peroxidase activity with 3,3'-diaminobenzidine as substrate for a maximum incubation time of 10 min.

RESULTS

Gel electrophoresis

The proteins and glycoproteins of *N. fischeri* and *A. fumigatus* WS mycelium preparations separated electrophoretically by SDS-PAGE and stained for proteins with Coomassie Blue (Figure 1) contained an array of molecules which ranged in apparent molecular weight from 10 to 100 kDa. Most of the protein fractions ranged between 30 and 95 kDa. All samples of *N. fischeri* showed two major fractions, a high molecular weight fraction, around 90 kDa, and a lower molecular weight fraction at 30 kDa. These proteins were also present in *A. fumigatus* samples, but the 30 kDa band was less important. A protein band of approximately 50 kDa was present in most of *N. fischeri* samples and in samples of *A. fumigatus*. The protein profiles of the WS from *N. fischeri* varieties were too heterogeneous to conclusively differentiate between them. However, the presence of similar profiles such as in *N. fischeri* var. *fischeri* F12 and *N. fischeri* var. *spinosa* S5 is indicative of either a possible misidentification or a very close relationship between them.

Ascospore extracts stained for proteins with silver nitrate exhibited fewer bands than mycelial extracts (Figure 2). In *N. fischeri* var. *spinosa*, two fractions were predominant (55 and 45 kDa for isolate S7, and 50 and 40 kDa for isolate S3). Samples from G6 and G8 also showed very similar profiles with a major band at approximately 45 kDa, whereas the profile of G9 resembled that of F9. This would again suggest a possible misidentification or close relationship between *N. fischeri* var. *glabra* G9 and *N. fischeri* var. *fischeri* F9. A 30 kDa band was common to all ascospore profiles, as seen in mycelial extracts.

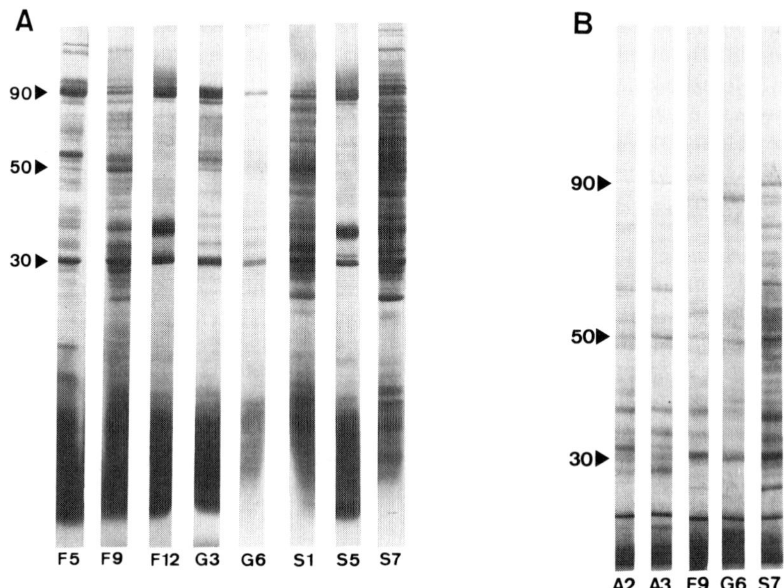

Figure 1. SDS-PAGE on (a) 12% and on (b) 7.5% gels of WS mycelium fractions from *N. fischeri* and *A. fumigatus* isolates stained for proteins with Coomassie blue. Molecular weights (kDa) of proteins are shown on the left margin.

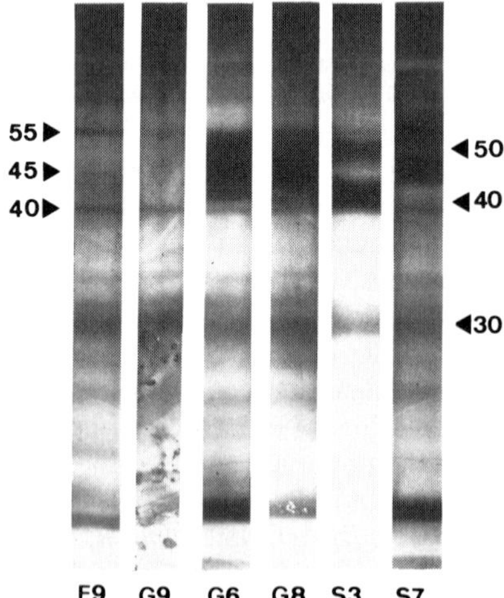

Figure 2. SDS-PAGE on 7.5% minigel of ascospore WS fractions from *N. fischeri* isolates stained for protein with silver nitrate. Molecular weights of proteins are expressed in kDa.

ELISA

Direct ELISA. Rabbit anti-mycelium and mouse anti-ascospore antisera were tested against mycelium WS and ascospore WS antigens. Each anti-mycelium antiserum cross-reacted with all mycelium extracts of *N. fischeri* and *A. fumigatus* (Table 3). These anti-mycelium antisera also gave positive responses with all the ascospore extracts tested. Each mouse anti-ascospore antiserum reacted positively with all ascospore extracts from the three varieties of *N. fischeri*. Anti-F9a and anti-G6a reacted exclusively with their homologous mycelium WS, whereas the anti-S7a gave a negative response with all the mycelium WS tested. These results suggest the presence of species specific antigens common to mycelium and ascospores.

ELISA inhibition. Only anti-S7m was significantly inhibited by *A. fumigatus* WS (Figure 3). This suggests that a close relationship exists between A2 and S7 antigens, whereas G6 and F9 antigens are distant from A2 antigens. No significant inhibition was observed when anti-A2m was incubated with *N. fischeri* WS (from isolates F9, G6 and S7).

Table 3. Cross-reactions in direct ELISA between the different sera raised against *N. fischeri* or *A. fumigatus* isolates[1]

Coated WS fractions[2]	Anti-mycelium				Anti-ascospore		
	F9m	G6m	S7m	A2m	F9a	G6a	S7a
F9m	+	+	+	+	+	-	-
G6m	+	+	+	+	-	+	-
S7m	+	+	+	+	-	-	-
A2m	+	+	+	+	-	-	-
F9a	+	+	+	+	+	+	+
G6a	+	+	+	+	+	+	+
S7a	+	+	+	+	+	+	+

[1] +, O.D. >0.6 with antiserum dilution of 1:1000; -, O.D. <0.6 with antiserum dilution of 1:1000
[2] m, mycelium; a, ascospore.

Immunoblots

The binding patterns of representative isolates of the three *N. fischeri* varieties and *A. fumigatus* to anti-F9m, anti-S7m and anti-A2m are shown in Figures 4-6. Most of the low molecular weight proteins and glycoproteins detected by Coomassie blue were not detected in blots, suggesting that they were poor antigens (Figure 4a). Washing the blots with TBS reduced the number of antigens detected (Figure 4b). This washing buffer was used to investigate the major antigens present in the different extracts. In all the *N. fischeri* and *A. fumigatus* samples tested, a major band of approximately 90 kDa was detected, although other bands appeared in A2 and S7 WS when probed with anti-A2m. The immunoblots obtained with anti-S7m and anti-A2m were similar (Figure 4b, 5). When probed with the anti-F9m (Figure 6), the WS preparation of F9 showed the best reactivity. The other samples incubated with this antiserum exhibited the same major bands as seen in the

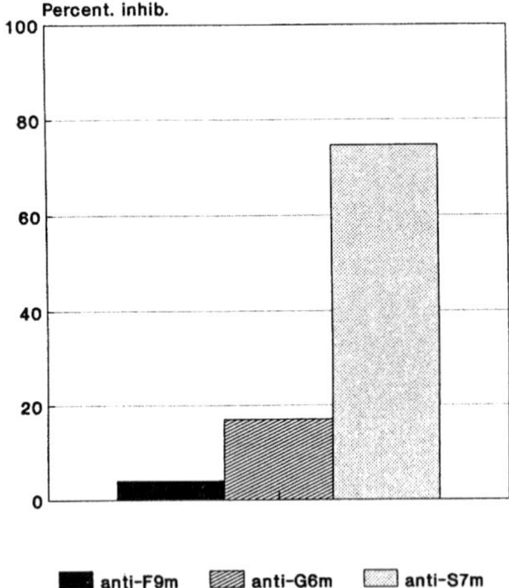

Figure 3. ELISA inhibition of anti-mycelium of *N. fischeri* antisera by WS of *A. fumigatus* (A2). Percent inhibition was calculated as: inhibition with *A. fumigatus* A2-antiserum complexes/inhibition with *N. fischeri*-antiserum complexes

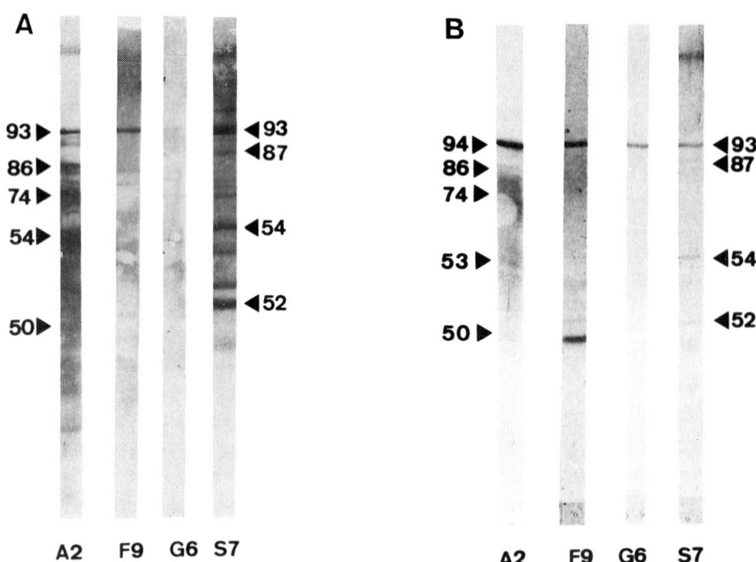

Figure 4. WS fractions from mycelia of *N. fischeri* and *A. fumigatus* isolates separated by SDS-PAGE (7.5% gel), transferred to nitrocellulose and probed with anti-A2m. PBS-T (A) or TBS containing EDTA and Tween (B) were used as washing buffers. Molecular weights of protein fractions are expressed in kDa.

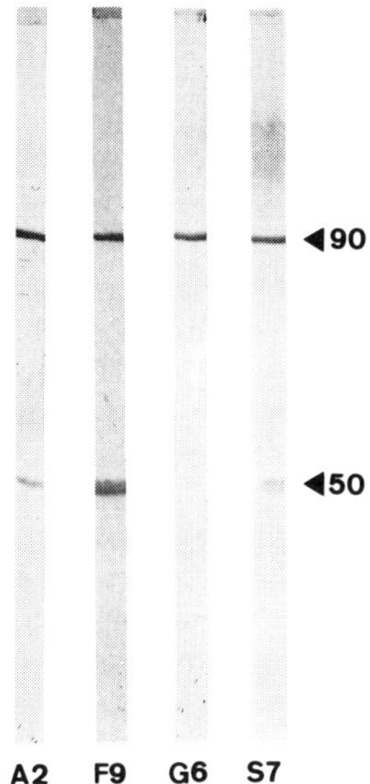

Figure 5. WS fractions from mycelia of *N. fischeri* and *A. fumigatus* isolates separated by SDS-PAGE (7.5% gel), transferred to nitrocellulose and probed with anti-S7m. Molecular weights of protein fractions are expressed in kDa.

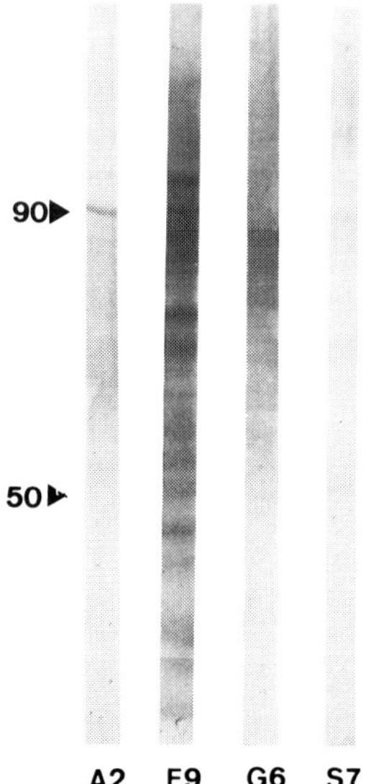

Figure 6. WS fractions from mycelia of *N. fischeri* and *A. fumigatus* isolates separated by SDS-PAGE (7.5% gel), transferred to nitrocellulose and probed with anti-F9m. Molecular weights of protein fractions are expressed in kDa.

anti-A2m and anti-S7m profiles. These results are in agreement with inhibition ELISA, presuming a close immunological relationship between A2 and S7 mycelial extracts and the distance between F9 and A2. However no specific major antigens from any variety were revealed in immunoblots.

DISCUSSION

As already shown by Hearn *et al.* (1990), the mycelial WS preparations of *N. fischeri* and *A. fumigatus* species by SDS-PAGE exhibit very complex and heterogeneous protein patterns when visualised with Coomassie blue. Consequently, the use of protein profiles for the separation of *A. fumigatus* from *N. fischeri* or *N. fischeri* varieties is extremely difficult. Because of this heterogeneity, the presence of similar patterns in two isolates, such as *N. fischeri* var. *fischeri* F12 and *N. fischeri* var. *spinosa* S5, must indicate a very high degree of homology between the two isolates. This similarity is confirmed by immunoblot patterns. No apparent homologies between the mycelial WS patterns are detectable, because the number of bands is too high. Fractionation of the protein extracts, or the selection of particular antigens by monoclonal antibodies, may contribute to a better interpretation of mycelial WS patterns. Antigenic preparations other than mycelial WS may be more suitable for discriminating between *N. fischeri* and *A. fumigatus* and between the three varieties of *N. fischeri*.

Some of the rabbit polyclonal antibodies prepared during this study produced heavy smears in immunoblotting experiments. These smears may be due to carbohydrate immunodominant antigens (Bennett *et al.*, 1985; Notermans *et al.*, 1988; Debeaupuis *et al.*, 1990) in response to immunisation of the total mycelium, including a very rich carbohydrate substrate like the wall. Immunisation of rabbits with WS extracts may overcome this problem and increase the resolution of the immunoblots obtained.

The SDS-PAGE profiles obtained with ascospore WS were less complex and less heterogeneous than those obtained with mycelial WS. Moreover, mouse anti-ascospore antisera did not cross-react with non homologous mycelial WS extracts. Therefore the study of protein patterns and the antigenic capacity of ascospore preparations may allow a better discrimination between *N. fischeri* varieties. Investigations in these areas will continue. Immunoblotting experiments with ascospore WS extracts and anti-ascospore antisera are currently being undertaken.

ACKNOWLEDGEMENTS

The authors wish to express their thanks to Dr. M.F. Roquebert, Laboratoire de Cryptogamie-Mycologie, Museum National d'Histoire Naturelle, Paris, and Dr. R.A. Samson, Centraalbureau voor Schimmelcultures, Baarn, for providing the *N. fischeri* isolates used in this study.

REFERENCES

BENNETT, J.E., BHATTACHARJEE, A.K. and GLAUDEMANS, C.P.J. 1985. Galactofuranosyl groups are immunodominant in *Aspergillus fumigatus* galactomannan. *Mol. Immunol.* **22**: 251-254.

DEBEAUPUIS, J.P., SARFATI, J., GORIS, A., STYNEN, D., DIAQUIN, M. and LATGE, J.P. 1990. Exocellular polysaccharides from *Aspergillus fumigatus* and related taxa. *In* Modern Concepts in

Penicillium and *Aspergillus* Classification, eds R.A. Samson and J.I. Pitt, pp. 209-223. New York: Plenum Press.

GARFIN, D.E. 1990. One dimensional gel electrophoresis. *In* Guide to Protein Purification, ed. M.P. Deutscher, *Methods in Enzymology* **182**: 425-441. New York: Academic Press.

HEARN, V.M., GRIFFITHS, B.L. and GORIN, P.A. 1989. Structural analysis of water-soluble fractions obtained from *Aspergillus fumigatus* mycelium. *Glycoconjugate* **6**: 85-100.

HEARN, V.M., MOUTAOUAKIL, M. and LATGÉ, J.P. 1990. Analysis of components of *Aspergillus* and *Neosartorya* mycelial preparations by gel electrophoresis and Western blotting procedures. *In* Modern Concepts in *Penicillium* and *Aspergillus* Classification, eds R.A. Samson and J.I. Pitt, pp. 235-245. New York: Plenum Press.

HOCKING, A.D., and PITT, J.I. 1984. Food spoilage fungi. II. Heat-resistant fungi. *CSIRO Food Res. Q.* **44**: 73-82.

HORIE, Y., and YAMAZAKI, M. 1981. Productivity of the tremogenic mycotoxins, fumitremorgins A and B in *Aspergillus fumigatus* and allied species. *Nippon Kingakkai Kaiko* **22**: 113-119.

JESENKA, Z., HAVRANEKOVA, D. and SAJBIDOROVA, I. 1984. On some problems of moulds on some products of canning industry. *Cesk. Hyg.* **24**: 102-109.

JESENSKA, Z., and PETRIKOVA, D. 1985. Microscopic fungi as an agent causing spoilage of preserved fruits. *Cesk. Hyg.* **30**: 175-177.

LAEMMLI, U.K. 1970. Cleavage of structural proteins during the assembly of the head of bacteriophage T4. *Nature, London* **227**: 680-685.

NOTERMANS, S., VEENEMAN, G.H., van ZUYLEN, C.W.E.M., HOOGERHOUT, P. and van BOOM, J.H. 1988. (1-5)-linked ß-D-galactofuranosides are immunodominant in extracellular polysaccharides of *Penicillium* and *Aspergillus* species. *Mol. Immunol.* **25**: 975-979.

PATTERSON, D.S.P., SHREEVE, B.J., ROBERTS, B.A. and McDONALD, S.M. 1981. Verruculogen produced by soil fungi in England and Wales. *Appl. Environ. Microbiol.* **42**:916-917.

RAPER, K.B., and FENNELL, D.I. 1965. The Genus *Aspergillus*. Baltimore, Maryland: Williams and Wilkins.

SCOTT, V.N., and BERNARD, D.T. 1987. Heat resistance of *Talaromyces flavus* and *Neosartorya fischeri* isolated from commercial fruit juices. *J. Food Prot.* **50**: 18-20.

SPLITTSTOESSER, D.F. and SPLITTSTOESSER, C.M. 1977. Ascospores of *Byssochlamys fulva* compared with those of a heat-resistant *Aspergillus*. *J. Food Sci.* **42**: 685-688.

TOWBIN, H., STACHELIN, T. and GORDON, J. 1979. Electrophoretic transfer of proteins from polyacrylamide gels to nitrocellulose sheets: procedure and some applications. *Proc. Nat. Acad. Sci.* **76**: 4350-4354.

WILSON, E.V. and HEARN, V.M. 1982. A comparison of surface and cytoplasmic antigens of *Aspergillus fumigatus* in an enzyme-linked immunosorbent assay (ELISA). *Mykosen* **25**: 653-661.

FACTORS CONTRIBUTING TO VARIATIONS IN D VALUES OBTAINED FOR ASCOSPORES OF *NEOSARTORYA FISCHERI*

D. E. Conner

Department of Poultry Science
Alabama Agricultural Experiment Station
Auburn University,
Auburn, AL 36849-5416, USA

SUMMARY

Ascospores of *Neosartorya fischeri* are exceptionally heat resistant by comparison with other fungal ascospores. However many factors, including heating menstruum, the presence of organic acids, pH and ascospore age influence the response of ascospores to heat. Ascospores of *N. fischeri* are more heat resistant in apple juice than in grape juice or phosphate buffer. Fumaric, acetic, citric and tartaric acids can increase the rate of heat inactivation, while pH and acid concentration act synergistically with heat. Ascospores develop greater heat resistance as they mature. Increasing incubation time from 10 to 13 days at 30°C increased $D_{80°C}$ values for ascospores from 27.0 to 66.7 min in apple juice, 28.5 to 33.3 min in grape juice and 18.2 to 50.0 min in phosphate buffer respectively. The rate of development of heat resistance is dependent upon incubation temperature at which ascospores are produced. These factors should be considered when developing standardised D values for *N. fischeri*.

INTRODUCTION

Ascospores of *Neosartorya fischeri* are exceptionally heat resistant, but many factors can ultimately influence the response of these ascospores to heat. Therefore, reference (standard) D values for *N. fischeri* have been difficult to determine. As an initial step in developing reliable quantitative heat resistance data, factors that can influence the ascospore heat resistance, usually quoted as D values, must be considered. This short review will focus on the results of a series of previously reported experiments that were conducted to elucidate intrinsic and extrinsic influences on heat resistance of ascospores of several strains of *N. fischeri* (Conner, 1987; Conner and Beuchat, 1987a,b; Conner *et al.*, 1987). In these studies, the effects of heating medium and pH, presence of organic acids and age of ascospores were investigated. In considering these data, this review is intended to aid in the development of reference or standard D values for ascospores of *N. fischeri* in foods and beverages.

EFFECT OF HEATING MEDIUM

Conner and Beuchat (1987b) found that ascospores of three strains of *N. fischeri* were exceptionally resistant to inactivation when heated in apple juice (ph 3.8), grape juice (pH 3.4) and phosphate buffer (pH 7.0). Treatment of an ascospore population of *N. fischeri* 110483 (University of Georgia, USA) with a viable count of $5x10^5$/mL at 84°C for 30 and

Table 1. Populations ($\log_{10}$ cfu/mL) of ascospores of *N. fischeri* recovered after heating at 84, 90 and 95°C in apple juice (pH 3.8, 12.3°Brix), Concord grape juice (pH 3.4, 16.6°Brix) and 0.1 m potassium phosphate buffer (pH 7.0) for 30, 60, 90 and 120 minutes. Population ($\log_{10}$ cfu/mL) of unheated ascospores was 5.7

Strain	Heating time (min)	Apple juice			Grape juice			Phosphate buffer		
		84°C	90°C	95°C	84°C	90°C	95°C	84°C	90°C	95°C
110483	30	6.4	3.3	2.6	6.7	<1	<1	6.2	2.5	<1
	60	6.4	<1	<1	6.6	<1	<1	6.1	<1	<1
	90	6.2	<1	<1	6.4	<1	<1	6.2	<1	<1
	120	6.4	<1	<1	6.2	<1	<1	6.5	<1	<1
FRR 2334	30	5.7	4.7	2.8	5.8	3.5	<1	3.5	3.8	
	60	5.8	2.9	<1	6.0	<1	<1	4.7	<1	
	90	5.7	<1	<1	5.9	<1	<1	5.5	<1	
	120	5.8	<1	<1	5.9	<1	<1	5.4	<1	
FRR 1833	30	6.8	3.3	2.0	7.0	3.3	<1	6.0	4.3	<1
	60	6.8	<1	<1	7.0	<1	<1	6.8	<1	<1
	90	6.7	<1	<1	6.5	<1	<1	6.6	<1	<1
	120	6.5	<1	<1	6.1	<1	<1	6.5	<1	<1

60 min caused activation, resulting in an approximately ten fold increase in viable count compared to unheated ascospores (Table 1). Heating for an additional 90 min at 84°C resulted in little or no inactivation. Viable count was reduced to 5×10^2 to 5×10^3 in apple juice and phosphate buffer by treatment at 90°C for 30 min. Less than 10 ascospores/mL survived this treatment in grape juice. Less than 10 ascospores/mL survived 60 min at 90°C regardless of heating media.

Ascospores of FRR 2334 responded similarly (Table 1). However, ascospores of this strain were slightly more heat resistant than those of strain 110483. The activation at 84°C observed with 110483 was also observed; however, there was an initial heat inactivation when FRR 2334 was heated for 30 min at 84°C in phosphate buffer, followed by heat activation over the next 60 min. No inactivation of this strain was detected when ascospores were heated at 84°C in apple juice and grape juice. Ascospores survived 90°C for 60 min in apple juice, and for 30 min in grape juice and phosphate buffer. As with 110483, ascospores of FRR 2334 survived 95°C for 30 min in apple juice only.

Ascospores of FRR 1833 also responded similarly to those of 110483 and FRR 2334 (Table 1). The initial heat activation at 84°C was also observed; the rate of activation in phosphate buffer was greater than for FRR 2334, but less than for 110483. Ascospores of FRR 1833 showed a small decrease in viable count at 90 and 120 min at 84°C in grape juice. Ascospores survived for 30 min at 90°C in all heating media, but survived 30 min at 95°C in apple juice only.

The observation that some ascospores of all three strains survived 95°C for 30 min in apple juice but not in grape juice or phosphate buffer suggested that certain constituents of apple juice may afford protection, or at least are not deleterious, to ascospores during heat treatment. These observations indicate that the nature of the heating medium, i.e. pH, soluble solids and organic acid content, may influence the sensitivity of ascospores to elevated temperatures.

EFFECT OF pH AND ORGANIC ACIDS

Fumaric, citric, acetic and tartaric acids at 0.5, 1.0 or 2.0% enhanced heat inactivation of *N. fischeri* ascospores, whereas malic acid had little effect (Conner and Beuchat, 1987b). Fumaric acid (0.5%) at pH 2.5 and 3.0, and 1.0 and 2.0% at pH 2.5, 3.0 and 3.5, greatly increased the rate of heat inactivation of ascospores as compared to distilled water and phosphate buffer controls. The concentration of undissociated fumaric acid under these conditions ranged from 0.022 M (0.5%, pH 3.0) to 0.132 M (2.0%, pH 2.5). With acetic, citric, fumaric, malic and tartaric acids, heat inactivation was unaffected at pH greater than 3.5 compared to water and phosphate buffer controls. Citric acid (0.5, 1.0 and 2.0%) enhanced heat inactivation at pH 2.5 and 3.0, but not at pH 3.5. The enhancement of heat activation by citric acid was less than that observed with fumaric acid; however, at the same percentages and pH values, i.e., 0.5 and 2.0% at pH 3.0 and 2.5, respectively, the concentrations of undissociated citric acid were considerably lower (0.015 and 0.084 M) than the concentrations of fumaric acid (0.022 and 0.132 M). Acetic acid (2.0%) enhanced heat inactivation at pH 2.5 (0.331 M), pH 3.0 (0.327 M) and pH 3.5 (0.315 M); however, at concentrations of 1.0 and 0.5%, this acid did not influence heat inactivation at pH 2.5, 3.0 and 3.5. This suggests that an undissociated acetic acid concentration of equal to or greater than 0.315 M is necessary to enhance heat inactivation. Concentrations of 0.097 M (2.0%), 0.048 M (1.0%) and 0.024 M (0.5%) tartaric acid enhanced heat inactivation at pH 2.5 only. Interestingly, malic acid (≤2.0%) did not affect rates of heat inactivation of ascospores at pH 2.5, 3.0 or 3.5.

In addition to the amount of undissociated acid present in the heating medium, inactivation was also dependent in part upon the type and molarity of organic acid present. Moreover, rates of heat inactivation of ascospores may also be dependent upon the pH of the heating medium, as evidenced by the observation that ascospores were inactivated in the presence of 0.5% tartaric acid at pH 2.5 (0.024 M undissociated) but not in the presence of 2.0% tartaric acid at pH 3.0 (0.061 M undissociated). Thus, pH and the type, concentration and molarity of the undissociated form of organic acid act synergistically with heat to inactivate ascospores of *N. fischeri*.

The fact that enhancement of heat inactivation did not occur in organic acid solutions adjusted to pH greater than 3.5 confirms that the undissociated acid molecule is largely responsible for the lethality of these acids on ascospores.

EFFECT OF ASCOSPORE AGE

Conner and Beuchat (1987a) found that ascospores of *N. fischeri* var. *glaber* FRR 1833 rapidly increased in heat resistance as cultures aged from 8 to 21 days (Table 2). This rapid increase in heat resistance necessitated increases in heating temperature during the course of the investigation in order to obtain thermal inactivation rates from which D values could be determined. For example, treatment at 70 and 80°C inactivated 8 day old ascospores (10^5/mL) within 15 min; therefore, a 60°C heat treatment was used to determine D values. However, a 10 day old ascospores (10^5/mL) were not inactivated by

120 min heating at 60 or 70°C, 80°C was used for determination of D values. Similar considerations led to the use of 85°C for D value determinations on ascospores aged over 13 days (Table 2). At 8 days, the earliest time at which abundant physically mature ascospores could be harvested from grape juice agar, $D_{60°C}$ values were 17.0, 17.5 and 13.3 min in apple juice, grape juice and phosphate buffer, respectively. A heat treatment of 80°C was used to determine D values of 10 and 13 day old ascospores. From 10 to 13 days, $D_{80°C}$ values increased from 27.0 to 66.7 min in apple juice, 28.5 to 33.3 min in grape juice and 18.2 to 50.0 in phosphate buffer. From 16 to 21 days, $D_{85°C}$ values were observed to increase from 6.9 to 13.2 in apple juice, 5.8 to 10.1 min in grape juice and 6.1 to 10.4 min in phosphate buffer.

After physical maturity is reached, ascospores of *N. fischeri* continue to undergo changes as indicated by an increase in heat resistance (Table 2). Maturation apparently continues after ascospores appear fully mature upon microscopic examination. Data indicate that changes in ascospores which occur during maturation do so at lower rates when *N. fischeri* is incubated at lower incubation temperatures; therefore, ascospores of the same age cultivated at 18 and 21°C tend to be less heat resistant than those cultivated at 25 and 30°C. Due to temperature dependent biochemical reaction rates, ascospores formed at lower incubation temperatures are physiologically younger than those formed at higher incubation temperatures.

Table 2. D values of *N. fischeri* ascospores in apple juice, grape juice and buffer

Culture age (days)	Temperature for D values (°C)	D values (min)		
		Apple juice[1]	Grape juice[2]	Buffer[3]
8	60	17.0	17.5	13.3
10	80	27.0	28.5	18.2
13	80	66.7	33.3	50.0
16	85	6.9	5.8	6.1
21	85	13.2	10.1	10.4

[1] pH 3.8; 12.3° Brix
[2] pH 3.4; 16.6° Brix
[3] 0.1 m potassium phosphate, pH 7.0 (Conner and Beuchat, 1987a).

CONCLUSION

The above data indicate that many factors can influence the response of *N. fischeri* ascospore to heating, and thus may cause variability in D values obtained in independent thermal death time (TDT) studies. For standardisation of D values, variability in these factors should be minimised when conducting TDT studies. The nature of the heating medium should be carefully considered, and the production of ascospore crops should be done under consistent, well defined conditions of time and temperature.

REFERENCES

CONNER, D.E. 1987. Extrinsic and intrinsic factors affecting heat resistance of ascospores of *Neosartorya fischeri*. Ph.D. Dissertation, University of Georgia, Athens, GA USA. 111 pp.

CONNER, D.E. and BEUCHAT, L.R. 1987a. Efficacy of media for promoting ascospore formation by *Neosartorya fischeri*, and the influence of age and culture temperature on heat resistance of ascospores. *Food Microbiol.* **4**: 229-238.

CONNER, D.E. and BEUCHAT, L.R. 1987b. Heat resistance of ascospores of *Neosartorya fischeri* as affected by sporulation and heating medium. *Int. J. Food Microbiol.* **4**: 303-312.

CONNER, D.E., BEUCHAT, L.R. and CHANG, C.J. 1987. Age-related changes in ultrastructure and chemical composition associated with changes in heat resistance of *Neosartorya fischeri* ascospores. *Trans. Br. Mycol. Soc.* **89**: 539-550.

5

IMMUNOLOGICAL DETECTION OF FUNGI IN FOODS

DEVELOPMENT OF A TECHNIQUE FOR THE IMMUNOLOGICAL DETECTION OF FUNGI

H.J. Kamphuis[1] and S. Notermans[2]

[1]*Department of Food Science*
Wageningen Agricultural University
6703 HD Wageningen, The Netherlands

[2]*Laboratory for Water and Food Microbiology,*
National Institute of Public Health and Environmental Protection
3720 BA Bilthoven, The Netherlands

SUMMARY

In this paper, basic principles of the development of an immunoassay for the detection of *Aspergillus* and *Penicillium* are summarised. The assay is based on the detection of a heat stable extracellular polysaccharide with unique properties. The reliability of the assay is enhanced by using a synthetic epitope of the *Aspergillus/Penicillium* antigen. This epitope specifically blocks the immunologically active sites of the immunoglobulins, permitting discrimination between positive and false positive reactions. The technique has been proven to be succesful in the detection of fungi in foods and feeds.

INTRODUCTION

Fungi are important in a wide range of agriculture commodities, processed food and beverage products, causing spoilage and producing mycotoxins. Certain species also produce desirable changes in some types of foods e.g. mould ripening of cheese. Detection of fungi is therefore of great interest. Available mycological detection methods lack precision and have some disadvantages (Jarvis *et al.*, 1983). Cultural methods are only suitable for products which were not heat treated or irradiated. Microscopic methods (e.g. Howard Mould Count) also lack precision and require skill. Chemical methods such as the chitin assay are subject to error from intrinsic glucosamine from food and contamination by invertebrates which also contain chitin. The assay is also laborious.

Immunological detection of fungi is a promising alternative technique. For optimal immunological detection of fungi, however, the antigen to be detected should be genus specific, produced by all species within the genus, and extracellular, heat stable and water extractable. This paper describes procedures used to produce an immunological system for detection of the common food borne fungi *Aspergillus* and *Penicillium*.

METHODS AND RESULTS

Production and purification of the mould antigen

For production of antigens, fungi (Table 1) were grown in submerged culture in dialysed malt extract (Oxoid) 60 g/L, at 24°C for 14 days. After filtration, the culture fluid was freeze dried. The freeze dried product was then dissolved in 80% ammonium sulphate and gel permeation chromatography of the supernatant layer was carried out (Notermans *et al.*, 1987) (Figure 1).

One main immunogenic fraction was observed. As this fraction did not show absorption at 280 nm, such fractions from different fungal species without 280 nm absorption were chemically analysed. These fractions were extracellular polysaccharides (EPS) and consisted mainly of glucose, mannose and galactose (Notermans *et al.*, 1987) (Table 1). In the case of *Mucor hiemalis* and *Rhizopus stolonifer* EPS also contained fucose. In a later study (de Ruiter *et al.*, in press) it became clear that additionally glucuronic acid was also present (approximately 50 mol %) in Mucoraceous fungi.

Antigenicity of the extracellular polysaccharides

The immunogenic fractions without absorption at 280 nm produced by *Penicillium digitatum*, *Cladosporium cladosporioides*, *Mucor racemosus* and *Fusarium oxysporum* were used to immunise rabbits (Notermans and Heuvelman, 1985). High serum titres against the extracellular polysaccharides were obtained (Table 2) and with the purified antibodies a highly specific ELISA for the detection of fungal EPS was developed.

Genus specificity of the extracellular polysaccharides

The sandwich ELISA was carried out using antibodies raised against EPS of *P. cyclopium*, *P. digitatum*, *M. racemosus*, *C. cladosporioides*, *F. oxysporum* and *G. candidum*. Species

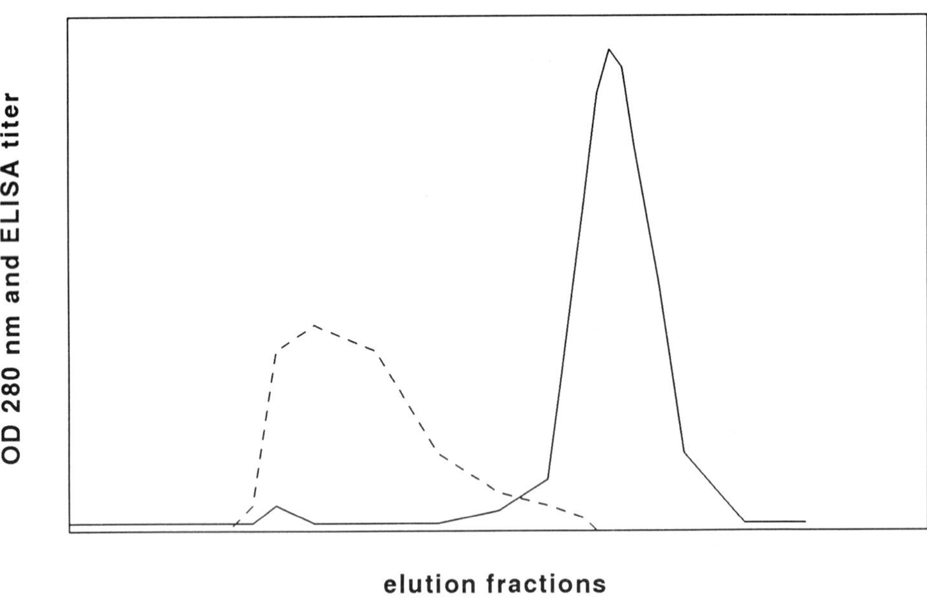

Figure 1. Chromatography of $(NH_4)_2SO_4$ treated freeze dried culture filtrate of *P. digitatum* on a Sepharose CL 4B column (Notermans *et al.*, 1987)

Table 1. Monosaccharide composition of extracellular fungal polysaccharides[1]

Species	polysaccharide fraction (%)	monosaccharides (mol %)			
		mannose	galactose	glucose	fucose
Penicillium cyclopium	70	76	17	7	0
P. digitatum	64	28	60	12	0
Aspergillus repens	71	30	66	2	0
Mucor hiemalis	57	17	12	4	18
Rhizopus stolonifer	61	9	11	22	14
Cladosporium cladosporioides	92	9	32	59	0
Fusarium oxysporum	60	38	17	46	0
Geotrichum candidum	68	56	33	8	0

[1] Notermans *et al.*, 1987.

Table 2. Antibody titre in rabbits immunised with purified extracellular polysaccharides (EPS) of some fungal species[1, 2, 3]

Day of test	*Penicillium digitatum*	*Cladosporium cladosporioides*	*Mucor racemosus*	*Fusarium oxysporum*
0	<50	<50	<50	<50
29	6000	20000	30000	NT
38	50000	78000	>50000	30000

[1] Notermans *et al.*, 1987.
[2] Titre is reciprocal of the highest serum dilution giving a positive result in the ELISA
[3] Rabbits were immunised subcutaneously on day 0 (0.25 mg EPS with Freund Complete Adjuvant) and on day 4 and 32 with 0.25 and 0.5 mg EPS, respectively, with Freund Incomplete Adjuvant.

of several genera which could be detected in the different sandwich ELISA systems are presented in Table 3.

The sandwich ELISA approached genus specificity. Closely related genera, e.g. *Aspergillus* and *Penicillium*, were detectable by the ELISA using antibodies raised against EPS of *P. digitatum* and *P. cyclopium*, but 3 species belonging to *Penicillium* Subgenus *Biverticillium* did not react in the sandwich ELISA carried out with antibodies raised against EPS of *P. digitatum* and *P. cyclopium*. Antibodies raised against *M. racemosus* were reactive with all tested species of *Mucor* and *Rhizopus*. Cross reactions were also observed with some species belonging to *Aspergillus*, *Penicillium* and *Trichothecium*.

In later experiments, carried out with IgG obtained from other immunised rabbits no cross-reactions were observed. Later it became clear that species of the closely related genera *Absidia* and *Syncephalastrum* also gave a positive reaction (De Ruiter *et al.*, 1991). For *F. oxysporum* antibodies some cross-reactivity was often observed with some species of *Aspergillus* and *Penicillium*, especially *Penicillium* Subgenus *Biverticillium*.

Table 3. Production of extracellular polysaccharide antigens by fungal species and their immunological relationship[1]

Tested genera	Sandwich ELISA reaction carried out with antibodies raised against EPS of					
	P. dig.[2]	*P. cycl.*	*M. rac.*	*C. clad.*	*F. oxy.*	*G. cand.*
Aspergillus	12/12[3]	12/12	8/12	0/12	3/12	0/12
Penicillium	41/44	41/44	11/44	0/44	7/44	0/44
Mucor	0/6	0/6	6/6	0/6	0/6	0/6
Rhizopus	0/4	0/4	4/4	0/4	0/4	0/4
Botrytis	0/3	0/3	0/3	0/3	0/3	0/3
Geotrichum	0/6	0/6	0/6	0/6	0/6	6/6
Cladosporium	0/4	0/4	0/4	4/4	0/4	0/4
Fusarium	0/6	0/6	0/6	0/6	6/6	0/6
Trichothecium	0/4	0/4	3/4	0/4	3/4	0/4
Scopulariopsis	0/3	0/3	0/3	0/3	0/3	0/3

[1] Notermans and Soentoro, 1986
[2] *P. dig.*, *Penicillium digitatum*; *P. cycl.*, *P. cyclopium*; *M. rac.*, *Mucor racemosus*; *C. clad.*, *Cladosporium cladosporioides*; *F. oxy.*, *Fusarium oxysporum*; *G. cand.*, *Geotrichum candidum*
[3] Number of species giving a positive ELISA reaction/total number of species tested

Relation between fungal growth and EPS production

For an effective immunological detection method based on fungal EPS in food it is of great importance that EPS production should be related to fungal growth. It was shown by Notermans *et al.* (1986) that EPS production was closely correlated with the mycelial dry weight. The production of EPS was not altered by growth on different media either in surface or submerged culture. No significant effects were observed on EPS production by changing the incubation temperature or water activity. Within the genus *Penicillium* all the strains tested produced comparable quantities of EPS with the exception of species in the Subgenus *Biverticillium*.

Elucidation of the immunodominant portion of *Aspergillus* and *Penicillium* EPS

As mentioned above, the EPS produced by species of *Aspergillus* and *Penicillium* consists mainly of glucose, mannose and galactose. Suzuki and Takeda (1975) showed indirectly that the immunological activity of galactomannans derived from *A. fumigatus* mycelium was due to galactofuranosyl residues. Bennet *et al.* (1985) confirmed that galactofuranosyl groups were immunodominant in *A. fumigatus* galactomannans.

For further elucidation of immunodominance in EPS of *Aspergillus* and *Penicillium* species, Notermans *et al.* (1988) carried out inhibition experiments with methylated mono- and oligosaccharides in a sandwich ELISA using antibodies against *P. digitatum*. It was demonstrated that ß-methyl-D-galactofuranoside was able to inhibit antibody binding to EPS of *P. digitatum* and *P. cyclopium* (Table 4).

Using differently linked dimers of ß-methyl-D-galactofuranosides it was concluded that the ß(1,5)-linked galactofuranoside possesses the greatest inhibitory effect (Table 4). The 50% inhibitory concentration of ß-methyl-D-galactofuranoside and dimer up to heptamer of ß(1,5)-linked-D-galactofuranosides between EPS of *P. digitatum* and *P. cyclopium*, respectively, and antibodies against *P. digitatum* EPS are shown in Figure 2.

Table 4. Inhibition of antibody binding to extracellular polysaccharide (EPS) of *P. digitatum* and *P. cyclopium*[1]

Inhibitor	50% inhibitory concentration (mg/ml)	
	EPS of *P. digitatum*	EPS of *P. cyclopium*
Methyl-α-D-galactopyranoside	>150	>150
Methyl-ß-D-galactopyranoside	>150	>150
Methyl-α-D-galactofuranoside	>150	>150
Methyl-ß-D-galactofuranoside	45	23
Methyl-α-D-glucopyranoside	>150	>150
Methyl-ß-D-glucopyranoside	>150	>150
Methyl-α-D-mannopyranoside	>150	>150
Methyl-ß-D-mannopyranoside	>150	>150
dimer of		
ß(1,2)-linked galactofuranoside	32	11
ß(1,3)-linked galactofuranoside	16	6
ß(1,5)-linked galactofuranoside	3	1
ß(1,6)-linked galactofuranoside	40	16

[1] Notermans *et al.*, 1988

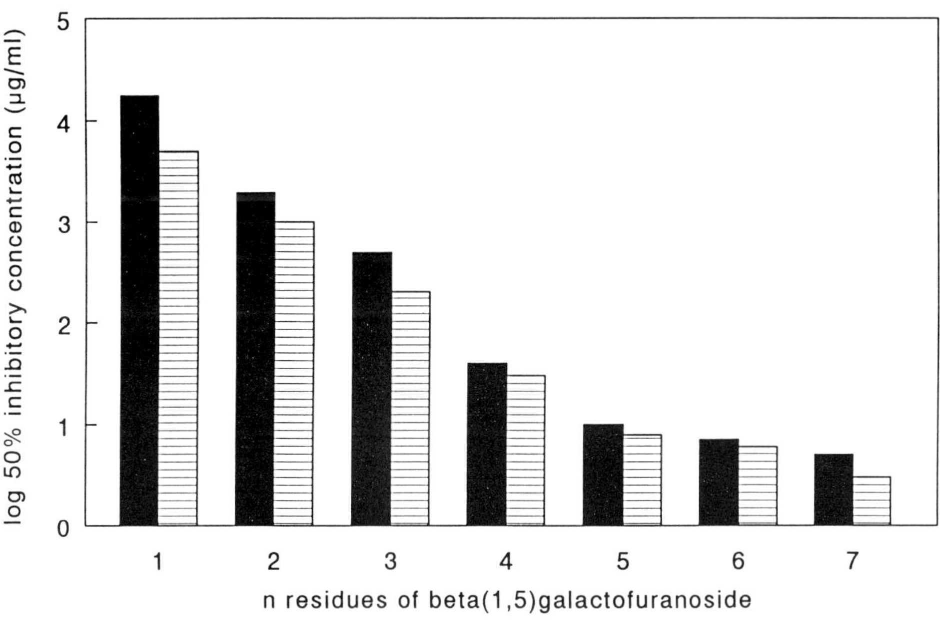

Figure 2. Inhibition of rabbit antibody binding to extracellular polysaccharides of *P. digitatum* and *P. cyclopium* by methyl-ß-D-galactofuranoside and dimer up to heptamer of ß-(1,5)-linked galactofuranosides (Notermans *et al.*, 1988).

The 50 % inhibitory concentration decreased with increasing chain length of the galactofuranosides. The pentamer up to the heptamer of ß(1,5)-linked-D-galactofuranosides interfered also but the 50% inhibitory concentration did not decrease as quickly as was observed with the monomer up to the tetramer. Complete inhibition of the reaction between EPS of *P. digitatum* and *P. cyclopium* and IgG anti *P. digitatum* EPS was observed with the trimer up to the heptamer of ß(1,5)-linked-D-galactofuranosides. These findings provided the opportunity to develop a very reliable immunoassay for the detection of EPS of *Aspergillus* and *Penicillium* in food and feed products (Kamphuis *et al.*, 1989). By using the synthetic tetramer of ß(1,5)-linked-D-galactofuranosides as a specific immunological inhibitor, the reliability of the immunoassay, such as latex agglutination, could be enhanced. Agglutination of the IgG anti *P. digitatum* EPS sensitised latex beads occurs with samples containing *Aspergillus* or *Penicillium* EPS. However, no reaction occurs with the sample if the sensitised latex beads are preincubated with the synthetic tetramer of ß(1,5)-linked-D-galactofuranosides. If agglutination did occur after this specific blocking it can be assumed that it is caused by nonspecific reactions of sample components. This principle is summarised in Figure 3.

Exo-ß-galactofuranosidase activity of moulds
Rietschel-Berst *et al.* (1977) reported that they isolated an exo-ß-galactofuranosidase from the culture filtrate of *P. charlesii*. This enzyme was able to degrade the galactofuranosyl residues of the extracellular glycopeptide (peptidophosphogalactomannan) of *P. charlesii*.

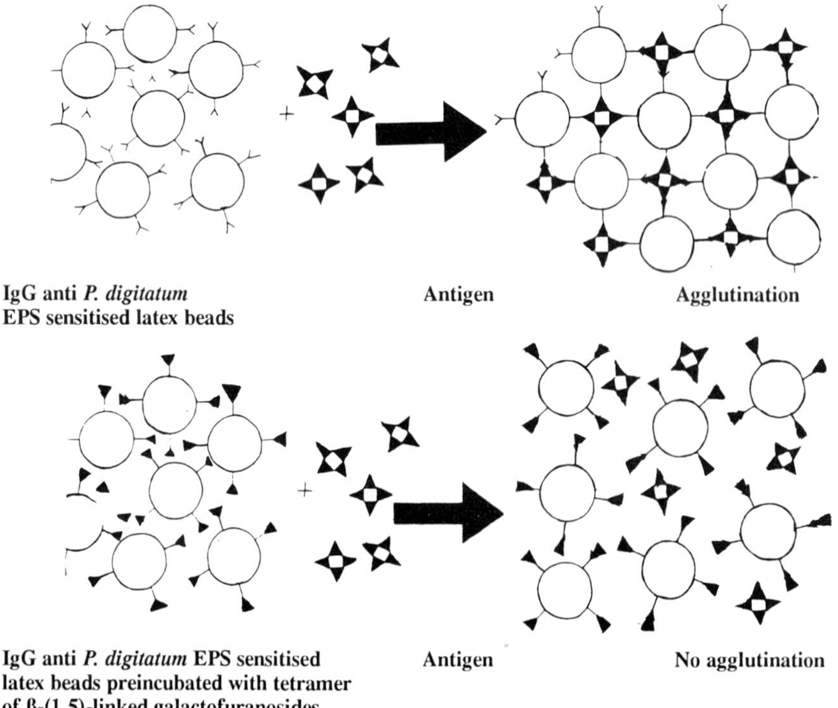

Figure 3. Principle of the latex agglutination test.

Enhanced production of the enzyme was noted after depletion of glucose in the medium (Pletcher *et al.* 1981). Because of the immunodominance of ß-galactofuranoside in *Aspergillus* and *Penicillium* EPS, Cousin *et al.* (1989) investigated the ability of several *Aspergillus* and *Penicillium* species to produce ß-galactofuranosidase. It was found that this enzyme was uncommon, except that biverticillate *Penicillium* species (*P. funiculosum, P. islandicum, P. rubrum* and *P. tardum*) produced substantial ß-galactofuranosidase activity. This enzyme digested the immunodominant galactofuranosyl units present on the EPS of these species preventing reaction with IgG anti *P. digitatum* EPS. This finding explains why some biverticillate *Penicillium* species do not show a positive reaction in the immunoassay for the detection of EPS produced by *Aspergillus* and *Penicillium* species.

REFERENCES

BENNET, J.E., BHATTACHARGEE, A.K. and GLAUDEMANS, C.P.J. 1985. Galactofuranosyl groups are immunodominant in *Aspergillus fumigatus* galactomannan. *Molec. Immunol.* **22**: 251-254.

COUSIN, M.A., NOTERMANS, S., HOOGERHOUT, P. and VAN BOOM, J.H. 1989. Detection of ß-galactofuranosidase production by *Penicillium* and *Aspergillus* species using 4-nitrophenyl ß-D-galactofuranoside. *J. Appl. Bacteriol.* **66**: 311-317.

DE RUITER, G.A., VAN DER LUGT, A.W., VORAGEN A.G.J., ROMBOUTS, F.M. and NOTERMANS, S.H.W. 1991. High performance size-exclusion chromatography and ELISA detection of extracellular polysaccharides from Mucorales. *Carbohydr. Res.* **215**: 47-57.

DE RUITER, G.A., JOSSO, S.L., COLQUHOUN, I.J., VORAGEN A.G.J. and ROMBOUTS, F.M. 1992. Isolation and characterization of ß(1-4)-D-glucuronans from extracellular polysaccharides of moulds belonging to Mucorales. *Carbohydr. Polym.* **18**: 1-7.

JARVIS, B., SEILER, D.A.L., OULD, A.J.L. and WILLIAMS A.P. 1983. Observations on the enumeration of moulds in food and feedingstuffs. *J. Appl. Bacteriol.* **55**: 325-336.

KAMPHUIS, H.J., NOTERMANS, S., VEENEMAN, G.H., VAN BOOM, J.H. and ROMBOUTS, F.M. 1989. A rapid and reliable method for the detection of moulds in foods: using a latex agglutination assay. *J. Food Prot.* **52**: 244-247

NOTERMANS, S. and HEUVELMAN, C.J. 1985. Immunological detection of moulds in food by using the enzyme-linked immunosorbent assay (ELISA): preparation of antigens. *Int. J. Food Microbiol.* **2**: 247-258.

NOTERMANS, S., HEUVELMAN, C.J., BEUMER, R.R. and MAAS, R. 1986. Immunological detection of moulds in food: relation between antigen production and growth. *Int. J. Food Microbiol.* **3**: 253-261.

NOTERMANS, S. and SOENTORO, P.S.S. 1986. Immunological relationship of extra-cellular polysaccharide antigens produced by different mould species. *Antonie van Leeuwenhoek* **52**: 393-401.

NOTERMANS, S., WIETEN, S., ENGEL, H.W., ROMBOUTS, F.M., HOOGERHOUT, P. and VAN BOOM, J.H. 1987. Purification and properties of extracellular polysaccharide (EPS) antigens produced by different mould species. *J. Appl. Bacteriol.* **62**: 157-166.

NOTERMANS, S., VEENEMAN, G.H., VAN ZUYLEN, C.W.E.M., HOOGERHOUT, P. and VAN BOOM, J.H. 1988. (1,5)-linked ß-D-galactofuranosides are immunodominant in extra-cellular polysaccharides of *Penicillium* and *Aspergillus* species. *Molec. Immunol.* **25**: 975-979.

PLETCHER, C.H., LOMAR, P.D. and GANDER, J.E. 1981. Factors affecting the accumulation of exocellular exo-ß-D-galactofuranosides and other enzymes from *Penicillium charlesii*. *Exp. Mycol.* **5**: 133-139.

RIETSCHEL-BERST, M., JENTOFT, N.H., RICK, P.D., PLETCHER, C., FANG, F. and GANDER, J.E. 1977. Extracellular exo-ß-galactofuranosidase from *Penicillium charlesii*. *J. Biol. Chem.* **252**: 3219-3226.

SUZUKI, S. and TAKEDA, N. 1975. Serologic cross-reactivity of the D-galacto-D-mannans isolated from several pathogenic fungi against anti *Hormodendrum pedrosoi* serum. *Carbohydrate Res.* **40**: 193-197.

DETECTION OF FUNGI IN FOODS BY LATEX AGGLUTINATION: A COLLABORATIVE STUDY

S. Notermans[1] and H. J. Kamphuis[2]

[1]*National Institute of Public Health and Environmental Protection*
P.O. Box 1
3720 BA Bilthoven, The Netherlands

[2]*Department of Food Science*
Bomenweg 2
Wageningen Agricultural University
6703 HD Wageningen, The Netherlands

In collaboration with: A.D. Hocking, V. Sanchis, M.A. Cousin, J. Krämer, S. Ewald, R. Steinman, D. Stynen and J.L. Cordier.

SUMMARY

The latex agglutination assay for detection of *Aspergillus* and *Penicillium* species in food products was collaboratively tested by nine different laboratories. The assay is a slide agglutination test using latex particles sensitised with immunoglobulins specific for extracellular polysaccharides (EPS) produced by species of *Aspergillus* and *Penicillium*. False positive results are recognised by use of sensitised latex particles, to which synthesised haptens have been added. These haptens, with a structure identical to the epitopes present on the EPS, specifically block the immunoglobulins present on the latex particles. False negative results are recognised by addition of EPS to test samples. Besides the latex agglutination assay, the collaborating laboratories used conventional methods for detection of fungi in the food products. Eight of the nine laboratories used colony counts for enumeration of fungi with a total of seven media. Eight laboratories were able to detect from 5 to 15 ng/ml of purified EPS. Cereals, animal feed and spices showed fair correlation between colony counts and latex agglutination titres. However, for other products, such as fruit juices, a correlation was not observed, while walnuts gave clearly false positive results. The collaborative study has shown that the latex agglutination test is a rapid, simple and reliable quantitive method for detection of *Penicillium* and *Aspergillus* in cereals, spices and animal feeds.

INTRODUCTION

Seeking alternatives to colony counts for enumeration of fungi in foods, Notermans and Heuvelman (1985) and Lin *et al.* (1986) reported the possibility of using immunological methods for the detection of heat stable extracellular polysaccharides (EPS) produced by moulds. The technique has proven to be useful for testing both heated or unheated food samples. EPS consists mainly of mannose, galactose and glucose: the ß(1,5)-linked D-galactofuranose part of the EPS produced by species of *Penicillium* and *Aspergillus* is immunodominant (Notermans *et al.*, 1988). Based on these findings, Kamphuis *et al.* (1989) developed a latex agglutination method for the detection of moulds. With this test EPS present in food can be detected within 10 minutes. The reliability of the test was enhanced by including a specific blocker in the assay, synthetic epitopes consisting of ß(1,5)-linked D-galactofuranosyl residues.

To evaluate the latex agglutination test for detection of *Penicillium* and *Aspergillus* a collaborative study was carried out, involving nine laboratories worldwide. This paper presents the results of this study.

MATERIALS AND METHODS

Samples and testing procedures

Participants were asked to test at least ten samples from the following foods: fruit juices, animal feeds, cereals, spices and walnuts. The samples were tested by the latex agglutination assay described below. Samples were also examined mycologically by the standard methods used by each participant. Where this method involved an enumeration such as a colony count, identification to genus level was requested.

Mould latex agglutination test

The mould latex agglutination test applied was essentially as described by Kamphuis *et al.* (1989). The test reagents were obtained from Hbt Holland Biotechnology bv, Leiden, The Netherlands. The assay is a slide agglutination test which uses latex particles sensitised with anti-EPS immunoglobulins (Reagent I). When the anti-EPS immunoglobulin sensitised latex particles come into contact with EPS, as produced by *Aspergillus* and *Penicillium* species in the test sample, a complex is formed resulting in agglutination. To prevent false positive readings, Reagent I is sensitised with the immunodominant hapten (a tetramer of ß(1,5)-linked D-galactofuranosyl residues). In this Reagent (II) the active sites of the immunoglobulins are blocked by the hapten preventing agglutination with EPS. If test samples dilutions show identical agglutination with Reagent I and Reagent II, this is the result of interference due to nonspecific factors in the specimen. Reagent III consists of EPS of *P. digitatum*. If a sample shows a negative result with Reagent I, then Reagent III has to be added to the test sample and retested using Reagent I. A sample is confirmed negative if agglutination now occurs with Reagent I. If, however, no agglutination occurs with Reagent I then the sample is to be regarded as unsuitable for testing.

Sample preparation and procedure for the latex agglutination used were as recommended by the manufacturer of the reagents. All participants were asked to determine the minimal quantity of EPS detectable by the latex agglutination assay. For this, participants tested quantities of 135, 45, 15, 5 and 1.7 ng/ml of EPS diluted in 0.07M phosphate buffered saline solution, pH 7.2 (PBS).

RESULTS

Methods used for mycological examination of samples

Eight of the nine collaborating laboratories used colony counts for estimating fungi (Table 1), using seven different media. Incubation temperature varied from room temperature to 26°C, and incubation times were usually 5 days. One laboratory used a latex agglutination assay which detected galactomannans produced by fungi (Eco-Bio Diagnostics Pasteur, Genk, Belgium). Intended for testing serum samples of patients with suspected aspergillosis, this test detected *Botrytis* and *Cladosporium* as well as *Aspergillus* and *Penicillium* species (results not presented).

Table 1. Methods used by collaborators for estimating fungi in foods

Participating laboratory	Method	Medium	Incubation temperature; time
1	Colony count	Wort agar + gentamycin and chloramphenicol;	25°C; 5 days
		Rose bengal chloramphenicol agar	25°C; 5 days
2	Colony count	Oxytetracycline glucose yeast extract agar	25°C; 5 days
3	Colony count	Not stated	Not stated
4	Colony count	Dichloran 18% glycerol agar	26°C; 5 days
5	Colony count	Potato dextrose agar + chloramphenicol and chlortetracycline	22°C; 5 days
6	Latex agglutination (Eco-Bio)		
7	Colony count	Sabouraud agar + chlortetracycline	22°C; 5 days
8	Colony count	Wort agar	Room temp.; 5-7 days
9	Colony count	Dichloran 18% glycerol agar and Dichloran rose bengal chloramphenicol agar	25°C; 5 days

Minimal quantity of EPS detectable by the latex agglutination assay.
Seven laboratories found that 15 ng/ml of EPS gave a positive test result whereas 5 ng/ml was negative (Table 2). One laboratory was able to detect 5 ng/ml of EPS and one was not able to detect 15 ng/ml of EPS.

Cereals, animal feed, spices and herbs
Each collaborating laboratory tested several samples of cereals, animal feeds, spices and herbs. Results obtained for cereals and animal feeds are summarised in Table 3 and for spices in Table 4. No false positive results were observed using the latex agglutination assay. This was demonstrated clearly by the use of Reagent II. However, for samples showing a high degree of mould contamination, blocking did not result in completely negative agglutination. False negative samples were not observed either. Samples that were negative in the latex agglutination assay were re-tested after adding Reagent III (EPS to give 100 ng/mL) and positive agglutination was obtained.

Fruit juices
Only four of the 17 fruit juice samples tested showed positive latex agglutination results (Table 5). Addition of Reagent III to the test samples showed that the other results were not false negatives. Viable fungi were not detected in most samples. The laboratory using the Eco-Bio test found two tested fruit juice samples positive whereas the latex agglutination test used in the collaborative study gave negative results.

Table 2. Limits for detection of *Penicillium digitatum* EPS by latex agglutination

Quantity of control reagent (ng/ml)	Participating laboratory								
	1	2	3	4	5	6	7	8	9
135	+	+	+	+	+	+	+	+	+
45	+	+	+	+	+	+	+	+	+
15	+	+	+	+	+	+	-	+	+
5	-	+	-	-	-	-	-	-	-
1.7	-	-	-	-	-	-	-	-	-

Table 3. Presence of moulds in decimal dilutions of cereals and animal feeds as determined by latex agglutination and colony counts[1]

Product tested	Lab	Mould latex agglutination									Colony count	
		Reagent I				Reagent II			Reagent III			
		10^{-1}	10^{-2}	10^{-3}	10^{-4}	10^{-1}	10^{-2}	10^{-3}	10^{-1}	10^{-2}	log cfu/g	Dominant genus
Wheat bran	5	+	-	-	-	-	-	-	NT		2.3	Pen./Asp.
Wheat bran	4	+	-	-	-	-	-	-	NT		2.7	Fus./Pen.
Wheat bran	1	+	+	-	-	-	-	-	NT		<3.0	-
Wheat bran	8	+	+	-	-	-	-	-	NT		2.0	Pen.
Wheat bran	3	+	+	+	+	+	+	-	NT		5.2	Pen./Asp.
Wheat flour	5	+	-	-	-	-	-	-	NT		2.7	Pen./Clad.
Wheat flour	8	+	-	-	-	-	-	-	NT		2.0	Pen.
Wheat flour	1	+	-	-	-	-	-	-	NT		<3.0	-
Wheat flour	4	+	-	-	-	-	-	-	NT		3.1	Asp./Fus.
Wheat flour	8	+	-	-	-	-	-	-	NT		4.0	Clad.
Wheat flour	6	+	+	-	-	-	-	-	NT		N.T.	-
Corn flour	8	-	-	-	-	-	-	-	NT		<2.0	-
Corn flour	7	-	-	-	-	-	-	-	+	+	<2.0	-
Corn flour	9	-	-	-	-	-	-	-	+	+	<2.0	-
Corn flour	9	-	-	-	-	-	-	-	+	+	3.7	Clad.
Corn flour	5	+	-	-	-	-	-	-	NT		4.0	Pen./Asp.
Corn flour	2	+	+	-	-	-	-	-	NT		3.7	Pen.
Oat flakes	1	+	+	+	-	+	+	-	NT		<3.0	-
Oat flakes	3	+	+	+	-	+	-	-	NT		4.6	-
Sesame seeds	9	+	+	-	-	-	-	-	NT		2.3	-
Sesame seeds	9	+	+	+	+	-	-	-	NT		<2.0	-
Animal feed	7	-	-	-	-	-	-	-	+	+	1.3	Fus.
Animal feed	3	+	+	-	-	-	-	-	NT		3.0	Asp.
Animal feed	1	+	+	-	-	-	-	-	NT		3.6	Asp.
Animal feed	4	+	+	+	+	+	+	-	NT		6.7	Eur./Pen.
Animal feed	8	+	+	+	+	+	-	-	NT		6.9	Pen.
Animal feed	3	+	+	+	+	+	+	-	NT		6.9	Pen./Asp.

[1] +, agglutination; -, no agglutination; NT, not tested; Pen., *Penicillium*; Asp., *Aspergillus*; Fus., *Fusarium*; Clad., *Cladosporium*; Eur., *Eurotium*.

Table 4. Presence of moulds in spices and herbs as determined by latex agglutination and colony count[1]

Product tested	Lab	Mould latex agglutination									Colony count	
		Reagent I				Reagent II			Reagent III			Dominant
		10^{-1}	10^{-2}	10^{-3}	10^{-4}	10^{-1}	10^{-2}	10^{-3}	10^{-1}	10^{-2}	$\log_{10}$ cfu/g	genus
Pepper	6	+	+	-	-	+	-	-	NT		N.T	-
Pepper	9	+	+	-	-	-	-	-	NT		4.8	Pen.
Pepper	9	+	+	-	-	-	-	-	NT		4.5	Asp./Pen.
Pepper	4	+	+	+	-	+	-	-	NT		2.5	Pen./Eur.
Pepper	1	+	+	+	-	+	-	-	NT		4.3	Asp.
Pepper	2	+	+	+	NT	+	-	-	NT		2.7	NT
Pepper	3	+	+	+	-	+	-	-	NT		4.4	Pen.
Pepper	3	+	+	+	+	+	+	-	NT		4.6	Pen./Asp.
Nutmeg	5	+	-	-	-	-	-	-	NT		<1.0	-
Nutmeg	6	+	+	+	-	+	-	-	NT		N.T	-
Paprika	2	+	+	+	NT	+	-	-	NT		3.5	NT
Paprika	2	+	+	+	NT	+	-	-	NT		4.3	NT
Oregano	8	-	-	-	-	-	-	-	NT		2.0	Pen.
Red pepper	7	+	-	-	-	-	-	-	+	+	<1.0	-
Curry	2	+	+	NT	NT	+	-	-	NT		2.2	NT
Zahatar	2	+	+	+	NT	+	-	-	NT		1.8	NT
Rosemary	6	+	+	+	-	+	-	-	NT		N.T	-

[1] +, agglutination; -, no agglutination; NT, not tested; fungal genera abbreviations, see Table 3.

Walnuts

Testing walnuts by latex agglutination may result in both false positive and false negative results (Table 6). Examples of false positives (agglutination with both Reagent I and Reagent II) are shown for laboratories 1 and 6. In all probability sample 5 is also false positive since no *Penicillium* or *Aspergillus* species were found to be present by counting. A false negative result was obtained by laboratory 2. Also the results obtained by laboratories 3 and 4 have to be regarded as false negative because the walnuts tested by these laboratories contained counts of more than 10^3 cfu/g of *Penicillium*. Inconsistent results from testing walnuts were also observed in a more detailed study carried out by us (results not presented).

Dominant genera detected

The dominant mould genera detected in cereals, animal feeds, spices and herbs and walnuts are summarised in Table 7. It is evident that *Penicillium* and *Aspergillus* are dominant in these products. In a few samples other genera were present in equal numbers to *Penicillium* and *Aspergillus*. Only three samples of the 29 tested showed dominant *Fusarium*, *Cladosporium* or *Rhizopus*.

Correlation between mould latex agglutination and colony count

Table 8 shows the correlation between mould latex agglutination titre and colony count, showing results of samples in which *Penicillium* and *Aspergillus* were dominant. In some

Table 5. Presence of moulds in fruit juices as determined by latex agglutination and colony counts[1]

Product tested	Lab	Mould latex agglutination					Colony count	
		Reagent I		Reagent II	Reagent III			
		10^{-1}	10^{-2}	10^{-1}	10^{-1}	10^{-2}	$\log_{10}$ cfu/g	Dominant genus
Orange juice	3	-	-	-	+	+	<2.0	-
Orange juice	8	-	-	-	NT		<2.0	-
Orange juice	1	-	-	-	+	+	<3.0	-
Orange juice	7	-	-	-	+	+	<1.0	-
Orange juice	4	-	-	-	+	+	<1.0	-
Orange juice	5	+	-	-	NT		2.8	Yeasts
Orange juice	3	+	-	-	NT		1.7	Pen.
Grapefruit juice	8	-	-	-	NT		<2.0	-
Grapefruit juice	1	-	-	-	+	+	<3.0	-
Grapefruit juice	7	-	-	-	+	+	1.3	Asp.
Grapefruit juice	6	-	-	-	+	+	N.T	-
Grapefruit juice	4	-	-	-	+	+	<1.0	-
Grapefruit juice	3	+	-	-	NT		1.0	Asp.
Grape juice	5	-	-	-	+	+	<1.0	-
Pineapple juice	8	-	-	-	+	+	4.0	Sporobol.
Lemon juice	7	-	-	-	+	+	<1.0	-
Lemon juice	4	+	-	-	NT		<1.0	-

[1] +, agglutination; -, no agglutination; NT, not tested; Pen., *Penicillium*; Asp., *Aspergillus*; Sporobol., *Sporobolomyces*

Table 6. Presence of moulds in walnuts as determined by the latex agglutination and colony counts[1]

Laboratory	Mould latex agglutination								Colony count	
	Reagent I			Reagent II			Reagent III			
	10^{-1}	10^{-2}	10^{-3}	10^{-1}	10^{-2}	10^{-3}	10^{-1}	10^{-2}	$\log_{10}$ cfu/g	Dominant genus
1	+	-	-	+	-	-	NT		<3.0	-
2	-	-	-	-	-	-	-	-	2.3	NT
3	+	-	-	-	-	-	NT		3.5	Pen.
4	-	-	-	-	-	-	+	+	3.4	Pen.
5	+	+	+	+	+	-	NT		3.3	Rhiz.
6	+	+	-	+	+	-	NT		N.T	-
7	-	-	-	-	-	-	+	+	<1.0	-
8	-	-	-	-	-	-	NT		<2.0	-
9	-	-	-	-	-	-	+	+	2.9	Asp./Pen.
9	+	+	+	-	-	-	NT		3.1	Eur./Asp./Pen.

[1] +, agglutination; -, no agglutination; NT, not tested; fungal genera abbreviations, see Tables 3 and 5; Rhiz., *Rhizopus*.

Table 7. Mould genera dominant in food and feed samples

Product tested	Number of samples with with a positive mould colony count	Dominant genus			
		Penicillium/ *Aspergillus*	*Fusarium*	*Cladosporium*	Others
Cereals	12	9	2	2	0
Animal feeds	6	5	1	0	1
Spices and herbs	7	7	0	0	1
Walnuts	5	4	0	0	2

Table 8. Correlation between latex agglutination test and colony count for analysis of cereals, animal feeds, spices and herbs where *Penicillium* and/or *Aspergillus* were dominant

Latex agglutination titre	Number of samples	Average $\log_{10}$ count plus standard deviation
< 10	3	< 2.0
≥ 10 - < 100	9	2.3 ± 0.90
≥ 100- < 1,000	6	3.6 ± 0.78
≥ 1,000 - < 10,000	4	4.0 ± 0.91
≥ 10,000	5	6.1 ± 1.06

samples, a high latex agglutination titre was observed in combination with a low colony count, e.g. the sesame seed tested by laboratory 9 (Table 3), that had been in domestic storage for a long time. The seeds were obviously mouldy, as they were clumping together, but the mould count was negative on both DG18 and DRBC.

DISCUSSION

The results obtained from this collaborative study indicate that the latex agglutination assay gives quantitive information about the presence of moulds. The minimal detectable quantity of purified EPS was estimated by the participants to be 5-15 ng/mL.

As the latex agglutination assay is a highly sensitive immunoassay, false positive reactions have to be taken into account. The reliability of the test is enhanced by using a control reagent, Reagent II, in which the immunoglobulins have been specifically blocked. Reagent II detected false positive results, for example, with walnuts. False negative results also have to be considered. False negative results may occur if the test procedure is not carried out correctly, or if the EPS becomes enzymatically destroyed or irreversibly bound to certain food components. Inactivation of the EPS by certain enzymes, such as ß-galactofuranosidase produced by *Penicillium* and *Aspergillus* species, has clearly been demonstrated (Cousin *et al.*, 1989). False negative results were only observed in this study for one sample of walnuts.

In this study, *Penicillium* and *Aspergillus* were the dominant genera present in cereals, animal feeds and spices and herbs. A clear correlation was found to exist between the colony count and the latex agglutination assay. In some samples, however, a high latex agglutination titre was observed in the absence of viable moulds. The viable mould count is not always a good reflection of the true mycological status of a food product (King and Schade, 1986), because, for example, mould viability may decline during storage. Also, heat treatment may render fungi nonviable, but the EPS can still be present (Notermans and Heuvelman, 1985).

Almost all fruit juices tested by the participants were definitely negative in the latex agglutination test. Presumably EPS of *Penicillium* and *Aspergillus* was present in only small quantities.

From the results obtained in this collaborative study it can be concluded that the latex agglutination assay is a rapid, reliable and quantitative method for the detection of *Aspergillus* and *Penicillium*, two genera common in animal feeds, cereals and spices. For fruit products, such as juices, the presence of EPS of *Aspergillus* and *Penicillium* may be less important.

REFERENCES

COUSIN, M.A., NOTERMANS, S., HOOGERHOUT, P. and VAN BOOM, J.H. 1989. Detection of ß-galactofuranosidase production by *Penicillium* and *Aspergillus* species using 4-nitrophenyl-ß-D-galactofuranoside. *J. Appl. Bacteriol.* **66**: 311-317.

KAMPHUIS, H.J., NOTERMANS, S., VEENEMAN, G.H., VAN BOOM, J.H.and ROMBOUTS, F.M. 1989. A rapid and reliable method for the detection of moulds in foods: using the latex agglutination assay. *J. Food Prot.* **52**: 244-247.

KING, A.D. and SCHADE, J.E. 1986. Influence of almond harvest, processing and storage on fungal population and flora. *J. Food Sci.* **51**: 202-205.

LIN, H.H., LISTER, R.M. and COUSIN, M.A. 1986. Enzyme-linked immunosorbent assay for the detection of mould in tomato puree. *J. Food Sci.* **51**: 180-192.

NOTERMANS, S. and HEUVELMAN, C.J. 1985. Immunological detection of moulds in foods by using the enzyme-linked immunosorbent assay (ELISA): preparation of antigens. *Int. J. Food Microbiol.* **2**: 247-258.

NOTERMANS, S., VEENEMAN, G.H., VAN ZUYLEN, C.W.E.M., HOOGERHOUT, P. and VAN BOOM, J.H. 1988. (1→5)-linked ß-D-galactofuranosides are immunodominant in extracellular polysaccharides of *Penicillium* and *Aspergillus* species. *Mol. Immunol.* **25**: 975-979.

CHARACTERISTICS OF A LATEX AGGLUTINATION TEST BASED ON MONOCLONAL ANTIBODIES FOR THE DETECTION OF FUNGAL ANTIGENS IN FOODS

D. Stynen[1], L. Meulemans[1], A. Goris[1], N. Braendlin[2] and N. Symons[1]

[1] *Eco-Bio Diagnostics Pasteur*
Woudstraat 25
B-3600 Genk, Belgium.

[2] *Nestec Ltd*
Quality Assurance Department
Avenue Nestlé 55
CH-1800 Vevey, Switzerland

SUMMARY

Sensitive immunological tests can rapidly detect microbiological contamination of foods without the need for cultural procedures. This paper describes the Mould Reveal Kit, an easy to use latex agglutination test for the rapid screening of foodstuffs for mould contamination. A rat monoclonal antibody (EB-A2) against *Aspergillus* galactomannan was coated onto latex beads. The sensitised latex beads detected this fungal extracellular polysaccharide at a concentration of 15 ng/mL after 5 minutes incubation. A negative control latex was developed to identify false positive results caused by non-specific agglutination of some types of foods. Of 35 common foodborne fungi tested, 27 gave positive reactions in the latex agglutination test. The test is rapid, easy to use, and can be semi-quantitative if titrations are carried out.

INTRODUCTION

Immunological detection systems have shown promise for overcoming some of the drawbacks of currently used methods for detecting fungi in foods, as they do not depend on the presence of living fungi. One effective approach is the latex agglutination test, in which antibodies are coupled to latex beads. In the presence of the antigen, these beads agglutinate to form macroscopically visible aggregates. Latex agglutination tests do not require any sophisticated apparatus, are very sensitive (at the level of ng/mL), very rapid (less than 15 min after sample preparation), and require little skill to carry out and interpret.

Kamphuis *et al.* (1989) published a latex agglutination test that detects galactomannan, the extracellular polysaccharide of *Aspergillus* and *Penicillium.* The polyclonal antibodies used to sensitise the latex beads react with the immunodominant epitope of the galactomannan, i.e. the ß(1,5)-linked D-galactofuranosyl side chains (Notermans *et al.*, 1988).

Eco-Bio developed the first latex agglutination test for the detection of *Aspergillus* galactomannan in the serum of patients with invasive aspergillosis (Dupont *et al.*, 1990;

Van Cutsem *et al.*, 1990a, b; Haynes *et al.*, 1990). The test, Pastorex *Aspergillus*, has been commercialised by Diagnostics Pasteur. The latex beads are sensitised with EB-A2, a rat monoclonal antibody that reacts with the same immunodominant epitope as the antibodies used in the test of Kamphuis *et al.* (1989) (Debeaupuis *et al.*,1990). This suggests that this latex might also be useful for mould detection in foods. We now describe the characteristics of the Mould Reveal Kit (Eco-Bio, Genk), a test kit for fungal antigen detection in foodstuffs, featuring EB-AL coated latex beads.

MATERIALS AND METHODS

Preparation of fungal culture filtrates

Most culture filtrates used in the experiments in Table 1 were obtained by growing fungal strains from the culture collection at Nestec, Vevey, in a liquid medium. The medium contained sucrose 100 g (400 g for xerophilic moulds); yeast extract, 0.5 g; $NaNO_3$, 2 g; KH_2PO_4, 2 g; KCl, 2 g; $MgSO_4$, 1.0 g; $FeSO_4$, 0.02 g per litre of distilled water. The medium was sterilised at 121°C, and the final pH was 4.6. Thirty millilitre flasks containing 10 mL of culture broth were inoculated with a dense suspension of conidia, previously harvested with sterile distilled water from an agar slant. The flasks were incubated for 3-5 days (depending on the growth rate) as stationary cultures at 26°C. After incubation, culture broths were filtered through a paper filter (Schleicher and Schüll, Keine). Other filtrates, from cultures grown in liquid Sabouraud medium were provided by the Centraalbureau voor Schimmelcultures (CBS), Baarn, The Netherlands.

Exoantigen preparation

Wallemia sebi, *Cladosporium cladosporioides*, *Cladosporium herbarum*, *Paecilomyces variotii*, *Trichoderma viride* and *Fusarium solani* culture filtrates were prepared by and obtained from M. van der Horst and R.A. Samson (CBS, Baarn). The fungi had been grown in malt extract (2%) broth (pH 7.0) for 11 days at 25°C on a rotary shaker (150 rpm). Cultures were then autoclaved for 60 min at 120°C and filtered on filter paper. Filtrates were mixed with four volumes of cold ethanol and allowed to stand overnight at 4°C. After 15 min centrifugation at 3000 rpm, pellets were resuspended in 40 mL of ethanol and centrifuged for 10 min at 48000 g. Supernatants were discarded and the pellets allowed to dry. The exoantigen pellets were redissolved in phosphate buffered saline (PBS; pH 7.2, 0.15 M) by sonication. Hexose concentrations were determined by the orcinol sulphuric acid method (Chandrasekaran and BeMiller, 1980) and proteins by the BCA reagent (Pierce Chemical Co., Rockford).

Exoantigens from *Cryptococcus neoformans* serotype A were prepared by precipitation of culture filtrates with ethanol acetate, followed by deproteination by extraction with chloroform and butyl alcohol (Dromer *et al.*, 1987). Mannan from *Candida albicans* was prepared essentially by precipitation with Fehling's solution (Kocourek and Ballou, 1969). Mannan from *Saccharomyces cerevisiae* was obtained from Sigma (St Louis, Missouri).

Exopolysaccharides from *Botrytis tulipae, Cladosporium cladosporioides* and *Penicillium digitatum* were provided by S. Notermans (National Institute of Public Health and Environmental Protection, Bilthoven). Exoantigen and purified galactomannan from *Aspergillus fumigatus* were provided by J.P. Latgé (Institut Pasteur, Paris).

Latex agglutination tests

The Mould Reveal Kit. Latex beads, coated with rat monoclonal antibody EB-A2 (IgM), were identical to those in the Pastorex *Aspergillus* kit. The negative control latex was prepared by covalently binding another rat monoclonal IgM, against an irrelevant antigen, to the latex beads.

Samples under investigation were dissolved and diluted in PBS. Samples (40 μL) were transferred to disposable plastic agglutination cards, included in the kit, and mixed with 10 μL of the latex reagent. After 5 min rotation at 170 rpm, samples showing agglutination visible to the naked eye were scored positive. Serial dilutions of the samples were tested to obtain semi-quantitative results. The samples were submitted to the same protocol with the negative control latex to identify non-specific agglutination reactions.

The Mould Latex Agglutination Test (MLAT) from Holland Biotechnology, Leiden was carried out according to the manufacturer's instructions with one modification: results were scored macroscopically, since microscopical observation was cumbersome and did not improve readability.

RESULTS

Sensitivity of the latex test

Figure 1 shows the minimal concentrations of *A. fumigatus* galactomannan detected after various incubation times with the Mould Reveal Kit and the MLAT. The sensitivity of the Mould Reveal Kit after 5 min was 15 ng/mL. Incubation for longer periods do not further increase the sensitivity. After 1.5 min, 30 ng/mL was detected; 60 ng/mL was detected after 1 min, 120 ng/mL after 30 sec, and 240 ng/mL after 15 sec. The MLAT reacted more slowly: 10 min for 120 ng/mL, 13 minutes for 60 ng/mL and 15 min for 30 ng/mL.

The negative control latex of the Mould Reveal Kit remained negative with galactomannan concentrations as high as 3.5 mg/mL, which was the most concentrated solution tested. The MLAT negative control latex was negative with 840 ng/mL galactomannan but yielded a positive result with the next concentration tested (4 μg/mL).

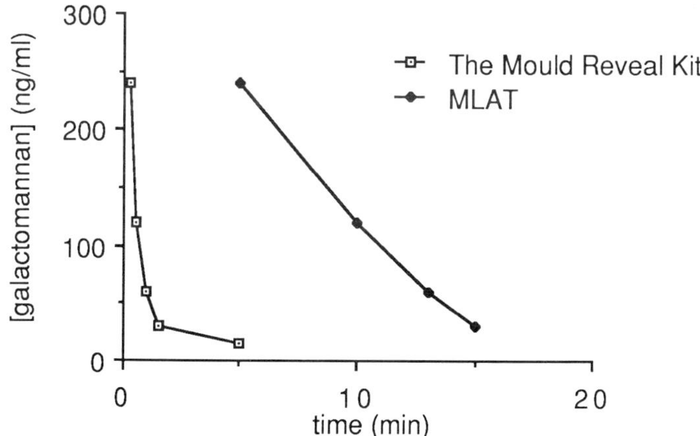

Figure 1. Time required to yield macroscopically visible agglutination of the Mould Reveal Kit and the Mould Latex Agglutination Test (MLAT). The latex suspensions were incubated with different concentrations of *Aspergillus fumigatus* galactomannan.

Agglutination of culture filtrates

A preliminary determination of the species specificity of the latex agglutination test was performed by testing culture filtrates (1:10 dilutions in PBS) of 39 fungal species (Table 1). All *Aspergillus* species (7) and *Penicillium* species (9) were positive. Positive reactions were also observed with *Eurotium repens*, *Eurotium herbariorum*, *Emericella nidulans*, *Neosartorya fischeri*, *Talaromyces* sp., *Paecilomyces variotii*, *Byssochlamys nivea*, *Trichoderma* sp., *Wallemia sebi*, *Eremascus albus*, *Aureobasidium pullulans*, *Botrytis cinerea*, *Cladosporium cladosporioides* and *Cladosporium herbarum*. Culture filtrates of *Alternaria* sp., *Epicoccum* sp., *Scopulariopsis* sp., *Geotrichum candidum*, *Fusarium solani*, *F. oxysporum*, *Mucor racemosus*, *M. plumbeus*, *Rhizopus oryzae* and *Syncephalastrum racemosum* did not cause agglutination reactions.

Agglutination of exoantigen preparations and purified fungal polysaccharides

Exoantigen preparations and purified exocellular polysaccharides of a selection of fungi were submitted to the latex agglutination test in order to obtain more direct information about their reactivity (Table 2).

Table 1. Screening of culture filtrates (1:10 dilutions) of different fungi with the Mould Reveal Kit

Positive cultures	Negative cultures
Aspergillus clavatus	*Alternaria* sp.
A. candidus	*Epicoccum* sp.
A. flavus	*Scopulariopsis* sp.
A. fumigatus	*Geotrichum candidum*
A. niger	*Fusarium solani*
A. ochraceus	*F. oxysporum*
A. versicolor	*Mucor racemosus*
Eurotium repens	*Mucor plumbeus*[1]
E. herbariorum[1]	*Syncephalastrum racemosum*
Emericella nidulans	*Rhizopus oryzae*[1]
Neosartorya fischeri	
Penicillium brevicompactum	
P. camemberti	
P. chrysogenum	
P. digitatum	
P. expansum	
P. roqueforti	
P. viridicatum	
P. glabrum	
P. spinulosum	
Talaromyces sp.	
Eremascus albus	
Paecilomyces variotii	
Trichoderma sp.	
Wallemia sebi	
Cladosporium cladosporioides	
C. herbarum	
Aureobasidium pullulans[1]	
Botrytis cinerea[1]	

[1] Fungi received from Centraalbureau voor Schimmelcultures, cultured in Sabouraud broth.

Table 2. The sensitivity of the Mould Reveal Kit towards ethanol precipitates of culture filtrates (exoantigens) and purified fungal polysaccharides

Species	Sensitivity (ng/mL)[1]
Exoantigens	
Aspergillus fumigatus	15
Penicillium digitatum	50
P. marneffei	100
Wallemia sebi	250
Cladosporium herbarium	500
C. cladosporioides	1000
Trichoderma viride	1000
Fusarium solani	>1000
Cryptococcus neoformans	>1000
Polysaccharides	
A. fumigatus	15
P. digitatum	15
C. cladosporioides	50
Botrytis tulipae	100
Candida albicans	>1000
Saccharomyces cerevisiae	>1000

[1] Hexose concentrations

Exoantigen preparations from fungi with positive culture filtrates were not equally agglutinated by the latex beads. The highest sensitivities were observed with *A. fumigatus* and both *Penicillium* species, *P. digitatum* and *P. marneffei* and an intermediate sensitivity with *Wallemia sebi*. The sensitivity for *C. herbarum* and *C. cladosporioides* exoantigens and *Trichoderma viride* was 10 to 100 times lower than for the *Aspergillus* exoantigen. *Fusarium solani* and *Cryptococcus neoformans* exoantigens were not agglutinated at the highest concentration tested (1 μg/mL).

Purified *A. fumigatus* and *P. digitatum* galactomannans were equally well detected (15 ng/mL). The sensitivity towards *C. cladosporioides* (50 ng/mL) and *Botrytis tulipae* (100 ng/mL) was lower. *Candida albicans* and *Saccharomyces cerevisiae* mannan remained negative at the highest concentration tested (1 μg/mL). The negative control latex remained negative in all cases.

DISCUSSION

The Mould Reveal Kit detects small concentrations of fungal polysaccharides after very short incubation times. The sensitivity is about 15 ng/mL of *Aspergillus* galactomannan if the sample and the latex beads are rotated for 5 min (Figure 1). The incubation time can be reduced considerably if somewhat lower sensitivities are acceptable. Galactomannan at 120 ng/mL, an antigen concentration commonly encountered in foodstuffs (Kamphuis *et al.*, 1989), will cause agglutination after as little as 30 sec. The MLAT test needs longer incubation times, ranging from 10 to 15 min for these detection limits.

Previous reports have shown that some foodstuffs contain ingredients that may cause nonspecific agglutination of the latex beads (Kamphuis *et al.*, 1989). These false positive reactions should be recognised with a negative control latex. The control latex in the MLAT test is identical to the positive latex but with a high concentration of a synthetic oligosaccharide added to the buffer that blocks specific, but not non-specific agglutination reactions. High concentrations of antigen in the sample, will successfully compete with the blocking agent and this leads to false, false positive results. Thus, the manufacturer defines real positive results as those where there is a 9 fold difference (i.e. two 3-fold dilutions) between the positive latex and the negative control latex. Since the Mould Reveal Kit uses monoclonal antibodies against galactomannan, immunological cross reactions with non-fungal antigens are unlikely. Other non specific reactions, such as those observed with walnuts (Kamphuis, 1989), are identified using a negative control latex reagent consisting of latex beads coated with a different rat monoclonal antibody of the same isotype. If a sample causes nonspecific agglutination of latex beads, both the positive latex and the negative control will agglutinate. If this is not the case, the negative control latex will remain negative, even with high galactomannan concentrations.

Screening of a large number of culture filtrates revealed the wide specificity of the Mould Reveal Kit (Table 1). Quantitative differences in affinity for the exopolysaccharides were studied by determining the sensitivity for semi-purified and purified polysaccharides produced by selection of fungal genera (Table 2). This experiment demonstrated that the reactivity varies from genus to genus. It is highest with *Aspergillus* and *Penicillium* galactomannan. This is not surprising since the monoclonal antibody EB-A2 was raised against *Aspergillus* (Goris *et al.*, 1989; Debeaupuis *et al.*, 1990) and *Penicillium* and *Aspergillus* galactomannans share their immunodominant epitopes (Notermans *et al.*, 1988). Previous characterisations of the antibody showed that it reacted with exoantigens from all nine *Aspergillus* species and all seven *Neosartorya* species studied (Goris *et al.*, 1989; Debeaupuis *et al.*, 1990). The reactivity of the Mould Reveal Kit with *Penicillium marneffei* antigens has also been shown in studies on an animal model of invasive mycoses: the latex beads not only detected antigen in plasma samples of guinea pigs with invasive aspergillosis, but also in those of animals infected with *P. marneffei* (Van Cutsem *et al.*, 1990a,b). Exopolysaccharides of other fungal species reacted less efficiently.

The lowest positive exoantigen concentrations of other species were 10 to 100 times higher. In the case of *C. cladosporioides*, the difference was smaller with polysaccharide preparations of higher purity. This observation suggests that, due to the complex composition of the exoantigens, the results with these preparations should be considered as estimates and interpreted with some caution. Nevertheless, the differences between these sensitivity estimates are large enough to conclude that the affinity of the latex beads may vary by one or two orders of magnitude, depending on the species under investigation. Whether these differences are due to a lower incidence of the epitope, a worse accessibility of the epitope or a somewhat different structure remains unknown.

Some fungal polysaccharides or culture filtrates did not react at all, including those of yeasts (both *Candida albicans* and *Cryptococcus neoformans*) and Zygomycetes (*Mucor plumbeus*, *M. racemosus*, *Rhizopus oryzae* and *Syncephalastrum racemosum*).

Our results indicate that the Mould Reveal Kit could be useful for the rapid detection of fungal contamination in foodstuffs. The presence of fungal antigens is revealed by strong agglutination patterns that become visible to the naked eye after a few seconds to five minutes. By testing serial dilutions of the sample, a semi-quantitative estimate of the degree of fungal contamination can be made. False positive agglutination reations can be identified by the use of the negative control latex. Due to the use of a monoclonal antibody, the frequency of this occurrence is reduced to a minimum. The use of monclonal antibodies also guarantees a constant quality without batch to batch variation. The wide specificity of the test and the fact that it can detect antigens in the absence of viable fungi can be advantageous or disadvantageous depending on the particular needs of the user. Further evaluations on foodstuffs will be necessary to confirm that the Mould Reveal Kit is useful in the routine laboratory.

ACKNOWLEDGEMENTS

We are grateful to Dr S. Notermans (National Institute of Public Health and Environmental Protection, Bilthoven) and Dr J.-P. Latgé (Institut Pasteur, Paris) for purified exopolysaccharides and to Ms M. van der Horst and Dr R.A. Samson (Centraalbureau voor Schimmelcultures, Baarn) for culture filtrates.

REFERENCES

CHANDRASEKARAN, E.V. and BeMILLER J.M. 1980. Constituent analysis of glycosaminoglycans. *In* Methods in Carbohydrate Chemistry, Vol. VIII., eds R.C. Whistler and J.N. BeMiller. pp. 89-96. New York: Academic Press.

DEBEAUPUIS, J.-P., SARFATI, J., GORIS, A., STYNEN, D., DIAQUIN, M. and LATGÉ, J.-P. 1990. Exocellular polysaccharides from *Aspergillus fumigatus* and related taxa. *In* Modern Methods in *Penicillium* and *Aspergillus* Classification, eds R.A. Samson and J.I. Pitt. pp. 209-223, New York: Plenum Press.

DROMER, F., SALAMERO, J., CONTREPOIS, A., CARBON, C. and YENI, P. 1987. Production, characterization, and antibody specificity of a mouse monoclonal antibody reactive with *Cryptococcus neoformans* capsular polysaccharide. *Infect. Immun.* **55**: 742-748.

DUPONT, B., IMPROVISI, L. and PROVOST, F. 1990. Détection de galactomannane dans les aspergilloses invasives humaines et animales avec un test au latex. *Bull. Soc. Fr. Mycol. Méd.* **19**: 27-34.

GORIS, A., SARFATI, J., DIAQUIN, M., DEBEAUPUIS, J.P., MOUTAOUAKIL, M., LATGÉ, J.-P. and STYNEN, D. 1989. Monoclonal antibodies against *Aspergillus* galactomannan: characterization and application in immunoelectron microscopy. Abstracts from 4th European Congress of Clinical Microbiology, Nice, p. 80.

HAYNES, K.A., PETROU, M.A. and ROGERS, T.R. 1990. Evaluation of a latex agglutination test for the detection of circulating *Aspergillus fumigatus* galactomannan in neutropenic patients. Abstract, 6th International Symposium on Infections in the Immunocompromised Host, Peebles, Scotland.

KAMPHUIS, H.J., NOTERMANS, S., VEENEMAN, G.H., VAN BOOM, J.H. and ROMBOUTS, F.M. 1989. A rapid and reliable method for the detection of molds in foods: using the latex agglutination assay. *J. Food Prot.* **52**: 244-247.

KOCOUREK, J. and BALLOU, C.E. 1969. Method for fingerprinting yeast cell wall mannan. *J. Bacteriol.* **100**: 1175-1181.

NOTERMANS, S., VEENEMAN, G.H., VAN ZUYLEN, C.W.E.M., HOOGERHOUT, P. and VAN BOOM, J.H. 1988. (1→5)-linked ß-D-galactofuranosides are immunodominant in extracellular polysaccharides of *Penicillium* and *Aspergillus* species. *Molec. Immunol.* **25**: 975-979.

VAN CUTSEM, J., MEULEMANS, L., VAN GERVEN, F. and STYNEN, D. 1990a. Detection of circulating galactomannan by Pastorex *Aspergillus* in invasive aspergillosis. *Mycoses* **33**: 61-69.

VAN CUTSEM, J., MEULEMANS, L., VAN GERVEN, F. and STYNEN, D. 1990b. Détection de galactomannane dans le plasma, à l'aide d'un test au latex dans l'aspergillose invasive du cobaye. *Bull. Soc. Fr. Mycol. Méd.* **19**: 35-42.

IMMUNOCHEMICAL DETECTION OF MUCORALES SPECIES IN FOODS

G.A. De Ruiter[1], T. Hoopman[1], A.W. Van der Lugt[1], S.H.W. Notermans[2] and M.J.R. Nout[1]

[1]*Department of Food Science*
Wageningen Agricultural University
6703 HD Wageningen, The Netherlands

[2]*Laboratory for Water and Food Microbiology*
National Institute of Public Health and Environmental Protection
3720 BA Bilthoven, The Netherlands

SUMMARY

In this paper, immunological detection of fungi of the order Mucorales in food is reviewed. The Enzyme Linked Immunosorbent Assay (ELISA), based on the immunochemical properties of the extracellular polysaccharides (EPS), has been found to be specific for species belonging to Mucorales. Testing of 161 food samples demonstrated that this method has several advantages compared with other detection methods, including colony count. The ELISA method was found to be sensitive, rapid and reliable for a number of food products. Checks for false positive reactions are necessary for samples of walnuts and jams. In general, the presence in food at some time of species belonging to Mucorales can be established using this ELISA assay, even if the mould itself is inactivated or removed by filtration.

INTRODUCTION

Mycological detection of species belonging to the order Mucorales in food is important because of their wide distribution in stored grain, fruits and vegetables, which leads in many cases to food spoilage (Pitt and Hocking, 1985). Medically, species belonging to Mucorales are important for their ability to cause mucormycosis in man (Kaufman *et al.*, 1989; De Ruiter *et al.*, 1991a). Testing of processed foods is often necessary to verify if the food was prepared from high quality raw materials. Conventional detection methods have been summarised by Jarvis *et al.* (1983). The mould colony count is used in many countries as a standard method prescribed in food legislation. However, this method cannot be used for processed foods in which the mould itself has been killed or removed. In addition, the mould colony count method lacks precision, is time consuming and is not specific for species belonging to Mucorales.

To overcome the disadvantages of conventional methods, immunological detection methods based on the immunochemically active extracellular polysaccharides (EPS) have been developed for various genera of fungi.

Aspergillus and *Penicillium* species could be detected either by an enzyme-linked immunosorbent assay (ELISA) or a latex agglutination assay (Notermans and Soentoro, 1986; Kamphuis *et al.*, 1989). It was found that EPS antigens involved have ß(1,5)-linked galactofuranose oligomeric epitopes, and that they are heat stable and water extractable, enabling detection of these moulds in food even after heat treatment or thermal processing (Kamphuis and Notermans, 1992).

A successful immunoassay for species of Mucorales requires that the antigens are easily extractable from food and that they be stable towards heating and other processes which inactivate fungi. Furthermore, the excreted immunologically active EPS must also be genus, or at least order, specific. For quantitative purposes the amount of excreted EPS and the amount of fungal biomass as well as the ELISA titre should be related. Also, it must be possible to recognise any false positive reaction which may occur as a result of the complexity of the food samples.

This paper reviews the features important for the immunological detection in food of fungi in Mucorales, including the genera *Mucor*, *Rhizopus*, *Rhizomucor*, *Absidia*, *Syncephalastrum* and *Thamnidium*. An ELISA assay developed for detection of these fungi was tested using a variety of food samples.

The extracellular polysaccharides of Mucorales

The immunochemical properties of polysaccharides excreted by species of Mucorales have been established partly by Miyazaki and Irino (1972) and Miyazaki *et al.* (1980). EPS are mainly composed of glucuronic acid, mannose, fucose, glucose, galactose and minor amounts of protein, and they are water extractable and resistant towards heating (Hough and Perry, 1955; Martin and Adams, 1956; De Ruiter *et al.*, 1991a, 1992). Immunological detection methods based on the immunochemically active EPS have been developed for detection of these moulds (Jones and Kaufman, 1978; Notermans and Heuvelman, 1985; Lin and Cousin, 1987; Kaufman *et al.*, 1989). Specific blocking of the antibody, which enables detection of false positive reactions, requires structural information about the epitopes. Although no detailed knowledge about the epitopes of *Mucor* EPS is available, inhibition using high amounts of methyl-α-mannosides can be performed as described by Notermans *et al.* (1988).

METHODS

Determination of EPS and biomass production

Moulds used in this study were grown at their optimal temperature in shaken submerged cultures for 7 days in a basal medium of yeast nitrogen base (YNB, Difco Laboratories, Detroit, USA) supplemented with 30 g/L glucose as carbon source. Extracellular polysaccharides (EPS) of the moulds were isolated from the culture fluid after filtration of the mycelium and purified by ethanol precipitation as described by De Ruiter *et al.* (1991b). The yield of biomass was measured after drying the extracted mycelium in an oven at 80°C overnight.

The *Mucor* sandwich ELISA

A sandwich ELISA assay with polyclonal IgG antibodies raised in rabbits against extracellular polysaccharides from *Mucor racemosus* was performed as described by Notermans and Heuvelman (1985) and De Ruiter *et al.* (1991b). The ELISA reactivity is expressed as the titre, which is the reciprocal dilution of the sample just giving a positive reaction, i.e. an extinction ≥0.1.

Specificity of the ELISA for species belonging to Mucorales

Notermans and Heuvelman (1985) and Notermans *et al.* (1986) established a complete cross reactivity in a sandwich ELISA between species of the genera *Mucor* and *Rhizopus* with antibodies raised against EPS of *Mucor racemosus*. No cross reactivity occurred with the major species of *Penicillium* and *Aspergillus* (Notermans and Soentoro, 1986). Antigenic similarity between some species of *Rhizopus* was also established by Polonelli *et al.* (1988). Species of other genera belonging to Mucorales were tested for their ability to react with antibodies raised against EPS of *Mucor racemosus*.

The use of the *Mucor* ELISA in food samples

In order to study the effectiveness of the *Mucor hiemalis* ELISA assay for detecting species belonging to Mucorales, 161 food samples were tested. To 10 g of sample 90 mL of a 0.07 M phosphate buffered saline (0.15 M) solution, pH 7.2, was added. The mixture was homogenised in a Colworth 400 Stomacher for 1 min, followed by centrifugation (15 min at 3000 rpm). The supernatant was used for all experiments.

The results of the ELISA were roughly compared with colony counts performed by the Netherlands Food Inspection Laboratories of Rotterdam, Zutphen and Utrecht. Fungal colonies were enumerated on Dichloran 18% Glycerol agar (DG18) as described by Hocking and Pitt (1980).

RESULTS

Specificity of the ELISA for species belonging to Mucorales

As shown in Table 1, EPS from 10 species belonging to six genera of Mucorales reacted to the ELISA system using EPS of *Mucor racemosus*. It appears that some EPS fractions are more reactive than others.

Table 1. ELISA reactivity of isolated EPS of various species of Mucorales

Species	Strain	EPS yield (mg/l)[1]	ELISA titre[2]
Mucor racemosus	RIVM H473-R5	190	400
Mucor hiemalis	CBS 201.28	378	220
Mucor circinelloides	RIVM 40	498	560
Rhizopus oryzae	LUW 581	626	50
Rhizopus stolonifer	CBS 609.82	300	140
Rhizomucor miehei	CBS 371.71	324	610
Rhizomucor pusillus	CBS 432.78	220	730
Absidia corymbifera	LUW 017	449	50
Syncephalastrum racemosum	CBS 443.59	504	240
Thamnidium elegans	CBS 342.55	46	70

[1] The yield of EPS is expressed as the isolated amount in milligrams per litre of culture medium.
[2] The ELISA titre is expressed as the reciprocal dilution of a 10 μg/ml solution of EPS in distilled water, just giving a positive reaction, compared to a blank, in the ELISA as described in Notermans *et al.* (1986). It can be used to determine the immunochemical reactivity of the isolated EPS fractions.

Relation between EPS production, biomass and ELISA titre

It was demonstrated that the excreted amount of EPS can be considered as a marker for the amount of biomass of *Mucor hiemalis*. For *M. hiemalis* grown on YNB medium with glucose, the ratio between the amount of excreted EPS, both determined by isolation and ELISA, and the fungal biomass varied between 0.1 and 0.4 (mg/mg) depending on carbon source and growth phase. This was valid also for the EPS of the 10 species listed in Table 1 (De Ruiter *et al.*, 1991a). In addition to the immunochemically active EPS it was found that other neutral polysaccharides could be excreted depending on species and carbon source used for growth (De Ruiter *et al.*, 1991b).

Table 2. *Mucor* ELISA titres on food samples with fungal counts of less than 200 cfu/g

Food	Samples tested	*Mucor* ELISA titre[1]			
		-[2]	+	++	>+++
Juices					
Grape juice	6	-	-	5	1
Apple juice	7	7	-	-	-
Pineapple juice	5	5	-	-	-
Blackcurrant juice	6	1	1	3	1
Orange juice	5	5	-	-	-
Mixed juices	7	6	-	1	-
Tomato juice	3	3	-	-	-
Spices					
Nutmeg	3	-	-	3	-
Paprika powder	2	-	-	-	2
Mixed spices	2	1	1	-	-
Nuts					
Pistachio	2	2	-	-	-
Pistachio[3]	2	1	-	1	-
Hazelnuts	6	6	-	-	-
Roasted peanuts	9	9	-	-	-
Peanut butter	11	9	-	-	2
Raw peanuts	3	3	-	-	-
Flour					
Maize flour	3	3	-	-	-
Buckwheat	1	1	-	-	-
Wheat flour	3	3	-	-	-
Dried fruits					
Dates	9	9	-	-	-
Figs	7	7	-	-	-
Apricots	4	4	-	-	-
Raisins	1	1	-	-	-

[1] *Mucor* ELISA titre is expressed as reciprocal dilution just giving a positive ELISA reaction, compared to the blank as described in Notermans *et al.* (1986).
[2] -, lower than 20; +, between 20 and 100; ++, between 100 and 1000; >+++, higher than 1000
[3] Results taken from Notermans *et al.* (1988).

The use of *Mucor* ELISA in food samples

The comparisons between ELISA results, expressed as the ELISA titre, and counts on DG18 are given in Tables 2 and 3. Samples with low counts, i.e. less than 200 cfu/g, are summarised in Table 2, while Table 3 summarises those with higher counts. Only samples with no false positive ELISA reactions are included in the tables: jams, which frequently showed false positive reactions, have been excluded.

As shown in Table 2, some pasteurised juices, mainly grape juices and blackcurrant juices, gave high *Mucor* ELISA titres. It is to be expected that the fruits used as raw materials for these juices would be contaminated with moulds belonging to Mucorales. Juices from fruits in which no contamination with these moulds is to be expected, e.g. orange, pineapple and tomato, did not give a positive ELISA reaction.

Spices, in which enumeration detected no fungal contamination, gave rise to high ELISA titres, indicating the previous presence of moulds. Most spices entering Europe are fumigated. Nuts, flours and dried fruits show low *Mucor* ELISA titres.

Table 3 shows results for samples which gave high colony counts on DG18. Counts are for total fungi, not specifically Mucorales: the number of samples in which Mucorales were detected on DG18 is given in the last column.

Table 3. *Mucor* ELISA titres on various food samples with relatively high mould counts

Food	Samples tested	*Mucor* ELISA titre[1]				Log_{10} cfu/g			Samples containing Mucorales[4]
		-[2]	+	++	>+++	2-<3	3-<4	>4	
Spices									
Nutmeg	8	-	-	2	6	2	4	2	-
Nutmeg[3]	1	-	-	-	1	-	-	1	-
Paprika powder	5	-	-	-	5	-	3	2	1
Mixed spices	2	2	-	-	-	-	-	2	-
Mixed spices[3]	4	-	-	-	4	-	-	4	2
Nuts									
Pistachios	2	1	1	-	-	-	-	2	-
Pistachios[3]	1	-	-	1	-	1	-	-	-
Hazelnuts	12	10	1	-	1	9	3	-	4
Hazelnuts[3]	3	1	2	-	-	-	3	-	-
Roasted peanuts	1	1	-	-	-	1	-	-	-
Raw peanuts	2	2	-	-	-	1	1	-	1
Flour									
Maize flour	5	1	-	3	1	1	4	-	1
Oatmeal	2	-	2	-	-	1	1	-	-
Barley meal	2	1	-	-	1	-	2	-	1
Buckwheat	1	-	-	-	1	-	-	1	1
Dried fruits									
Figs	1	1	-	-	-	1	-	-	-
Currants	2	1	-	-	1	1	1	-	-

[1, 2, 3] see Table 2.

[4] Number of samples in which species belonging to Mucorales were detected on DG18.

Most spices in Table 3 were fairly heavily contaminated with fungi, and had very high *Mucor* ELISA titres. However, in only 3 of these 20 samples were species of Mucorales found on DG18. Eighteen samples of walnuts had high false positive ELISA titres and are not included. Notermans *et al.* (1988) and Kamphuis *et al.* (1989) reported similar problems with walnuts in studies on immunological detection of *Penicillium* and *Aspergillus* species. On the other hand samples of pistachios, hazelnuts and peanuts showed low ELISA titres in the presence of relatively high total mould counts, including 5 samples in which species of Mucorales could be detected.

Samples of flour showed a good correlation between colony count and ELISA titre for species of Mucorales. In the three flour samples with high (>+++) *Mucor* ELISA titres, species of Mucorales were detected on DG18.

DISCUSSION

In this study, the possibility of using an ELISA technique to detect mould species belonging to Mucorales in food was tested on 161 food samples. The *Mucor heimalis* ELISA was shown to react with 11 species in six genera of Mucorales. Colony counts did not always correlate well with the ELISA test, but DG18 is not well suited to the isolation of Mucorales.

This ELISA method appears to have several advantages compared with other mould detection methods, including the colony count. It was found to be sensitive, rapid and reliable for a number of food products. The earlier occurrence of species belonging to Mucorales can be established using this ELISA assay, even if the mould itself is inactivated or removed by filtration. This was shown in pasteurised juices, in which the *Mucor* ELISA can be used to determine the quality of fruit used as raw materials for the production of these juices. Therefore the ELISA provides information on the history of mould contamination of raw materials, which is important in the food industry.

In flour samples there was a good correlation between the colony counts and the *Mucor* ELISA titre. The use of this ELISA test for samples of walnuts and jams should be further examined, because of the high number of false positive reactions that occurred.

The reliability of the *Mucor* ELISA could be improved by elucidation of the structure of epitopes involved in the immunochemical reaction. Research is in progress to gain more detailed structural information about the immunogenic properties of EPS produced by species of Mucorales.

It can be speculated that the specificity can be extended to a large number of species belonging to Mucorales. It has been assumed in this study that EPS derived from species of Mucorales have common antigenic determinants, which are specific and characteristic for this order.

ACKNOWLEDGEMENTS

Most of the fungi were kindly provided by Dr R.A. Samson, Centraalbureau voor Schimmelcultures, Baarn, the Netherlands. We gratefully acknowledge the Food Inspection Services of Utrecht, Zutphen and Rotterdam, the Netherlands for supplying the food samples tested. The investigations were supported by the Netherlands Foundation for Chemical Research (SON) with financial aid from the Netherlands Technology Foundation (STW).

REFERENCES

DE RUITER, G.A., SMID, P., VAN DER LUGT, A.W., VAN BOOM, J.H., NOTERMANS, S.H.W., and ROMBOUTS, F.M. 1991a. Immunogenic extracellular polysaccharides of Mucorales. *In* Fungal Cell Wall and Immune Responses, eds J.-P.Latgé and D. Boucias, pp. 169-180. NATO ASI Ser. **H 53**. Berlin: Springer Verlag.

DE RUITER, G.A., VAN DER LUGT, A.W., VORAGEN A.G.J., ROMBOUTS, F.M. and NOTERMANS, S.H.W. 1991b. High performance size-exclusion chromatography and ELISA detection of extracellular polysaccharides from Mucorales. *Carbohydr. Res.* **215**: 47-57.

DE RUITER, G.A., JOSSO, S.L., COLQUHOUN., I.J. VORAGEN A.G.J. and ROMBOUTS, F.M. 1992. Isolation and characterization of ß(1-4)-d-glucuronans from extracellular polysaccharides of moulds belonging to Mucorales. *Carbohydr. Polym.* **18**: 1-7.

HOCKING, A.D. and PITT, J.I. 1980. Dichloran-glycerol medium for enumeration of xerophilic fungi from low moisture foods. *Appl. Environ. Microbiol.* **39**: 488-492.

HOUGH, L. and PERRY, M.B. 1955. An extracellular polysaccharide from *Mucor racemosus*. *Biochem. J.* **61**: viii-ix.

JARVIS, B., SEILER, D.A.L., OULD, A.J.L. and WILLIAMS, A.P. 1983. Observations of the enumeration of moulds in food and feedingstuffs. *J. Appl. Bacteriol.* **55**: 325-336.

JONES, K.W. and KAUFMAN, L. 1978. Development and evaluation of an immunodiffusion test for diagnosis of systemic zygomycosis (Mucormycosis). *J. Clin. Microbiol.* **7**: 97-103.

KAMPHUIS, H.J., NOTERMANS, S., VEENEMAN, G.H., VAN BOOM, J.H. and ROMBOUTS F.M. 1989. A rapid and reliable method for the detection of molds in foods: using the latex agglutination assay. *J. Food Prot.* **52**: 244-247.

KAMPHUIS, H.J. and NOTERMANS, S. 1992. Development of a technique for the immunological detection of fungi. *In* Modern Methods in Food Mycology, eds R.A. Samson, A.D. Hocking, J.I. Pitt and A.D. King. pp. 197-203. Amsterdam: Elsevier.

KAUFMAN, L., TURNER, L.F. and McLAUGHLIN, D.W. 1989. Indirect enzyme-linked immunosorbent assay for zygomycosis. *J. Clin. Microbiol.* **27**: 1979-1982.

LIN, H.H. and COUSIN, M.A. 1987. Evaluation of enzyme-linked immunosorbent assay for detection of molds in foods. *J. Food Sci.* **52**: 1089-1096.

MARTIN, S.M. and ADAMS, G.A. 1956. A survey of fungal polysaccharides. *Can. J. Microbiol.* **34**: 715-721.

MIYAZAKI, T. and IRINO, T. 1972. Studies on fungal polysaccharides. X. Extracellular heteroglycans of *Absidia cylindrospora* and *Mucor mucedo*. *Chem. Pharm. Bull.* **20**: 330-335.

MIYAZAKI, T., YADOMAE, T., YAMADA, H., HAYASHI, O., SUZUKI, I. and OSHIMA, Y. 1980. Immunochemical examination of the polysaccharides of Mucorales. *In* Fungal Polysaccharides, eds P.A. Sandford and K. Matsuda, pp. 81-94. *ACS Symp. Ser.* **126**. Washington D.C.: American Chemical Society.

NOTERMANS, S. and HEUVELMAN, C.J. 1985. Immunological detection of moulds in food by using the enzyme-linked immunosorbent assay (ELISA); preparation of antigens. *Int. J. Food Microbiol.* **2**: 247-258.

NOTERMANS, S., HEUVELMAN, C.J., BEUMER, R.R. and MAAS, R. 1986. Immunological detection of moulds in food: relation between antigen production and growth. *Int. J. Food Microbiol.* **3**: 253-261.

NOTERMANS, S. and SOENTORO, P.S.S. 1986. Immunological relationship of extra-cellular polysaccharide antigens produced by different mould species. *Antonie van Leeuwenhoek* **52**: 393-401.

NOTERMANS, S., DUFRENNE, J. and SOENTORO, P.S. 1988. Detection of molds in nuts and spices: the mold colony count versus the enzyme linked immunosorbent assay (ELISA). *J. Food Sci.* **53**: 1831-1833, 1843.

PITT, J.I. and HOCKING A.D. 1985. Fungi and Food Spoilage. Sydney: Academic Press.

POLONELLI, L., DETTORI, G., MORACE, G., ROSA, R., CASTAGNOLA, M. and SCHIPPER, M.A.A. 1988. Antigenic studies on *Rhizopus microsporus*, *Rh. rhizopodiformis*, progeny and intermediates (*Rh. chinensis*). *Antonie van Leeuwenhoek* **54**: 5-18.

EVALUATION OF AN IMMUNOLOGICAL MOULD LATEX DETECTION TEST: A COLLABORATIVE STUDY

Hetty Karman[1] and R. A. Samson[2]

[1]*Food Inspection Services*
Nijennoord 6
3552 AS Utrecht
The Netherlands

[2]*Centraalbureau voor Schimmelcultures*
P.O. Box 273
3740 AG Baarn
The Netherlands

SUMMARY

A latex agglutination test for the detection of *Aspergillus* and *Penicillium* was compared with the standard plate count technique. Seven laboratories tested 500 food samples belonging to eight product groups: fruit juices (117 samples); flour (78); nuts (58); spices (49); low moisture products (53); peanut products (31); marmalades and fruit pulps (84); and cheese (39).

The collaborative study showed that the mould latex agglutination test can be used for a range of food products. With some experience the latex agglutination test can be carried and interpreted without too much difficulty. A relatively good correlation between colony count and the titre of the latex test could be seen in products such as unpasteurised fruit juices and pulp, marmalades and flour. In heat treated products, however, there was no correlation between the latex test and the colony count but with the latex test high titres of non-viable moulds could be detected. For cheese products and walnuts the available latex test was apparently not suitable. Further research is needed for the application of the test kit for products like spices, peanuts and low moisture products.

INTRODUCTION

At the Food Inspection Services in the Netherlands, numerous food samples are tested for fungal contamination, using colony counts obtained by the dilution plating method. Disadvantages of this method are that only viable moulds can be detected, and that the method cannot be used to detect the use of contaminated ingredients in a heated product. Recently a mould latex agglutination test was developed which was a serological method based on detection of an extracellular polysaccharide (EPS) produced by various moulds (Kamphuis *et al.*, 1989). In this test, when latex particles coated with specific antibodies raised against EPS of *Aspergillus* and *Penicillium* are mixed with a sample containing mould EPS, large aggregates will form, resulting in visible clumping. A control for false positives is performed by mixing the sample with latex particles coated with antibodies of which the specific binding places are blocked. With the latex test viable and nonviable moulds are detected whereas the plate count technique only detects viable moulds. The aim of this study was to compare the results achieved by the latex test and those of the standard plate count technique in an interlaboratory study using 500 food samples.

MATERIALS AND METHODS

Seven laboratories tested 500 food samples belonging to eight product groups: fruit juices (117 samples); flour (78); nuts (58); spices (49); low moisture products (53); peanut products (31); marmalades and fruit pulp (84); and cheese (39). Samples were collected at local stores in the Netherlands: fruit juices and flour came from health food stores and the other samples were bought in supermarkets.

For the tests, a 10 g sample was weighed into a sterile Stomacher bag, diluted 10 fold with Phosphate Buffered Saline (PBS; pH 7.4) and homogenised for 1 minute. From this dilution (10^{-1}) further decimal dilutions (10^{-2} and 10^{-3}) were prepared. The latex test kit was obtained from Holland Biotechnology BV, Leiden, Netherlands, and was carried out by the procedure described in the manufacture's instructions. Colony counts were carried out using DG18 agar (Pitt and Hocking, 1985), using 0.1 mL aliquots of the 10^{-1} and 10^{-2} dilution, surface plated. Plates were incubated at 24°C for 3 to 5 days, then colonies were counted, and species types determined.

RESULTS AND DISCUSSION

Correlations between colony counts and latex agglutination titres are set out in Table 1.

Fruit juices
The 117 fruit juice samples showed mainly negative results with the colony count technique, to be expected for pasteurised products. Some apple juices had a titre of 10, while a blackcurrant, and a grape and raspberry had high titres. These fruits are easily damaged during harvest and transport and therefore likely to become mouldy.

Nuts
No correlation was found between the results of the latex agglutination test and colony counts of 58 nut samples. Nine samples with a titre of <10 and a high colony count (10^3-10^4) were all walnuts. These products gave false negative results (see also Kamphuis *et al.*, 1989) and therefore the latex test is not suitable for this product group.

Flour
The 78 flour samples included wheat, buckwheat, corn, rice, oatmeal and barley products. Overall a fair correlation was found between the results of the latex test and the colony count. High titres and low colony counts, however, may occur with products that have received a heat treatment. Some samples obtained from health food stores showed a titre of <10 and colony count $>10^3$. One sample of wheat flour had a colony count between 10^2 and 10^3 but was negative with the latex test.

Spices
Thirty four of 49 spice samples showed titres of 100 or higher so it may be concluded that spices are heavily contaminated products. Mould colony counts were variable because spices are often treated by gamma irradiation. Seven samples had a high *Penicillium* and *Aspergillus* colony count and no titre. These were samples of nutmeg (2), paprika (2), pepper (2) and cinnamon (1).

Low moisture products
Most of the 53 low moisture samples were dried fruit (e.g. figs, dates, raisins and apricots) and had a low count and no titre. The results were similar to those obtained with

Table 1. Correlation between colony count and latex agglutination titres for various food products

Food	Count/g	Titres			
		<10	10	100	1000
Fruit juices	$<10^2$	81	24	10	1
	10^2-10^3	0	0	1	0
	10^3-10^4	0	0	0	0
	$>10^4$	0	0	0	0
Nuts	$<10^2$	26	1	0	0
	10^2-10^3	5	9	0	1
	10^3-10^4	9	1	3	1
	$>10^4$	0	0	1	1
Flour	$<10^2$	8	7	6	1
	10^2-10^3	4	19	5	2
	10^3-10^4	1	7	9	3
	$>10^4$	0	1	2	3
Spices	$<10^2$	5	3	8	7
	10^2-10^3	0	0	2	1
	10^3-10^4	5	0	3	7
	$>10^4$	2	0	2	4
Low moisture products	$<10^2$	43	1	1	2
	10^2-10^3	2	1	1	2
	10^3-10^4	0	0	0	0
	$>10^4$	0	0	0	0
Peanuts; peanut products	$<10^2$	17	7	4	0
	10^2-10^3	1	0	1	0
	10^3-10^4	0	1	0	0
	$>10^4$	0	0	0	0
Marmalades	$<10^2$	55	23	5	0
	10^2-10^3	0	0	1	0
	10^3-10^4	0	0	0	0
	$>10^4$	0	0	0	0
Cheese; cheese products	$<10^2$	21	10	1	2
	10^2-10^3	2	0	1	0
	10^3-10^4	1	1	0	0
	$>10^4$	0	0	0	0

fruit juices and marmalades. Two samples of plums and raisins showed a colony count of $<10^2$ and a titre of 1000.

Peanuts and peanut products

No viable moulds were detected in 31 samples of peanuts and peanut products, except for one sample of unroasted peanuts.

Marmalades and fruit pulp

Sterilised products like marmalades contained no viable moulds. In fruit pulp, low numbers of moulds, mainly *Neosartorya* and *Eupenicillium* species were present, which were detected by both methods. Counts between 10 and 100 and titres of 10 were found.

Cheese

Most of the 39 cheese samples were grated hard cheese. In contrast to what might be expected, low titres and low colony counts were found. In these experiments cheese samples proved to be a difficult product, for using the latex test. In preliminary experiments carried out in one laboratory, 10 samples of heavily contaminated cheese were tested with a mould ELISA test using the same type of EPS antibody described here. Mould counts varied from 10^5 to 10^7 and only low titres (100) were found.

CONCLUSIONS

The mould latex agglutination test can be used to detect moulds in a wide variety of food products. With some experience the test can be carried out and interpreted without difficulty. A relatively good correlation between colony count and the titre of the latex test were found in products such as unpasteurised fruit juices and pulp, marmalades and flour. In heat treated products, however, nonviable remnants of *Penicillium* and *Aspergillus* can be detected. The latex test does not appear to be suitable for cheese products and nuts, and this is true to a lesser extent for spices, peanuts and low moisture products. Further research on the use of the test for these products will be necessary.

ACKNOWLEDGEMENTS

This interlaboratory study was a collaborative survey of the Netherlands Working Group on Moulds. The following members participated: Food Inspection Services at Amsterdam, Utrecht (H. Karman), Groningen (A. Dolfing), Enschede (M. van de Broek), Rotterdam (J. Besling), Zutphen (E. de Boer), Agricultural University at Wageningen (R. Nout), CIVO-TNO (B. Hartog) and Centraalbureau voor Schimmelcultures (M.I. van der Horst). Holland Biotechnology (Leiden, Netherlands) kindly donated some of the test kits used in this study.

REFERENCES

KAMPHUIS, H.J., NOTERMANS, S., VEENEMAN, G.H., VAN BOOM, J.H. and ROMBOUTS, F.M. 1989. A rapid and reliable method for the detection of molds in foods: using the latex agglutination assay. *J. Food Prot.* **52**: 244-247.

PITT, J.I. and HOCKING, A.D. 1985. Fungi and Food Spoilage. Sydney: Academic Press.

IMMUNOAGGLUTINATION ASSAY FOR RAPID DETECTION OF *ASPERGILLUS* and *PENICILLIUM* CONTAMINATION IN FOOD

N. Braendlin and L. Cox

Quality Assurance Department, Nestec Limited
Avenue Nestlé 55
CH-1800 Vevey
Switzerland

SUMMARY

This study was undertaken to evaluate a Mould Latex Agglutination Test kit (MLAT) (Holland Biotechnology B.V. Leiden, the Netherlands), for the analysis of various raw and processed dry foods. MLAT agglutination titres of 160 samples were compared with fungal plate counts on 118 samples. A second immunoassay, the Pastorex *Aspergillus* test (PAT; Eco-Bio Diagnostic Pasteur, Genk, Belgium) was tested on 50 samples. Aflatoxins B and G were measured chemically in 50 samples. Thirty six of the processed food samples with counts of less than 10 cfu/g were examined. In 14 of these no agglutination was observed but in the remaining 22, agglutination titres in the range of 1/10 to 1/7290 were detected. This shows that the agglutination test may detect samples that could have had significant contamination prior to processing. Sixty eight samples of raw materials showed an approximate positive correlation of MLAT titre to mould count. In eight samples counts of 20-1000 cfu/g did not give detectable agglutination titres even though *Aspergillus* and *Penicillium* were the dominant genera isolated. A relation was observed between both MLAT and PAT agglutination titres and aflatoxin levels. In samples of corn, peanuts and cottonseed, an MLAT agglutination titre of 1/90 corresponded to a maximum aflatoxin contamination of 3.0 μg/kg, while a titre of 1/810 corresponded to a maximum of 72 μg/kg. Positive agglutination titres for PAT were usually 3 times higher than those of MLAT. Because latex agglutination tests indirectly determine both dead and viable fungal biomass, they have a very useful application in the general examination of foods for *Aspergillus* and *Penicillium* contamination and in preliminary screening for the presence of aflatoxin.

INTRODUCTION

Aspergillus and *Penicillium* species are implicated in both food spoilage and mycotoxin formation (Pitt and Hocking, 1985; Smith and Moss, 1985). This is mainly due to inadequate drying and storage of raw materials or products. Traditionally, examination for fungal contamination in foods is performed by the plate count method. This has some drawbacks, e.g. the necessity for incubation for 5 days, and the preparation of media used in counts (King *et al.*, 1986). The mould plate count evaluates viable propagules at some point in time. However, propagules may lose their viability during food storage or after minimal food processes, so a viable propagule count can give no indication of the content of dead fungal biomass. The evaluation of both dead and viable fungal biomass gives a better indication of the extent to which growth may have occurred and thus the possibility of mycotoxin contamination, off flavour or other defects.

Immunological methods developed recently have provided a specific and sensitive way of estimating fungal biomass contamination (Notermans *et al.*, 1986a; 1986b; 1988a). The Mould Latex Agglutination Test (MLAT) (Kamphuis *et al.*, 1989a) and the Pastorex *Aspergillus* Test (PAT), first developed for medical purposes (Dupont *et al.*, 1990) are the first kits to be commercialy available. They are both based on the detection, by immuno-agglutination, of the same antigenic determinant, a specific part of galactomannan, an extracellular polysaccharide (EPS) common to both *Aspergillus* and *Penicillium* species (Notermans *et al.*, 1988b; Kamphuis *et al.*, 1989b). EPS are water soluble and heat-stable and levels in these fungi correlate well with mycelium weight (Notermans *et al.*, 1986b; 1987).

In previous studies (unpublished results; Stynen *et al.*, 1992), it was noticed that culture filtrates of various fungi isolated from food also gave an agglutination with the two kits. These were teleomorph species closely related to *Aspergillus* and *Penicillium* (e.g. *Eurotium*, *Emericella*, *Neosartorya*, *Talaromyces*, *Byssochlamys* species), *Paecilomyces* and *Wallemia sebi*. Other genera agglutinated only PAT, e.g. *Cladosporium*, *Trichoderma*, and *Aureobasidium pullulans*. The antibody sensitivity to EPS of these species is different from that of *Aspergillus* and *Penicillium* (Stynen *et al.*, 1992), indicating that the kits can be used for the general examination of foods for storage fungi.

The relation between EPS titres (mesured by ELISA), growth and mycotoxin production for two *Penicillium* species has been studied (Notermans *et al.*, 1986b), as have EPS titres and aflatoxin B_1 levels in naturally contaminated samples (Notermans *et al.*, 1986a). The aim of this study was to interpret the agglutination titres obtained with the MLAT kit in food products compared with the PAT agglutination kit, the mould plate count method, and the level of aflatoxin in naturally contaminated foods.

MATERIALS AND METHODS

Aspergillus and *Penicillium* Latex Immuno-agglutination tests

The Mould Latex Agglutination Test (MLAT), was obtained from Holland Biotechnology B.V. Leiden, the Netherlands, and the Pastorex *Aspergillus* Test (PAT) was obtained from Eco-Bio Diagnostics Pasteur, Genk, Belgium. The PAT was first developed for medical purposes. The sensitivity of each kit is about 15 ng/mL of EPS. With MLAT 10 μL of sample are tested with 10 μL of reagent and for PAT 40 μL of sample with 10 μL of reagent.

The kits were used according to the suppliers' instructions. Food samples (10 g) were blended in 90 mL PBS (0.06M, pH 7.2) and then centrifuged for 10 min at 10,000 g. If necessary, the supernatant was filtered on a paper filter (Schleicher and Schull, Dassle, Germany). Series of 3-fold dilutions from 1/10 to 1/7290 were then prepared from each supernatant and tested with anti-EPS coated latex and negative controls to find the highest dilution that gave an agglutination result. Negative samples were retested after the addition of a positive control (a solution of EPS) to ensure that sample constituents did not interfere with the agglutination reaction.

Food products tested

A total of 160 food samples of low a_w and from various origins were analysed (Table 1). Fungal counts, MLAT titres and PAT (only at a dilution of 1/10 to provide qualitative results) were performed on 118 of these samples. Fifty samples were analysed for aflatoxin levels and MLAT and PAT titres.

Enumeration of fungi

Colony counts were performed on DG18 medium (CM729, Oxoid Ltd, Basingstoke). Samples (10 g) were soaked in 90 mL of Tryptone Salt (tryptone 1 g/L, L42, Oxoid plus 8.5 g/L NaCl) for 30 min. They were then blended for 2 minutes using a Stomacher, or a Waring blender for solid samples. Plates were inoculated by spreading, examined after 5 days at 26°C and re-examined after 7 days. Mould colonies were identified directly under the stereomicroscope to genus level if possible or after isolation and growth on standard identification media (Pitt and Hocking, 1985). *Aspergillus* and *Penicillium* were considered to be the dominant genera if their count exceeded other genera on plates with between 10 and 50 colonies.

Aflatoxin determination

Aflatoxins B_1, B_2, G_1 and G_2 were determinated by the official Swiss method (Anon., 1982) using 50 g samples.

RESULTS

MLAT agglutination titres

The distribution of agglutination titres for the samples tested is summarised in Table 1. Agglutinations were clearly detected in 132 samples of the 160 tested. The highest agglutination titres observed corresponded to EPS levels >109 μg/mL. The reproducibility of the test was good and no samples showed non-specific agglutination reactions.

Table 1. Distribution of food samples showing a positive Mould Latex Agglutination Test at given dilutions

Foods	Total samples	nd[1]	1/10	1/30	1/90	1/270	1/810	1/2430	1/7290
					Highest positive dilution				
Cereals	49	5	5	11	9	5	9	3	2
Pulses	11			3	2	3	3		
Oil seeds and nuts	30	9	5	5	1	4	1	3	2
Spices	10			1	3	2	2	2	
Herbs	5	1	2	1	1				
Dried mixed prep'ns	12	7	1	2		1		1	
Dry fruits and jams	10	3		1	1	2		3	
Edible mushrooms	4	2	1					1	
Cocoa products	6		1	1		4			
Green coffee beans	10	1	3			2	3	1	
Cottonseed presscake	13				1	1	3	7	1
Total	160	28	18	25	18	24	21	21	5

[1] nd, not detected

Comparison of agglutination by MLAT and PAT at 1/10 dilutions
Of the 160 samples tested, 132 were positive with both the kits. Of 28 samples negative with MLAT, 24 were positive using PAT. Four samples were negative by both tests: one each of buckwheat, rice, milk powder and strawberry jam.

Comparison of MLAT with colony counts
The data for MLAT versus colony counts on DG18 are presented in Figure 1. Samples were distributed in 4 areas as indicated in the figure. Areas 1 and 4 included 82 samples with fungal counts greater than 10 cfu/g. These were mainly raw material samples. *Aspergillus* and *Penicillium* were the dominant genera in 74 of these samples. Yeasts were dominant in three samples and *Fusarium* spp. in five. Areas 2 and 3 included samples with a fungal count less than 10 cfu/g. These 36 samples were processed foods.

Areas 1 and 2 included samples that gave an agglutination for the 1/10 dilution and higher. Areas 3 and 4 included samples without agglutination at a 1/10 dilution.

Comparison of aflatoxin levels with MLAT and PAT titres
The results of aflatoxin determinations, MLAT and PAT agglutination titres are presented in Table 2. Aflatoxins were detected in 36 samples. Levels of aflatoxin compared to agglutination titres for the 36 samples contaminated with detectable levels of aflatoxins are reported in Figure 2 for MLAT and Figure 3 for PAT. For all samples, the highest positive dilution was greater for PAT than MLAT and in most samples the PAT titres were 3 times higher.

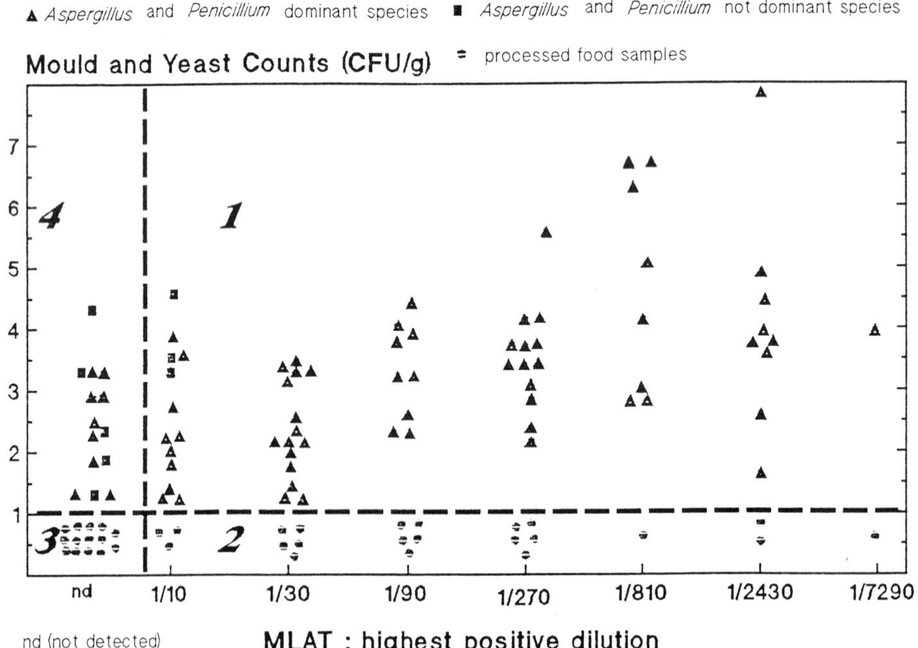

Figure 1. Fungal count vs Mould Latex Agglutination Test

Table 2. Comparison of aflatoxin assays with immuno-agglutination titres

Foods	Total aflatoxins, μg/kg[1]	Highest positive dilution MLAT	PAT
Cottonseed presscake (13 samples)	2.6	1/270	1/810
	2.9	1/90	1/810
	10.1	1/810	1/7230
	10.5	1/2430	>1/7230
	18.4	1/2430	1/7230
	18.8	1/2430	1/7230
	21.9	1/2430	1/7230
	26.8	1/2430	1/7230
	27.1	1/2430	1/7230
	28.2	1/2430	1/7230
	28.6	1/2430	1/7230
	38.7	1/2430	1/7230
	82.0	1/7290	>1/7230
Corn (12 samples)	n.d.	1/810	1/2430[2]
	0.6	1/270	1/810
	0.8	1/270	1/810
	0.9	1/270	1/810
	8.4	1/810	1/2430
	10.3	1/810	1/2430
	10.5	1/2430	1/7290
	33.0	1/810	>1/7230
	38.0	1/810	1/2430
	72.0	1/810	1/2430
	75.0	1/7230	>1/7230
Peanuts (7 samples)	n.d.	1/10	1/30
	n.d.	1/30	1/90[2]
	60.0	1/2430	1/7290
	70.0	1/2430	1/7290
	102.0	1/7290	>1/7290
	112.0	1/7290	>1/7290
Soy beans	n.d.	1/270	1/810
	0.5	1/810	1/2430
	1.6	1/270	1/810
	2.9	1/810	1/7290
Rice	0.8	1/2430	1/7290
	1.0	1/810	1/2430
	3.0	1/810	1/2430
Wheat	n.d.	1/30	1/270[3]
	n.d.	1/90	1/810
	0.1	1/90	1/810
Pistachios	n.d.	1/30	1/270
	n.d.	1/90	1/270
	60.7	1/7290	>1/7290
Hazelnuts	n.d.	1/10	1/90
Almonds	n.d.	1/10	1/270
Carob flour	0.1	1/10	1/30

[1] n.d., not detected
[2] Two samples gave this result
[3] Three samples gave this result

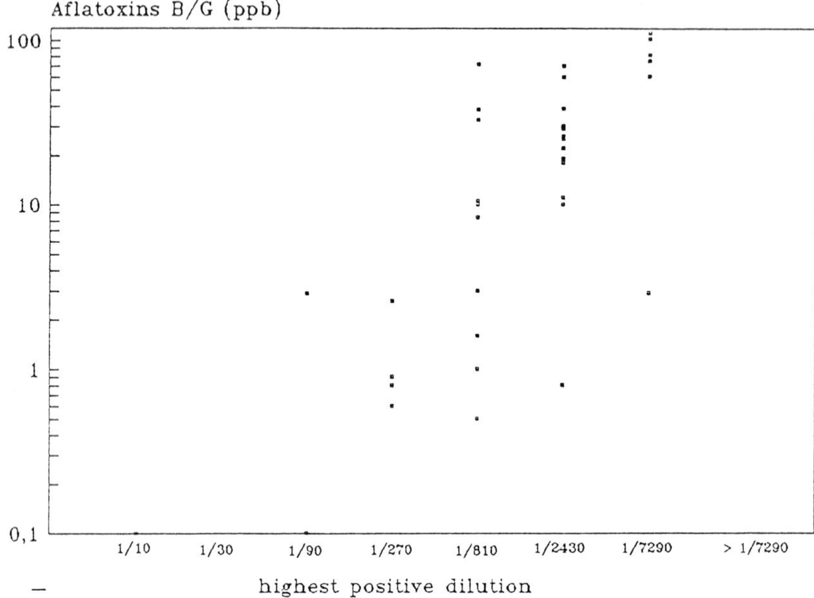

Figure 2. Aflatoxin levels vs MLAT titres in naturally contaminated samples

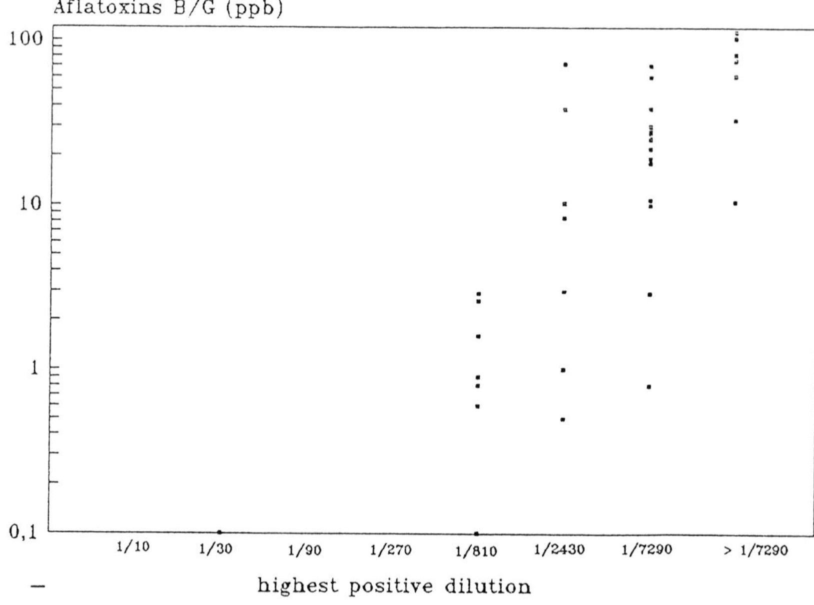

Figure 3. Aflatoxin levels vs PAT titres in naturally contaminated samples

DISCUSSION

Results reported here show that EPS levels can vary considerably in foods. In areas 1 and 3 of Figure 1, counts and agglutination titres are to some extent positively correlated, but in other areas these methods of estimating fungal growth exhibit no relationship.

In area 4, some samples gave no agglutination with MLAT despite the fact that approximately 1000 cfu/g were detected. These samples (walnut pieces, ground hazelnuts and green coffee beans mainly contaminated with *Penicillium* spp., and ground almonds contaminated with *Aspergillus niger*), gave an agglutination with PAT. These results are probably due to the fact that more species agglutinate PAT than MLAT (Stynen *et al.*, 1992). Although differences have been shown here between the MLAT and PAT kits, the aim of this work was not to compare them. We included the PAT kit in order to give some indication of how to interprete the results from MLAT.

Area 2 shows the stability of EPS in products which have undergone heat treatment, irradiation (five samples of spices) or other processes which have killed the fungi. These results show that agglutination titres and fungal counts measure two different properties and cannot be used for the same purpose. MLAT is particularly suitable for checking product history. Data pertaining to the normal levels of contamination expected in foods would be required before the test could be applied rationally.

The results of the comparison of aflatoxin levels to MLAT and PAT agglutination titres show that samples with high agglutination titres can be suspected to contain aflatoxins, particularly in susceptible crops like peanuts, corn and cottonseed. Samples positive for the 1/90 dilution appear to be at the critical point. When the 1 /810 dilution was positive, as high as 72 μg/kg aflatoxin were detected (Figure 2). With PAT the critical range of agglutination titres indicative of aflatoxin presence was between 1/810 and 1/2430 (Figure 3). However, high agglutination titres can also be attributed to contamination by *Aspergillus flavus* and *Aspergillus parasiticus* under conditions not necessarily conducive to the production of aflatoxin, or to contamination by other species, e.g. high agglutination titres in corn, soya and rice (Table 2). Overall, MLAT titres were lower than those of PAT, but the results are comparable as trends are similar for both systems.

These results show that high EPS agglutination titres in the samples indicate that high fungal counts are likely when samples have not been treated by a preservation process, heat, irradiation or chemicals. This high agglutination titre may also indicate the presence of mycotoxins produced by *Aspergillus* and *Penicillium* species. No agglutination reaction, or low agglutination titres in the range 1/10 to 1/30, indicate a low *Aspergillus* and *Penicillium* biomass, suggesting good handling practice and high quality products.

The MLAT method has been shown to be rapid and easy to use. For low a_w products, raw or processed samples such as cereals, nuts and oil seeds, agglutination titres can be used as an initial semi-quantitative method to check product quality. The MLAT and PAT kits can also be used for screening raw materials for presumptive mycotoxin contamination so that only suspect samples need to be further analysed chemically. This will probably be the most important application of agglutination kits.

ACKNOWLEDGEMENTS

We would like to thank A. Pittet and his collaborators of Nestec Quality Assurance Laboratory for the aflatoxin analyses. The Pastorex *Aspergillus* test was provided by D. Stynen, Eco-Bio Diagnostics Pasteur.

REFERENCES

ANON. 1982. Bestimmung der Aflatoxine B_1, B_2, G_1 und G_2. Arbeitsgruppe "Toxine" 2 der Eidgenossischen Lebensmittelbuch-Kommission. *Mitt. Geb. Lebensm. Hyg.* **73**: 362-367.

DUPONT, B., IMPROVISI, L. and PROVOST, F. 1990. Detection de galactomannane dans les aspergilloses invasives humaines et animales avec un test au latex. *Bull. Soc. Fr. Mycol. Med.* **1**: 35-42.

KAMPHUIS, H.J., NOTERMANS, S., VEENEMAN, G.H., VAN BOOM, J.H. and ROMBOUTS, F.M. 1989a. A rapid and reliable method for the detection of molds in foods: using the latex agglutination assay. *J. Food Prot.* **52**: 244-247.

KAMPHUIS, H.J., VEENEMAN, G.H., ROMBOUTS, F.M., VAN BOOM, J.H. and NOTERMANS, S. 1989b. Antibodies against synthetic oligosaccharide antigens reactive with extracellular polysaccharides produced by moulds. *Food Agric. Immunol.* **1**: 235-242.

KING, A.D., PITT, J.I., BEUCHAT, L.R. and CORRY, J.E.L., eds. 1986. Methods for the Mycological Examination of Food. New York: Plenum Press.

NOTERMANS, S., HEUVELMAN, C.J., VAN EGMOND, H.P., PAULSCH, W.E. and BESLING, J.R. 1986a. Detection of mould in food by enzyme linked immunosorbent assay. *J. Food Prot.* **49**: 786-791.

NOTERMANS, S., HEUVELMAN, C.J., BEUMER, R.R. and MAAS, R. 1986b. Immunological detection of moulds in food: relation between antigen production and growth. *Int. J. Food Microbiol.* **3**: 253-261.

NOTERMANS, S., WIETENS, S., ENGEL, H.W.B., ROMBOUTS, F.M., HOOGERHOUT, P. and VAN BOOM, J.H. 1987. Purification and properties of extracellular polysaccharide (EPS) antigens produced by different mould species. *J. Appl. Bacteriol.* **62**: 157-166.

NOTERMANS, S., DUFRENNE, J. and SOENTORO, P.S. 1988a. Detection of molds in nuts and spices: the mold colony count versus the enzyme linked immunosorbent assay (ELISA). *J. Food Sci.* **55**: 1831-1833.

NOTERMANS, S., VEENEMAN, G.H., VAN ZUYLEN, C.W.E., HOOGERHOUT, P. and VAN BOOM, J.H. 1988b. (1-5)-linked ß-D-Galactofuranosides are immunodominant in extracellular polysaccharides of *Penicillium* and *Aspergillus* species. *Molec. Immunol.* **25**: 975-979.

PITT, J.I. and HOCKING, A.D. 1985. Fungi and Food Spoilage. Sydney: Academic Press.

SMITH, J.E. and MOSS, M.O. 1985. Mycotoxins - Formation, Analysis and Significance. Chichester: John Wiley and Sons.

STYNEN, D., MEULEMANS, L., GORIS A., BRAENDLIN, N. and SYMONS, N. 1992. Characteristics of a latex agglutination test based on monoclonal antibodies for the detection of fungal antigen in foods. *In* Modern Methods in Food Mycology, eds R.A. Samson, A.D. Hocking, J.I. Pitt and A.D. King. pp. 213-219. Amsterdam: Elsevier.

COMPARISON OF TWO COMMERCIAL KITS TO DETECT MOULDS BY LATEX AGGLUTINATION

Marjolein van der Horst[1], R. A. Samson[1] and Hetty Karman[2]

[1]*Centraalbureau voor Schimmelcultures*
P. O. Box 273
3740 AG Baarn
The Netherlands

[2]*Food Inspection Service*
Nijenoord 6
3552 AS Utrecht
The Netherlands

SUMMARY

Two commercial latex agglutination kits were used for detecting *Penicillium* and *Aspergillus* antigens in 59 food samples including spices, herbs, flour and tomato products selected at local stores. For comparison viable moulds were detected qualitatively on Dichloran 18% Glycerol (DG18) agar plates. Agglutination could be observed effectively by the naked eye, but a dissecting microscope is recommended for visualisation of the agglutination of one kit. The Mould Reveal Kit (Eco-Bio, Genk, Belgium) showed reactions with *Cladosporium*, *Botrytis* and *Wallemia* as well as with *Penicillium* and *Aspergillus*. For processed and mostly pasteurised tomato products, the Mould Reveal Kit gave positive results whereas the HBT kit (Holland Biotechnology, Leiden, Netherlands) was negative.

INTRODUCTION

This study was undertaken to compare two commercial latex agglutination kits for detecting moulds in foods. The HBT kit (Holland Biotechnology, Leiden, Netherlands) uses latex beads sensitised with polyclonal antibody raised against the extracellular polysaccharide (EPS) of *Penicillium digitatum* (Kamphuis *et al.*, 1989; Kamphuis and Notermans, 1992), while the latex beads of the Mould Reveal Kit (Eco-Bio, Genk, Belgium) are sensitised with monoclonal antibody raised against EPS of *Aspergillus fumigatus* (Stynen *et al.*, 1992). This kit was initially developed for medical purposes, for detecting aspergillosis. Both test kits include a control system which detects false positive results, based on blocking the reactive sites of the antibody so that no reaction can occur with the specific antigen. This study compared these kits in terms of sensitivity and ease of handling.

MATERIALS AND METHODS

Fifty nine food samples including spices, herbs, flour and tomato products were selected at local stores. Sensitivity of the HBT kits was 10 ng/ml *Aspergillus*/*Penicillium* galactomannan while the Eco-Bio sensitivity 15-20 ng/ml *Aspergillus*/*Penicillium* galactomannan.

For both kits, samples were prepared for testing as follows: sample (10 g) was weighed into a sterile Stomacher bag, diluted 10 fold with 0.07 M phosphate buffered saline (PBS, pH 7.4) and homogenised for 1 minute. Where necessary samples were centrifuged and the liquid fraction tested. From this dilution (10^{-1}) further decimal dilutions (10^{-2} and 10^{-3}) were prepared, also in PBS. All dilutions were made with ultra pure water. To avoid false negative reactions, 2 % bovine serum albumin was added to the 10^{-1} dilution. False negative reactions can occur when samples, especially cheese and nuts, contain a lot of free fatty acids or proteins. The kits and samples were equilibrated at room temperature before testing.

The latex test was carried out according to the manufacturer's instructions. The incubation time of the HBT test was 15 minutes and the Eco-Bio test 5 minutes. The results can be visualised with the naked eye but a dissecting microscope is recommended with the HBT kit. Negative samples were retested with the positive control (EPS solution) supplied with the HBT kit.

For mould counts, samples (10 g) were weighed into a Stomacher bag, homogenised for 1 minute and diluted 10 fold with peptone (0.1%) sodium choride (0.85%) solution. Aliquots (0.1 mL) were surface plated on Dichloran 18% Glycerol (DG18) agar plates and incubated at 25°C for 5 d. Colony counts were then carried out, and moulds were identified at least to genus level.

Table 1. Latex agglutination titres and detection of *Aspergillus* and *Penicillium* in tomato products

Products	Titre HBT[1]	Titre MRK[1]	*Penicillium* and *Aspergillus* on DG18
Tomato juice 1	<10	(±)100	-
Tomato juice 2	<10	100	-
Tomato juice 3	<10	10	-
Tomato juice 4	<10	10	+
Tomato juice 5	<10	10	+
Tomato juice 6	<10	<10	-
Tomato juice 7	10	10	-
Tomato pulp 1	10	(±)100	-
Tomato pulp 2	<10	100	-
Peeled tomatoes 1	<10	10	-
Peeled tomatoes 2	<10	10	-
Peeled tomatoes 3	<10	(±)10	-
Peeled tomatoes 4	<10	10	+
Peeled tomatoes 5	<10	1000	-
Tomato purée 1	10	100	-
Tomato purée 2	10	100	-
Tomato purée 3	10	10	-

[1]HBT = Holland Biotechnology BV, Leiden, Netherlands; MRK = Mould Reveal Kit (Eco-Bio, Genk, Belgium)

RESULTS AND DISCUSSION

The results of the latex tests and mould detection on DG18 are summarised in Tables 1-3. The Mould Reveal kit gave positive results for 57 of the 58 food samples tested while HBT was positive for 42. The HBT kit was mostly negative for the tomato products. Positives with the Mould Reveal kit may be due to cross reaction with fungi other than *Aspergillus* or *Penicillium*. Colony counts were negative for most of the tomato products as these are usually heat treated or irradiated.

Table 2. Latex agglutination titres and detection of *Aspergillus* and *Penicillium* in various products

Product	Titre		*Penicillium* and *Aspergillus* on DG18
	HBT[1]	MRK[1]	
Basil	1000	1000	+
Celery leaf	100	10	-
Chives	10-1000	10	-
Coriander	100	1000	+
Curry	1000	1000	+
Curry mix	100	1000	+
Ginger root	1000	1000	+
Bay leaf	< 10	100	ND[2]
Bay leaf	1000	1000	+
Mace	1000	10000	+
Mace	1000	10000	+
Marjoram	100	100	+
Nutmeg	1000	1000	+
Nutmeg	1000	1000	ND
Nutmeg	1000	1000	+
Nutmeg (whole)	1000	10000	+
Oregano	100	1000	+
Parsley	10	10	-
Savory	10	100	+
Chicken spice	1000	1000	+
Salad spices	100	1000	+
Cloves	1000	1000	ND
Cloves	1000	1000	+
Mixed spice	100	100	+
Sweet pepper	1000	10000	-
Thyme	1000	1000	+

[1]HBT = Holland Biotechnology BV, Leiden, Netherlands; MRK = Mould Reveal Kit (Eco-Bio, Genk, Belgium)
[2]ND = no moulds detected

Both test kits give high titres with spices and herbs. For some samples, the HBT kit showed agglutination in the negative control (see also Notermans and Kamphuis, 1992). The herbs and spices contained a variety of fungi belonging to *Aspergillus, Penicillium, Eurotium*, and even *Fusarium*, Zygomycetes and yeasts.

In conclusion, both kits performed satisfactorily for various food products. Technicans working in laboratories for food quality control can readily carry out the tests.

Table 3. Latex agglutination titres and detection of *Aspergillus* and *Penicillium* in various products

Product	Titre		*Penicillium* and *Aspergillus* on DG18
	HBT[1]	MRK[1]	
Peanut product	100	100	ND[2]
Peanut product	100	100	ND
Peanut product	100	100	ND
Peanut product	100	100	ND
Peanut product	100	100	ND
Peanut product	100	100	ND
Peanut product	100	100	ND
Peanut product	100	100	ND
Hazelnuts	10	100	+
Apple sauce	<10	10	ND
Apple sauce	<10	10	ND
Apple sauce	<10	10	ND
Flour	10	10	+
Flour	100	100	+
Flour	10	100	+

[1]HBT = Holland Biotechnology BV, Leiden, the Netherlands; MRK = Mould Reveal Kit (Eco-Bio, Genk, Belgium)
[2]ND = no moulds detected

ACKNOWLEDGEMENT

Kits were kindly provided by Holland Biotechnology, (Leiden, Netherlands) and Eco-Bio (Genk, Belgium).

REFERENCES

KAMPHUIS, H. AND NOTERMANS, S. 1992. Development of a technique for the immunological detection of fungi. *In* Modern Methods in Food Mycology, eds R.A. Samson, A.D. Hocking, J.I. Pitt and A.D. King. pp. 197-203. Amsterdam: Elsevier.

KAMPHUIS, H.J., NOTERMANS, S., VEENEMAN, G.H., VAN BOOM, J.H. and ROMBOUTS, F.M.

1989. A rapid and reliable method for the detection of molds in foods: using the latex agglutination assay. *J. Food Prot.* **52**: 244-247.

NOTERMANS S. and KAMPHUIS, H. 1992. Detection of fungi in foods by latex agglutination: a collaborative study. *In* Modern Methods in Food Mycology, eds R.A. Samson, A.D. Hocking, J.I. Pitt and A.D. King. pp. 205-212. Amsterdam: Elsevier.

STYNEN, D., MEULEMANS, L., GORIS, A., BRAENDLIN, N. AND SYMONS, N. 1992. Characteristics of a latex agglutination test, based on monoclonal antibodies, for the detection of fungal antigen in foods. *In* Modern Methods in Food Mycology, eds R.A. Samson, A.D. Hocking, J.I. Pitt and A.D. King. pp. 213-219. Amsterdam: Elsevier.

TOWARDS THE IMMUNOLOGICAL DETECTION OF FIELD AND STORAGE FUNGI

J. N. Banks, Sarah J. Cox, J. H. Clarke, R. H. Shamsi and Beverley J. Northway

MAFF Central Science Laboratory
London Road
Slough, Berks., SL3 7HJ
UK

SUMMARY

Polyclonal antibodies were raised in rabbits using as antigen the field fungus *Fusarium culmorum* and the storage fungi *Eurotium amstelodami* and *Penicillium aurantiogriseum* var. *melanoconidium*. Sera taken 14 weeks after the first immunisations were tested for cross-reactivity by means of an indirect ELISA. The antigen comprised pure cultures of 10 field fungi and 27 storage fungi.

Some field fungi, such as *Aureobasidium pullulans, Cladosporium herbarum, Alternaria alternata* and *Hyalodendron* sp. showed little or no reaction with any of the antibodies. However, the four *Fusarium* species showed strong cross reactivity towards all three antisera.

Of the storage fungi, *Wallemia sebi* was notable in not reacting with any of the three antisera, while others such as *Apiospora montagnei, Aspergillus flavus, A. terreus, Byssochlamys nivea, Mucor racemosus* and *Scopulariopsis brevicaulis* gave low to moderate levels of cross reactivity. However, other storage fungi in *Aspergillus* and *Penicillium*, the four common species of *Eurotium, Monascus ruber* and *Trichothecium roseum* all cross-reacted strongly with at least one but often two or three of the antisera.

The wide cross reactivity shown by these polyclonal antibodies makes them suitable for use in an ELISA to estimate total fungal biomass. These antibodies could also be applied, possibly in conjunction with monoclonal antibodies also being produced at this laboratory, in more specific immunoassays.

INTRODUCTION

Mycological examination of foods and feeds, still requires the well recognised techniques of direct plating or dilution plating, followed by incubation for 5 to 7 days or longer until the fungi present have formed sporulating colonies that can be examined and identified by eye or by microscope. This is currently the only approach if a wide range of fungi have to be identified to species. However, since about 1964 (Beno and Allen, 1964) workers have attempted an immunological approach in which the fungi are used as antigens to induce an immune response in an animal and antibodies are produced. These antifungal antibodies can then be used in a number of different ways to estimate or identify the original antigenic fungus in the commodity itself.

In a review of the immunological detection of fungi in plants, including stored cereals, Clarke *et al.*, (1986) reported that no standardisation existed for producing antigens, polyclonal antibodies far outweighed monoclonals (e.g. Kohler and Milstein, 1975), immunofluorescence was still popular and many of the studies did not attempt to detect the antigenic fungus in the plant itself.

More recently, a simple culture washing technique for producing antigens was used by Dewey *et al.* (1990) but very sophisticated extracellular polysaccharide antigen preparations were used in the Netherlands, e.g. Kamphuis *et al.* (1989). A mycelial preparation was used in this study, because mycelium is the part of the fungus needing to be detected, as it produces spoilage enzymes and mycotoxins. Monoclonal antibodies are now rapidly gaining favour though polyclonal antibodies are still widely used. Enzyme immunoassays, first used about 20 years ago (Engvall and Pearlman, 1971), have almost completely supplanted immunofluorescence; however, rapid latex agglutination techniques are favoured at present in the Netherlands. In the last few years several papers on immunological studies on fungi in foods have appeared (Notermans *et al.*, 1985; 1986; 1987).

This study had two main aims. First, to try to develop enzyme immunoassays as a replacement for dilution plating, especially for total counts; and second, to develop assays for individual spoilage or toxigenic species. This paper reports on polyclonal antibodies produced to a field fungus and two storage fungi.

METHODS

Fungi

Fusarium culmorum PIL 234 was grown on potato sucrose agar (Booth, 1971), *Eurotium amstelodami* PIL 218 on 2% malt + 10% salt agar (Christensen and Lopez, 1963) which favours the formation of the anamorph, and *Penicillium aurantiogriseum* var. *melanoconidium* PIL 333 on malt agar. PIL numbers indicate the former name of Pest Infestation Laboratory. Spore suspensions were made in 0.01% Tween 80 (British Drug Houses) and after 3 washings by centrifugation, 1 mL of a spore suspension of 10^6/mL was inoculated into Erlenmeyer flasks containlng 100 mL of liquid medium (Burrell *et al.*, 1966), for *E. amstelodami* supplemented with 100g/L NaCl. This medium is of value in immunological studies on fungi as it does not contain large molecules, such as peptone, that could be immunogenic. Flasks were shaken on a rotary shaker (L.H. Engineering Ltd., Slough) at 140 rpm, for 7 d in the dark at 25°C. Typical pellets were formed by *Eurotium* and *Penicillium* species, but a mycelial sludge by the *Fusarium*.

The isolates chosen had given low or zero levels of mortality in a brine shrimp assay (Harwig and Scott, 1971) and were negative for mycotoxins by thin layer chromatography, to ensure healthy animals during the intended long immunisation period.

The above and the following operations were performed in a sterile Class II (BS 5726:1979) biohazard cabinet to prevent contamination wlth other microorganisms.

Antigen preparation

Mycelia were filtered into a sintered glass funnel (Grade 3, Gallenkamp) and washed with sterile distilled water then sterile phosphate buffered saline. The filtrate was discarded. Mycelia were frozen overnight at -20°C, then thawed, distributed between several sterile centrifuge tubes and dried in a vacuum dryer (Edwards). Mycelia were pooled and snap frozen with 50 mL of liquid nitrogen then comminuted at full speed for 1 min in an OmniMixer blender (Camlab) until the nitrogen had evaporated. The material was immediately ground in a sterile pestle and mortar to create a very fine powder. The powdered mycelium was suspended in PBS (200 mg of fungus in 10 mL PBS) and centrifuged at 4500 rpm (3000 G) for 10 min in an MSE Chilspin centrifuge precooled to 4°C. The supernatant representing the soluble antigen was divided into 0.5mL aliquots and stored at -20°C.

Protein assays were carried out on the antigenic material to give some form of standardisation. Diluted samples were compared with bovine serum albumin (BSA) standards using a Coomassie Blue protein reagent (Pierce). For spectrophotometry, two wavelengths were used: 630 nm as primary wavelength and 410 nm as reference wavelength. Samples were placed in 96 well microplates and read with a MR600 microplate reader (Dynatech). Results were previously shown to be comparable with the method of Bradford (1976) which measures protein absorption at 595 nm.

Immunisation

The soluble antigen preparations were first mixed with equal quantities of Freund's Complete Adjuvant (Difco)(Freund and Thomson, 1948) so that the final protein concentration of the mixture was about 1 mg/mL.

After testing to show the absence of native antibodies to the three proposed antigens, duplicate Simone Noir rabbits were given an intramuscular injection (1 mL) of the cell wall or soluble antigen. Subsequent injections of the soluble preparation, given in Freund's Incomplete Adjuvant were made 1, 4, 8 and 12 weeks after the first, a schedule relatively similar to that of Notermans and Heuvelman (1985). Blood samples were taken 4 weeks after the first injections and at 14, 16, and 18 weeks and sera prepared from these by clotting followed by centrifugation. The test animals remained alive allowing further booster injections for antibody production.

Enzyme linked immunosorbent assay (ELISA)

These techniques are similar to the indirect ELISA described in Voller *et al.* (1979). The fungi listed above were all prepared and processed as for the homologous antigens.

The soluble antigen preparations were diluted in sodium hydrogen carbonate/sodium carbonate buffer (pH 9.6) to give a protein concentration of about 1 μg/mL. Aliquots (100 μL) were then pipetted into 96 wall microplates (Dynatech) and kept at 4°C in a moist chamber overnight, washed three times with phosphate buffered saline and Tween 20 (PBST) and blotted dry. Blocking of antigens was then carried out using 250 μL of a 3% bovine serum albumin (Sigma) in PSBT for 1 h at 25°C. After two further washings the coated microplates were stored at -20°C until required. No loss of antigenic activity occurred. Before use, microplates were then washed once with PBST. Antisera were diluted 1:1000 in PBST, applied in triplicate in 50 μL aliquots and incubated in a moist chamber at 25°C for 1 h. Appropriate positive and negative control sera and PBST were included. Microplates were then washed three times in PBST.

A 1:3000 dilution of horseradish peroxidase conjugated swine anti-rabbit immunoglobulin (Dako) in 200 μL aliquots was applied to all wells. Plates were again incubated for 1 h at 25°C, then washed 3 times in PBST. An enzyme substrate of ortho-phenylene diamine (OPD) was prepared from a stock solution of 0.4 g OPD in 100 mL of phosphate cltrate buffer, pH5, 0.15 M (Voller *et al.*, 1979), which had been frozen. For use, aliquots (2.5 mL) were diluted 1:10 with PCB. Hydrogen peroxide (10 μL) was added to 25 mL of the OPD immediately before use, and 200 μL then added to the microplate wells. Plates were incubated for 15 min in the dark at room temperature before the reaction was stopped by adding 50 μL 2.5 M sulphuric acid to each well.

Absorbances were read at two wavelengths (490nm/630nm) on a Dynatech MR600 microplate reader linked to an Apple IIe microcomputer and Epson FX80 printer.

Table 1. Cross-reactivity of antisera with field and storage fungi

		anti-*Penicillium aurantiogriseum* var. *melanoconidium* antisera		anti-*Eurotium amstelodami* antisera		anti-*Fusarium culmorum* antisera	
	PIL No.	A	B	C	D	E	F
FIELD FUNGI							
Alternaria alternata	764	++[1]	-	+	-	-	-
Aureobasidium pullulans	202	+	-	-	-	-	-
Cladosporium cladosporioides	182	++	++	+	-	-	-
C. herbarum	183	+	+	+	-	-	+
Fusarium avenaceum	569	+++	+	+++	+++	+++	+++
F. culmorum	234	+++	-	+++	++	+++	+++
F. graminearum	451	+++	-	+++	+++	+++	+++
F. moniliforme	450	+++	-	+++	+++	+++	+++
Hyalodendron sp.	204	-	-	-	-	-	-
Verticillium lecanii	188	+++	+	+++	+	++	++
STORAGE FUNGI							
Apiospora montagnei	184	++	+	+	-	-	-
Aspergillus candidus	129	+++	+	++	-	-	-
A. flavus	295	++	+	+	-	-	-
A. nidulans	730	+++	+	+++	+	-	-
A. ochraceus	253	+++	+	++	-	++	+
A. restrictus	300	+++	++	+++	+++	+++	+++
A. terreus	422	++	-	+	-	-	-
A. versicolor	399	+++	++	+++	+	++	-
Byssochlamys nivea	203	++	+	+	-	-	-
Eurotium amstelodami	218	+++	++	+++	+++	++	++
E. chevalieri	280	+++	++	+++	+++	+++	+++
E. repens	364	+++	++	+++	+++	+++	+++
E. rubrum	219	++	-	+++	+++	+++	+++
Monascus ruber	213	+++	-	+++	++	+++	++
Mucor racemosus	48	-	++	-	++	++	-
Paecilomyces variotii	195	+++	+	++	+	-	+
Penicillium aurantiogriseum	563	+++	++	++	++	+	-
P. aurantiogriseum var. *melanoconidium*	333	+++	++	++	+	+	-
P. cyclopium	118	+++	++	++	+	+	-
P. hordei	351	+++	++	++	+	+	-
P. martensii	9	+++	++	+	+	-	-
P. notatum	645	+++	+	+++	+++	+++	++
P. roqueforti	186	+++	++	++	+	+	-
P. rubrum	162	+++	+	+	-	+	-
Scopulariopsis brevicaulis	222	-	-	+	-	++	+
Trichothecium roseum	80	+++	+	+++	+++	+++	+++
Wallemia sebi	152	-	-	-	-	-	-

[1] +++ strong positive, ++ medium positive, + weak positive, - negative

Cross reactivity tests
Antisera were tested for cross reactivity against the fungi listed in Table 1, by the ELISA described above. All fungi were prepared in the same way as the homologous antigens.

RESULTS

Of the 10 field fungi, *Alternaria alternata*, *Aureobasidium pullulans*, *Cladosporium herbarum* and *Hyalodendron* sp. all gave negative or low absorbancies with all six antisera (Table 1). *Cladosporium cladosporioides* gave moderate absorbances with the anti-*Penicillium* serum only. However, the four *Fusarium* species and *Verticillium lecanii* gave high absorbances with most antisera.

Of the storage fungi *Wallemia sebi* was notable in not reacting positively with any of the antisera. *Apiospora montagnei* (*Arthrinium*), *Aspergillus flavus*, *A. terreus*, *Byssochlamys nivea*, *Mucor racemosus* and *Scopulariopsis brevicaulis* gave a range of reactivities ranging from negative to moderate with all the antisera. Moderate absorbances were mostly observed with the anti-*Penicillium* antiserum A (Table 1). Other *Aspergillus* and *Penicillium* species, the four *Eurotium* species, *Monascus ruber*, *Paecilomyces variotii* and *Trichothecium roseum* gave high absorbances with usually one of the two pairs of antisera from each genus. *Eurotium* species gave the most consistently high absorbances of any of the storage fungi.

DISCUSSION

The polyclonal antibodies were originally intended to be the first antibody stage in a double antibody sandwich (DAS) ELISA. For this stage, a highly cross reacting antibody, or antibodies, was considered desirable. The antibodies produced, especially the anti-*Penicillium* antiserum A, had those characteristics as it reacted with 33 of the 37 fungi tested. Perhaps a purely polyclonal DAS ELISA could be constructed to achieve detection of total fungi in a sample.

Notermans and Heuvelmann (1985) used a soluble mycelial antigen like that described here, but more recent studies (Notermans and Soentoro, 1986; Notermans *et al.*, 1987) have concentrated on using extracellular polysaccharides (EPS) obtained after column chromatography of culture filtrates. Some very interesting results have been obtained, with indications that these are the most important antigenic determinants. The questions remains whether these substances, produced in abundance in liquid culture, are common in the natural environment. Dewey *et al.* (1990) used surface washings from sporulating *Penicillium islandicum* cultures, while we have used washings of the disrupted mycelium.

We have also tried the mycelium itself after washing without adjuvant as its molecular weight is large enough to be immunogenic, but it gave poor responses (unpublished data). The soluble preparations however, needed adjuvant and gave greater immune response.

ACKNOWLEDGEMENT

We wish to thank Mr J.A. Vaughan, Miss L.M. Lelliot and Mr M. Hetmanski for their assistance with this study.

REFERENCES

BENO, D.W. and ALLEN, O.N. 1946. Immunofluorescent staining for the identification of *Puccinia sorghina* germinated uredospores. *Phytopathology* **54**: 872-873.

BOOTH, C. 1971. The Genus *Fusarium*. Kew, Surrey: Commonwealth Mycological Institute.

BRADFORD, M.M. 1976. A rapid and sensitive method for the quantitation of microgram quantities of protein, utilizing the principle of protein-dye binding. *Anal. Biochem.* **72**: 248-254.

BURRELL, R.G., CLAYTON, C.W., GALLEGLY, M.E. and LILLY, V.G. 1966. Factors affecting the antigenicity of the mycelium in three species of *Phytophthora*. *Phytopathology* **56**: 422-426.

CHRISTENSEN, C.M. and LOPEZ, L.C. 1963. Pathology of stored seeds. *Proc. Int. Seed Testing Assoc.* **28**: 701-711.

CLARKE, J.H., MACNICOLL, A.D. and NORMAN, J.A. 1986. Immunological detection of fungi in plants, including stored cereals. *Int. Biodet.* **22** (Supplement): 123-130.

DEWEY, F.M., MACDONALD, M.M., PHILLIPS, S.I., and PRIESTLY, R.A. 1990. Development of monoclonal-antibody-ELISA and Dip-Stick immunoassays for *Penicillium islandicum* in rice grains. *J. Gen. Microbiol.* **136**: 753-760.

ENGVALL, E. and PERLMANN, P. 1971. Enzyme-Linked-Immunosorbent-Assay (ELISA). Quantitative assay of immunoglobulin G. *Immunochemistry* **8**: 871-874.

FREUND, J. and THOMSON, H.J. 1948. Antibody formation and sensitization with the aid of adjuvants. *J. Immunol.* **60**: 383-.

HARWIG, J. and SCOTT, P.M. 1971. Brine shrimp *Artemia salina* L. larvae as a screening system for fungal toxins. *Appl. Microbiol.* **21**: 1011-1016.

KAMPHUIS, H.J., VEENEMAN, G.H., ROMBOUTS, F.M., VAN BOOM, J.H. and NOTERMANS, S. 1989. Antibodies against synthetic oligosaccharide antigens reactive with extracellular polysaccharides produced by moulds. *Food Agric. Immunol.* **1**: 235-242.

KOHLER, G. and MILSTEIN, C. 1975. Continuous culture of fused cells secreting antibody of predefined specificity. *Nature, London* **256**: 495-497.

NOTERMANS, S. and HEUVELMAN, C.J. 1985. Immunological detection of moulds in foods by using the enzyme-linked immunosorbent assay (ELISA); preparation of antigens. *Int. J. Food Microbiol.* **2**: 247-258.

NOTERMANS, S. and SOENTORO, P.S.S. 1986. Immunological relationship of extracellular polysaccharide antigens produced by different mould species. *Antonie van Leeuwenhoek* **52**: 393-401.

NOTERMANS, S., WIETEN, G., ENGEL, H.U.B., ROMBOUTS, F.M., HOOGERHOUT, P. and VAN BOOM, J.H. 1987. Purification and properties of extracellular polysaccharide (EPS) antigens produced by different mould species. *J. Appl. Bacteriol.* **62**: 157-166.

VOLLER, A., BIDWELL, D.E. and BARTLETT, A. 1979. The enzyme-linked-immunosorbent assay. A guide with abstract of microplate applications. *Guernsey Dynatec Europe*.

6

MEDIA AND METHODS FOR MYCOTOXIGENIC FUNGI

COLLABORATIVE STUDY ON MEDIA FOR DETECTING AND ENUMERATING TOXIGENIC *PENICILLIUM* AND *ASPERGILLUS* SPECIES

J. C. Frisvad, O. Filtenborg, U. Thrane and P. V. Nielsen

Department of Biotechnology,
Technical University of Denmark,
DK-2800 Lyngby, Denmark

In collaboration with: L. R. Beuchat, Annette Lillie, F. Boisen and Susanne Elmholt

SUMMARY

A selective medium for toxigenic fungi was compared with three standard media in a collaborative study to examine its efficacy for enumerating the toxigenic mycoflora of stored foods. Originally designed for *Penicillium verrucosum*, DRYES (Dichloran Rose bengal Yeast Extract Sucrose agar) may also be used for a more general determination of toxigenic *Penicillium* and *Aspergillus* species. DRYES was compared with other media in five laboratories. The number of *P. verrucosum* and *P. aurantiogriseum* was very low or nil in most samples tested in the five laboratories. In general the counts on DRYES were not significantly different from counts on DG18 (Dichloran 18% Glycerol agar), PCACC (Plate Count Agar with antibiotics) or DRBC (Dichloran Rose Bengal Chloramphenicol agar), despite the higher recovery of xerophilic species on DG18.

INTRODUCTION

Pentachloronitrobenzene Rose bengal Yeast Extract Sucrose agar (PRYES) and Dichloran Rose bengal Yeast Extract Sucrose agar (DRYES) were developed primarily for the selective enumeration of *Penicillium verrucosum*, the producer of the carcinogenic nephrotoxin ochratoxin A (Frisvad, 1983). The selective principles were two bacteriocides and one of two fungicides, with a somewhat lowered water activity (0.96-0.97) and incubation at 20°C. The indicative principle was a consistently produced (unidentified) non-diffusible violet brown secondary metabolite in the reverse of *P. verrucosum*. DRYES was also of value in the detection of other cereal fungi, especially *P. aurantiogriseum*. Further experience has shown that DRYES is valuable for more general detection of toxigenic fungi in foods and feeds, because the selectivity of the medium is good and because several diagnostic mycotoxins are produced in this medium and can be detected directly by the agar plug method (Filtenborg and Frisvad, 1990). As a collaborative study, DRYES was compared with Dichloran Rose bengal Chloramphenicol agar (DRBC, King *et al.*, 1979), Dichloran 18% Glycerol agar (DG18, Hocking and Pitt, 1980) and Plate Count Agar with Chlortetracycline and Chloramphenicol (PCACC, King *et al.*, 1986, p. 299) for enumerating *P. verrucosum* in foods. Unfortunately this species was not detected by any collaborator in any of the food samples tested. For this reason, as an alternative the collaborative study tested the value of DRYES as a more general enumeration medium for toxigenic fungi.

MATERIALS AND METHODS

A total of 51 samples were collaboratively tested in the study. Samples were: wheat flour (21 samples), barley (4), rye (8), oats (7), mixed feed (1), corn meal (6), buckwheat (1), pepper (3), walnuts (2), fresh thyme (1), tarragon (1), sage (1), marjoram (1), curled mint (1), parsley (2), balm (1), spring onion (1), cress (1) and water cress (2).

Samples were dilution plated as recommended and described by King *et al.* (1986) on DRBC, DG18, PCACC and DRYES, modified from Frisvad (1983) by including 0.5 g/L $MgSO_4.7H_2O$. Chloramphenicol (50 mg/L) and chlortetracycline (50 mg/L) added after autoclaving) were incorporated in all media, based on the experience that plates with 100 ppm chloramphenicol were often completely overgrown with *Bacillus* species. Samples labelled USA in Table 1 were exceptional: chloramphenicol (100 mg/L) was used. $CuSO_4.5H_2O$ (50 mg/L) and $ZnSO_4.7H_2O$ (100 mg/L) were added to all media to compensate for loss of trace metals if media were made in distilled water. If tap water was used these trace metals were not added.

Typical cultures of *P. verrucosum* (4 isolates), *P. aurantiogriseum* and *P. viridicatum* (one isolate each) were sent to all participants as reference material. The results were compared statistically according to Niemela (1983).

RESULTS AND DISCUSSION

The results of the collaborative study are listed in Tables 1-4. Only three laboratories found a few colonies of *Penicillium verrucosum* or *P. aurantiogriseum*, and only in occasional samples, so this study failed to show the efficacy of DRYES to recover these important toxigenic Penicillia in comparison with DG18 or DRBC.

Pairwise t-tests of log total counts showed that the counts on the four media tested were not significantly different for the 65 samples (Table 5). However, when the xerophilic fungus *Wallemia sebi* was determined separately (Table 2), the counts of this species on DG18 were significantly higher than on the other media (Table 6, Figure 1). Other xerophiles, e.g. *Eurotium* spp., were not recovered in significantly different numbers on DG18 compared with the other media (Tables 3 and 6). It was unexpected that counts of *Cladosporium* species were higher on DG18 than on DRBC or DRYES (Table 4, herb samples). Several other authors have also found higher counts of field fungi on DG18 compared with high water activity media such as DRBC (King *et al.*, 1986) and suggested that DG18 is of more general applicability than it was originally designed to be. It is interesting that total counts completely mask the significant differences between counts of individual species. For the detection of individual species in food mycology, e.g. mycotoxin producers, new evaluation methods should be developed for the comparison of media or for development of media designed for the mycoflora of different foods.

For toxigenic *Penicillium* and *Aspergillus* species, differences in counts in this study appeared to be statistically insignificant on all media tested. The selective agents used in the different media clearly work in that significant differences in counts were obtained for particular groups of fungi. More work is needed in the area of the selective agents used in mycological media for important species. A combination of two or more different media, more replicate analyses or better sampling plans may result in good recovery of all species relevant for the food product examined. Also a combination of dilution plating with direct plating using surface disinfected food particles will also give a more complete picture of the toxigenic potential of the mycoflora actually growing in foods.

Table 1. Fungal counts (log_{10}/g) on four different media for various food samples

	DRYES	DG18	DRBC	PCACC
Corn meal, USA-1	5.56	5.63	5.59	5.57
Corn meal, USA-2	4.42	4.50	4.45	4.42
Corn meal, USA-3	4.54	4.76	4.75	4.73
Corn meal, USA-4	5.11	5.23	5.04	5.11
Corn meal, USA-5	4.43	4.46	4.53	4.54
Corn meal, DK-1	3.38	3.60	-	3.52
Wheat flour, USA-1	2.90	3.69	2.90	2.88
Wheat flour, USA-2	3.34	3.34	2.90	2.78
Wheat flour, USA-3	3.23	3.38	3.49	3.42
Wheat flour, USA-4	2.70	2.78	2.00	2.54
Wheat flour, USA-5	3.34	3.32	2.95	2.90
Wheat flour, DK-1	3.55	3.66	3.45	3.46
Wheat flour, DK-2	3.18	3.16	3.13	3.20
Wheat flour, DK-3	3.62	-	2.78	2.90
Wheat flour, DK-4	3.00	-	2.95	2.78
Wheat flour, DK-5	3.28	-	3.30	3.34
Wheat flour, DK-6	3.00	-	2.85	2.90
Wheat flour, DK-7	3.30	4.53	2.95	3.08
Wheat flour, DK-8	3.32	4.62	2.90	2.93
Wheat flour, DK-9	3.29	4.89	3.11	3.00
Wheat flour, DK-10	3.18	3.53	3.15	3.11
Wheat flour, DK-11[1]	3.34	4.75	-	2.70
Wheat flour, DK-12[1]	2.68	4.28	-	2.93
Wheat flour, DK-13[1]	3.72	4.19	-	3.16
Wheat flour, DK-14[1]	3.28	4.74	-	2.54
Wheat flour, DK-15[1]	3.19	4.28	-	2.88
Wheat flour, DK-16[1]	3.02	3.46	-	2.94
Barley flour, DK[1]	2.18	3.09	<2.00	2.24
Barley, DK-1	3.74	3.72	3.11	3.08
Barley, DK-2	3.08	-	3.04	3.10
Barley, DK-3	3.68	3.22	-	3.10
Barley, DK-3[1]	3.68	3.21	-	3.11
Rye, DK-1	3.13	-	3.08	3.08
Rye, DK-2	3.06	-	2.90	2.97
Rye, DK-3	3.56	3.74	3.43	3.37
Rye, DK-4	3.23	3.82	3.06	3.11
Rye, DK-5[1]	3.27	3.70	-	3.20
Rye, DK-6[1]	3.24	3.73	-	2.74
Rye, DK-7[1]	3.35	3.66	-	3.04
Rye, DK-8	2.81	3.37	-	2.57
Oats, DK-1	3.60	3.95	3.32	3.36
Oats, DK-2	3.42	3.30	3.18	3.19
Oats, DK-3	3.54	-	3.53	3.57
Oats, DK-4	3.24	3.24	2.85	3.16
Oats, DK-5[1]	3.45	3.89	-	3.69
Oats, DK-6[1]	3.59	3.73	3.58	3.65
Oats, DK-7	3.16	3.33	-	3.08
Mixed feed, DK	3.58	3.71	3.19	3.32
Buckwheat, DK	4.13	4.45	-	4.19
Black pepper	4.38	5.12	3.85	4.02
White pepper-1[1]	3.48	3.99	-	3.35
White pepper-2[1]	5.08	5.19	-	4.78

Table 1 (continued)

	DRYES	DG18	DRBC	PCACC
Walnuts-1	2.89	2.97	-	2.97
Walnuts-2[1]	3.26	3.22	-	3.27
Parsley, DK	4.15	4.66	4.11	(bacteria)
Parsley, DK	5.06	3.81	5.02	4.95
Water cress, DK-1	3.04	3.14	2.81	(bacteria)
Water cress, DK-2[2]	3.57	3.68	3.61	2.93
Tarragon, DK[2]	3.75	3.94	-	3.82
Thyme, DK[2]	3.54	3.28	2.89	3.26
Balm, DK[2]	-	3.75	2.85	3.23
Sage, DK[2]	3.38	3.07	2.54	2.98
Marjoram, DK[2]	3.30	3.44	3.09	3.41
Curled mint, DK[2]	3.87	4.10	3.38	3.76
Spring onion, DK	3.20	3.32	2.65	-
Cress, DK	4.86	4.91	4.88	-

[1] These samples contained a number of xerophilic fungi. Where possible, these were counted separately and are presented in Tables 2 and 3.
[2] These samples contained high numbers of *Cladosporium* species, which were counted separately and are presented in Table 4.

Table 2. Counts ($\log_{10}$/g) of *Wallemia sebi* on three different media

	DRYES	DG18	PCACC
Wheat flour, DK-11	2.70	4.75	<3.00
Wheat flour, DK-12	<3.00	4.24	<3.00
Wheat flour, DK-13	2.18	4.13	<3.00
Wheat flour, DK-14	2.70	4.73	<3.00
Wheat flour, DK-15	2.39	4.23	<3.00
Wheat flour, DK-16	1.88	3.30	<2.00
Barley flour, DK	<2.00	3.01	<2.00
Barley, DK-3	<2.00	2.54	<2.00
Rye, DK-5	<2.00	3.42	<2.00
Rye, DK-6	2.40	3.23	<2.00
Rye, DK-7	2.30	3.46	<2.00
Oats, DK-5	<2.00	2.18	<2.00
Oats, DK-6	<2.00	2.70	<2.00

Table 3. Counts ($\log_{10}$/g) of *Eurotium* species on three different media

	DRYES	DG18	PCACC
Rye, DK-5	2.74	3.22	2.85
Rye, DK-6	2.98	2.98	2.93
Oats, DK-5	2.00	3.58	2.98
White pepper-1	2.88	3.50	3.11
White pepper-2	4.85	4.88	<4.00
Walnuts-2	<2.00	2.35	<2.00

Table 4. Counts (log_{10}/g) of *Cladosporium* species on four different media

	DRYES	DG18	DRBC	PCACC
Water cress, DK-2	3.16	3.29	3.18	<2.00
Tarragon, DK	3.37	3.69	-	1.15
Thyme, DK	3.12	3.24	2.40	3.16
Balm, DK	-	3.55	2.10	3.03
Sage, DK	2.24	2.44	1.70	2.00
Marjoram, DK	3.17	3.20	2.57	3.24
Curled mint, DK	3.77	4.02	3.08	3.76

Table 5. Pairwise t-tests for differences between log total counts of filamentous fungi in 66 food samples (see Table 1)

Media	Average	Standard deviation	t[1]	n[2]
DRYES-DG18	-0.33	0.45	-0.73	56
DRYES-DRBC	0.22	0.29	0.77	43
DRYES-PCACC	0.21	0.36	0.58	61
DG18-DRBC	0.49	0.50	0.99	36
DG18-PCACC	0.55	0.61	0.90	54
DRBC-PCACC	-0.02	0.42	-0.04	41

[1] All values are less than the t-fractile value at 90% level for 30-60 samples: 1.30, i.e. differences are not significant.
[2] n = number of samples

Table 6. Pairwise t-test for *W. sebi*, *Eurotium* spp. and *Cladosporium* spp. on different media

Media	Average	Standard deviation	t	n[1]
Wallemia sebi				
DRYES-DG18	-1.86	0.56	-3.33[2]	13
DRYES-PCACC	0.64	0.68	0.93	13
DG18-DRBC	2.31	0.43	5.36[3]	2
DG18-PCACC	2.50	0.46	5.41[3]	13
Eurotium spp.				
DRYES-DG18	-0.77	0.81	-0.96	6
DRYES-PCACC	0.15	1.05	0.14	6
DG18-PCACC	0.92	0.91	1.02	6
Cladosporium spp.				
DRYES-DG18	-0.18	0.10	-1.85[4]	6
DRYES-DRBC	0.50	0.30	1.67[4]	5
DRYES-PCACC	0.40	1.02	0.40	6
DG18-DRBC	0.78	0.43	1.80[4]	6
DG18-PCACC	0.58	0.91	0.63	7
DRBC-PCACC	-0.14	1.30	-0.11	6

[1] Number of samples
[2] Signicant difference at the 99.5% level.
[3] Significant difference at the 98% level.
[4] Significant difference at the 90% level.

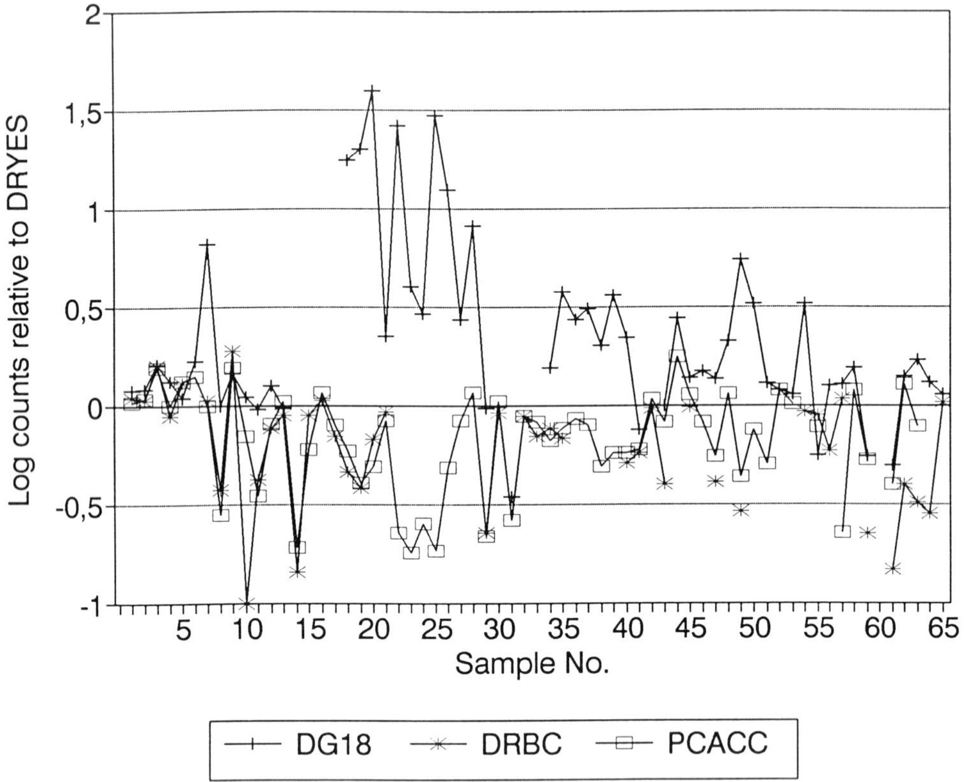

Figure 1. The $\log_{10}$ counts of filamentous fungi relative to the counts on DRYES are plotted (vertical axis) against the 65 samples (horizontal axis). Samples containing *Wallemia sebi* and other xerophilic fungi appear to be responsible for the tendency of higher counts on DG18.

Several results in this collaborative study indicated that rose bengal may be more toxic than previously assumed. Some counts on DRBC performed in our laboratory were not included in Table 1, because the counts were unusually low. The plates used had been protected against light during the whole process, so maybe other toxic effects of rose bengal may have resulted in the strong inhibition of the fungi. This occasional toxic effect was also noted by Madelin (1987) and it may not be due to the toxic effect of light induced singlet oxygen products of rose bengal alone (see Gams and van Laar, 1982). More work is needed to evaluate the production of singlet oxygen by exposure of rose bengal to light, or to find alternatives to rose bengal.

Mycotoxin production on the isolation plates was also tested on DRYES, DG18, DRBC and PCACC in our laboratory. PCACC did not support production of any known mycotoxins, while DG18 and DRBC supported the production of only a few (see Filtenborg and Frisvad, 1990). In general DG18 supported the production of secondary metabolites of *Eurotium* species and *A. versicolor*, while DRBC supported production of roquefortine C in several *Penicillium* species (Filtenborg and Frisvad, 1990). However ochratoxin A and aflatoxin were only produced in significant amounts on media based on Yeast Extract Sucrose (YES) agar with 0.5% $MgSO_4.7H_2O$ (Filtenborg *et al.*, 1990). More work is needed to develop media that give consistent recovery of important toxigenic fungal species and also support their mycotoxin production.

REFERENCES

FILTENBORG, O. and FRISVAD, J.C. 1990. Identification of *Penicillium* and *Aspergillus* species in mixed cultures in Petri dishes using secondary metabolite profiles. *In* Modern Concepts in *Penicillium* and *Aspergillus* Classification, eds R.A. Samson and J.I. Pitt, pp. 27-36. New York: Plenum Press.

FILTENBORG, O., FRISVAD, J.C. and U. THRANE. 1990. The significance of yeast extract composition on metabolite production in *Penicillium*. *In* Modern Concepts in *Penicillium* and *Aspergillus* Classification, eds. R.A. Samson and J.I. Pitt, pp. 433-440. New York: Plenum Press.

FRISVAD, J.C. 1983. A selective and indicative medium for groups of *Penicillium viridicatum* producing different mycotoxins in cereals. *J. Appl. Bacteriol.* **54**: 409-416.

GAMS, W. and van LAAR, W. 1982. The use of solacol (validamycin) as a growth retardant in the isolation of soil fungi. *Netherland J. Plant Pathol.* **88**: 39-45.

HOCKING, A.D. and PITT, J.I. 1980. Dichloran-glycerol medium for enumeration of xerophilic fungi from low moisture foods. *Appl. Environ. Microbiol.* **39**: 488-492.

KING, A.D., HOCKING, A.D. and PITT, J.I. 1979. Dichloran-rose bengal medium for enumeration and isolation of molds from foods. *Appl. Environ. Microbiol.* **37**: 959-964.

KING, A.D., PITT, J.I., BEUCHAT, L.R. and CORRY, J.E.L., eds. 1986. Methods for the Mycological Examination of Food. New York: Plenum Press.

MADELIN, T.M. 1987. The effect of surfactant in media for the enumeration, growth and identification of airborne fungi. *J. Appl. Bacteriol.* **63**: 47-52.

NIEMELA, S. 1983. Statistical evaluation of results from quantitative microbiological examinations. NMKL Report No. 1, 2nd edn. Uppsala: Nordic Committee on Food Analysis.

SIMPLE IDENTIFICATION PROCEDURE FOR SPOILAGE AND TOXIGENIC MYCOFLORA OF FOODS

O. Filtenborg, J. C. Frisvad, F. Lund and U. Thrane

Department of Biotechnology
Technical University of Denmark
DK-2800 Lyngby, Denmark

SUMMARY

In the quality management of food industries simple screening methods for important spoilage and toxigenic fungal species are needed. The following screening and identification procedure has been designed for food products in which species of *Penicillium*, *Aspergillus*, *Eurotium* and *Neosartorya* are the major spoilage fungi. The media used are dichloran rose bengal yeast extract sucrose agar and dichloran creatine sucrose bromocresol agar. In mixed cultures on dilution plates, the isolates are grouped and the important species are preliminarily identified and counted using macroscopic characters, i.e. colour of conidia, mycelium obverse and reverse, and extent of sporulation. Descriptions and photographs of standard cultures on the same media are used in this preliminary identification. Final identifications are then performed by direct thin layer chromatography of secondary metabolites of selected colonies. Metabolite profiles are identified by comparison with standard profiles. The procedure has been tested on samples of cereals and mouldy bread and air samples from bread factories. In cereal samples, all the important species could be identified. In mouldy bread and air samples two less important species could not be confirmed because no metabolites were detected. This procedure is currently being implemented in several Danish food industries permitting companies to perform an increasing part of their mycological control unaided. It is our experience that this procedure is of great benefit to the companies: implementation is simple, encompassing the few species of importance to each company.

INTRODUCTION

Mycological problems in the food and feed industry are numerous and whether caused by spoilage or mycotoxins are often very costly. Problems are often discovered too late to prevent economic losses or press coverage. A major contributory factor is lack of knowledge of mycological control in the food industry: food mycology is a new discipline, understood by experts only.

It is of vital importance that mycological control in any food company is performed by staff or consultants in close contact with the company. This is the only effective way to monitor mycological quality and prevent mycological problems. The introduction of Hazard Analysis: Critical Control Points (HACCP; ICMSF, 1988) based on understanding of the qualitative composition of the mycoflora in samples taken at carefully selected places will increasingly make it possible to prevent infection and growth of important toxigenic and spoilage fungi.

An important tool in mycological control is simple, specific identification methods. Identification to species level is necessary, to pinpoint important properties such as mycotoxin production, resistance to physical, chemical and biological factors, etc. Only two diagnostic media for important toxigenic fungi exist so far, Aspergillus flavus and parasiticus agar (AFPA; Pitt *et al.*, 1983) for *Aspergillus flavus* and *A. parasiticus* and Dichloran Rose bengal Yeast Extract Sucrose agar (DRYES; Frisvad, 1983) for *Penicillium verrucosum*. The development of media is very time consuming, so one vision for the future (Pitt and Samson, 1990) is towards direct identification in mixed cultures on effective isolation media. For this purpose we have suggested the use of secondary metabolites (Filtenborg and Frisvad, 1990), and this paper deals with further investigations in that area.

For use in industry, procedures should be simple and quick, and be able to detect all the important species. Investigation of all isolates present on the medium is impractical, so some kind of grouping has to take place. Substrates are therefore required which ensure similarity of isolates from the same species and differences in isolates of different species. This demand, electivity (Mossel, 1986), is met by several substrates for some Aspergilli, but the number of useful substrates is very limited for Penicillia. As this work is based on production of secondary metabolites in the substrate as well, the optimal choice of media for identification of *Penicillium*, *Aspergillus*, *Eurotium* and *Neosartorya* species so far are DRYES and Dichloran Creatine sucrose bromocresol agar (CREAD; Frisvad *et al.*, 1992).

DRYES allows growth of all species in these four genera, but some species are more competitive than others. The optimal choices of samples, inoculation technique and dilution level compensate for this, allowing the most important species to be detected on DRYES. CREAD on the other hand is selective, and only allows good growth of some Penicillia and a few other species, favouring certain slowly growing species.

The important toxigenic and spoilage fungi in barley malt are *Fusarium* and *Aspergillus* species (Flannigan *et al.*, 1986). *Penicillium verrucosum*, *P. aurantiogriseum* and *P. hirsutum* var. *hordei* (Frisvad and Filtenborg, 1988) are important in cereals used in food manufacture and feeds. This mycoflora is important in temperate climate cereals using drying as the only conservation method.

Identification of the Fusaria is described elsewhere (Thrane *et al.*, 1992). Identification of the *Aspergillus* and the *Penicillium* species is described below.

The most important spoilage fungus of bread to which preservatives have been added is *P. roqueforti* (Engel and Teuber, 1978). If preservatives have not been added and no other preservation process has been used, *Eurotium* species are also important. Other less important species are *P. commune*, *Aspergillus flavus*, *A. niger*, *P. corylophilum*, *Monascus ruber* and *Paecilomyces variotii*.

MATERIALS AND METHODS

Samples

Because this work was primarily directed towards the brewing and baking industries, barley kernels, barley flour, mouldy bread and air from bread factories were sampled.

Media

The media used were Dichloran Rose bengal Yeast Extract Sucrose agar (DRYES) and Dichloran Creatine sucrose bromocresol agar (CREAD)(Frisvad *et al.*, 1992).

Inoculation methods

Surface disinfected barley and wheat kernels were direct plated, seven per Petri dish.

Flour samples were diluted, homogenised and spread plated. Closely spaced dilutions were used e.g. $2x10^{-2}$, 10^{-2}, $2x10^{-3}$, 10^{-3} and so on, as it was important in the identification procedure that number of colonies per 9 cm Petri dish did not exceed about 30. To ensure the number of isolates is sufficient to include all important species in the sample, at least triplicates should be made from each dilution. Alternatively, single dilutions on 14 cm Petri dishes may be used. From mouldy bread samples isolates were streaked onto the medium surface or pieces of the infected bread (approx. 5x5x5 mm) were cut and direct plated. Media were incubated at 25°C for 7 days. For bread pieces, the incubation time may be reduced to 3-4 days.

Secondary metabolite analysis

After incubation, selected colonies were directly analysed for the production of secondary metabolites, using a simple thin layer chromatography (TLC) technique (Filtenborg *et al.*, 1983). For isolates from DRYES the TLC plates were eluted with TEF (toluene: ethyl acetate: formic acid, 5:4:1); isolates from CREAD were eluted with TEF and CAP (chloroform: acetone: propane-2-ol, 85:15:20). Elution time was 15 minutes, permitting 20 x 20 cm TLC plates to be used twice, from opposite sides. Anisaldehyde (after elution in TEF) and cerium (IV) sulphate spray (after elution in CAP) were used for identification of the spots. For detailed description of sprays see Filtenborg *et al.* (1983) and Filtenborg and Frisvad (1980). The identification of metabolite profiles was based on comparison with mixtures of metabolites extracted from standard reference cultures.

Identification procedure in mixed cultures

The identification procedure starts with a grouping of the colonies according to their macroscopic appearance, i.e., mainly on colour of conidia, colony obverse and reverse, and extent of sporulation. By comparing colonies with photographs and brief descriptions, a preliminary qualitative and quantitative description of the mycoflora can be established. The photographs and descriptions of the species are taken from streak inoculated standard cultures on DRYES or CREAD.

Table 1. Macroscopic characteristics after 7 days on DRYES of toxigenic fungi important in cereals.

Penicillium verrucosum	Brown red reverse, often small colonies (5-10 mm) with solid white mycelium. Very sparse sporulation concentrated in the centre of the colony.
P. aurantiogriseum var. *aurantiogriseum*	Yellow reverse, very dome-shaped colonies.
P. aurantiogriseum var. *polonicum*	Yellow reverse, very heavy sporulation with blue-green conidia. Flat colonies.
P. aurantiogriseum var. *melanoconidium*	Yellow reverse, heavy sporulation with dark green conidia. Flat colonies.
P. viridicatum	Yellow reverse, light yellow mycelium, very sparse sporulation concentrated in the centres or absent. Light green conidia. Umbonate colonies.
P. hirsutum var. *hordei*	Yellow reverse, very velvety colonies with yellow white mycelium. Sporulation sparse and concentrated in the centre.
Aspergillus flavus	White reverse, heavy sporulation with yellow green conidia. Large colonies (40-50 mm) with white mycelium.
Eurotium spp.	Cream reverse. Heavy sporulation, dominated by green conidia, yellow ascomata, or both.

Preliminary identifications must be confirmed, because macroscopic appearance may vary significantly within species, because many *Penicillium* species have macroscopic characteristics in common, and because the procedure is meant to be used in industry by persons with a specialised knowledge of only a few species.

Final identifications are based on the production of secondary metabolites by one or more isolates of each group. Using a simple TLC technique, the metabolite profile is identified by comparison with mixtures of standard metabolites produced by the important species. The number of isolates analysed must depend on the experience of the personnel.

Table 2. Macroscopic characteristics of toxigenic fungi important in bread, on bread, DRYES and CREAD.

Penicillium roqueforti	On **bread** white mycelium with a rhizoid margin. Blue green conidia. On **DRYES** dark green reverse, large colonies (often 40-50 mm) with a rhizoid margin and very sparse mycelium formation. Heavy sporulation, blue green conidia. On **CREAD** good growth. Other characteristics as on DRYES.
P. commune	On **bread** white mycelium with a distinct colony margin. Greyish green conidia. On **DRYES** white reverse, heavy to sparse sporulation of greyish green conidia and white mycelium. On **CREAD** good growth, base production. Sporulation as on DRYES.
P. corylophilum	On **bread** grey green conidia. On **DRYES** light yellow brown reverse and heavy sporulation of grey green conidia. On **CREAD** very poor growth.
Eurotium spp.	On **bread** blue green conidia and usually yellow ascomata. Without ascomata like *P. roqueforti*, but surface more floccose. On **DRYES** see Table 1. On **CREAD** no growth.
Aspergillus flavus	On **bread** yellow to cinnamon conidia. On **DRYES** see Table 1. On **CREAD** very weak growth.
A. niger	On **bread** white-yellowish mycelium and black conidia. On **DRYES** white mycelium and heavy to sparse sporulation of black conidia. On **CREAD** very weak growth.
Monascus ruber	On **bread** white, floccose mycelium, turning reddish brown in the centre with age. On **DRYES** white floccose mycelium and no sporulation. On **CREAD** very poor growth.
Paecilomyces variotii	On **bread** yellow brown conidia. On **DRYES** light yellow brown reverse, white floccose mycelium, heavy sporulation of light yellow brown conidia. On **CREAD** very poor growth.

Identification of fungi in cereals

The identification of *Aspergillus* and *Penicillium* isolates is carried out from cultures on DRYES. The isolates are grouped, based on descriptions in Table 1.

Figures 1 and 2 provide examples of fungi detected on DRYES from barley and wheat samples. As well as these species, *A. fumigatus*, *A. niger* and *Alternaria alternata* can also be identified. Confirmation of the identification of the fungi in Figures 1 and 2 by secondary metabolite profiles is shown in Figures 3 and 4. Most species produce several unidentified metabolites in addition to those listed in Table 3. From Figures 3 and 4 it appears that almost every isolate produces several metabolites on TLC plates in a characteristic profile.

Some of the profiles in the Figures are identical, perhaps indicating close relationships among species. A few profiles only consist of one metabolite. If this is penicillic acid, it is only possible to state that the isolate belongs to *P. aurantiogriseum* and that isolation and further investigation are needed to establish the identity to varietal level. For practical purposes, identification to *P. aurantiogriseum* is sufficient, indicating that the isolate belongs to the normal spoilage mycoflora of cereals and thus has the potential to produce mycotoxins.

Table 3. Secondary metabolite production by common toxigenic storage fungi on DRYES

Fungus	Secondary metabolites
Aspergillus candidus	Terphenyllin; xanthoascin
A. flavus	Aflatoxin B_1; cyclopiazonic acid; kojic acid
A. fumigatus	Verruculogen
A. niger	Napthoquinones
A. ochraceus	Ochratoxin A; penicillic acid
A. parasiticus	Aflatoxins B & G; kojic acid
A. sydowii	-
A. tamarii	Cyclopiazonic acid; kojic acid
A. terreus	Terrein; patulin; citreoviridin
A. versicolor	Sterigmatocystin
A. wentii	-
Emericella nidulans	Sterigmatocystin
Eurotium spp.	-
Monascus ruber	-
Paecilomyces variotii	Viriditoxin
Penicillium aurantiogriseum var. *aurantiogriseum*	Xanthomegnin; penicillic acid; viridicatin; terrestric acid; aurantiamin
Penicillium aurantiogriseum var. *polonicum*	Verrucosidin; penicillic acid
Penicillium aurantiogriseum var. *melanoconidium*	Penitrem A; verrucosidin; penicillic acid
P. viridicatum	Xanthomegnin; penicillic acid; brevianamide A
P. brevicompactum	Mycophenolic acid; Raistrick phenols; brevianamide A
P. chrysogenum	-
P. commune	Cyclopiazonic acid
P. crustosum	Penitrem A; viridicatin; terrestric acid
P. echinulatum	Penechins; viridicatin
P. expansum	Citrinin; patulin; chaetoglobosin C
P. griseofulvum	Griseofulvin; patulin
P. hirsutum var. *hordei*	Terrestric acid
P. roqueforti var. *roqueforti*	PR-toxin; mycophenolic acid
P. roqueforti var. *carneum*	Patulin; mycophenolic acid
P. solitum	Compactin; cyclopenin
P. verrucosum	Ochratoxin A; citrinin; verrucolone

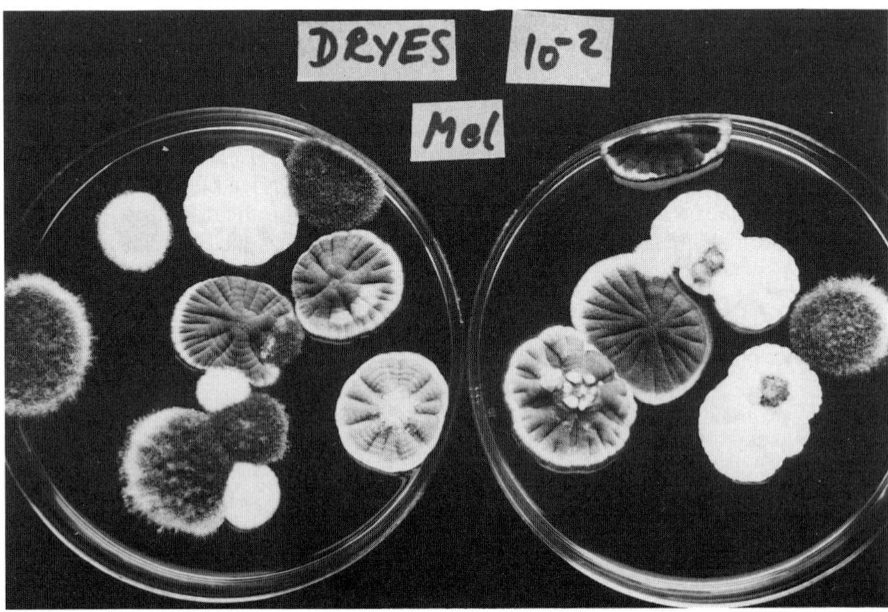

Figure 1. Spread plates on DRYES of barley flour dilution incubated at 25°C for 7 days

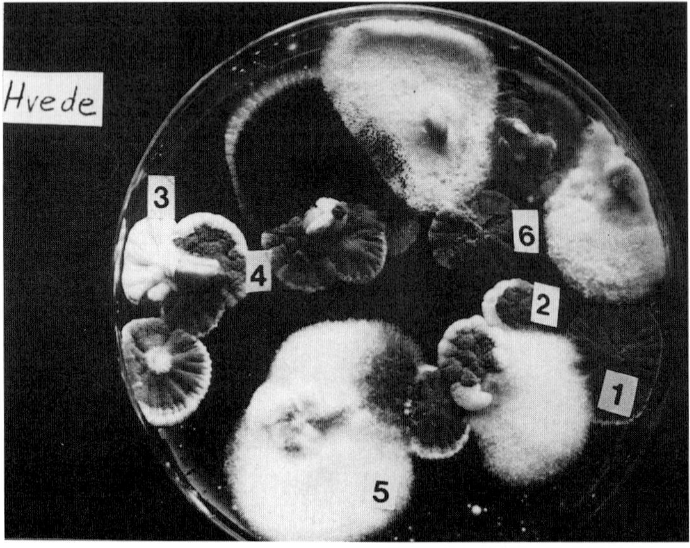

Figure 2. Surface sterilised wheat kernels direct plated on DRYES incubated at 25°C for 7 days

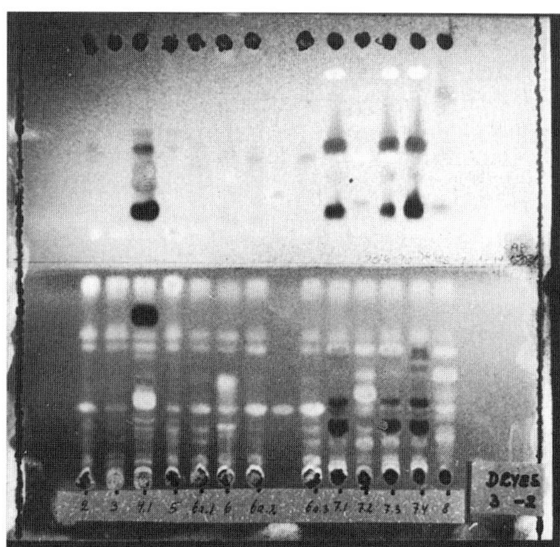

Figure 3. TLC metabolite profiles of isolates from a barley flour sample in mixed cultures on DRYES (see Fig. 1). TLC conditions: elution in TEF and CAP from opposite directions, spraying with anisaldehyde and $Ce(SO_4)_2$ respectively and 365 nm UV illumination. Profiles characteristic of *P. aurantiogriseum* varieties, *Eurotium* species and *P. verrucosum* are present

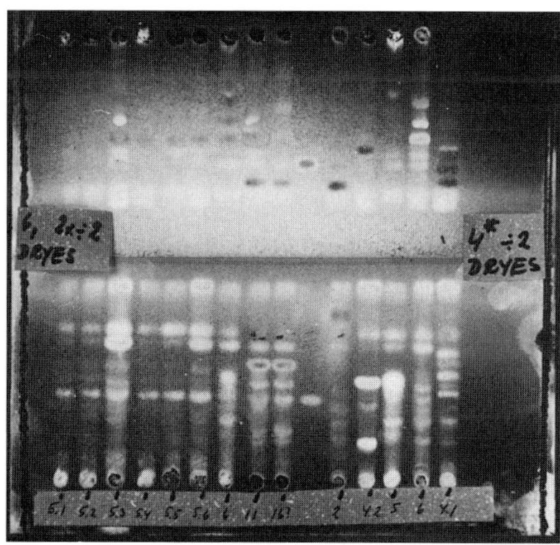

Figure 4. TLC metabolite profiles of isolates from a whole wheat samples in mixed cultures on DRYES (see Fig. 2). TLC conditions: elution in TEF and CAP from opposite directions, spraying with anisaldehyde and $Ce(SO_4)_2$ respectively and 365 nm UV illumination. Profiles characteristic of *P. aurantiogriseum* varieties, *P. verrucosum* and two chemotypes of *Alternaria alternata* are present

Identification of fungi in bread

Identifications from bread may be performed directly on the mouldy bread or in mixed culture on DRYES and CREAD using the procedure described above. Mixed cultures originate from samples taken in the bread production environment (air, machinery, hands and so on). The isolates may be grouped according to the description in Table 2. Examples of mouldy bread samples are shown in Figures 5 and 6.

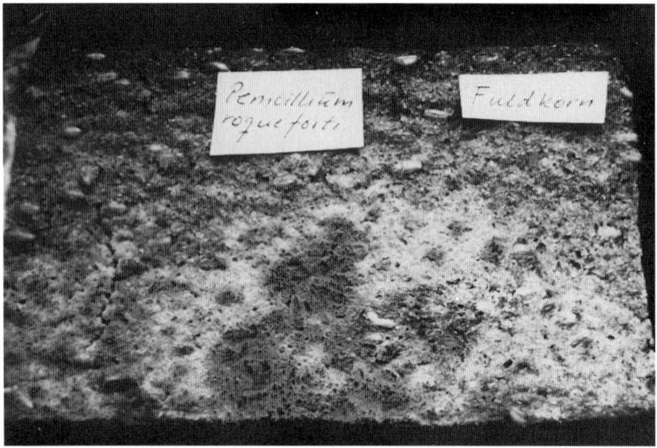

Figure 5. Whole rye bread infected with *Penicillium roqueforti*

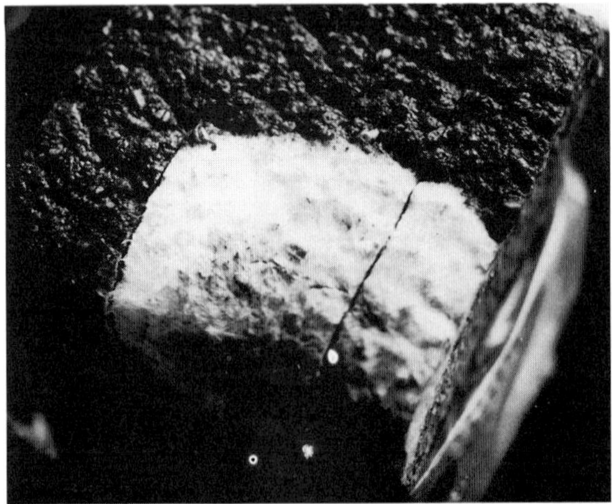

Figure 6. Slices of rye bread infected with *Monascus ruber*

Table 4. Production of secondary metabolites by food-borne fungi growing on CREAD

Fungus	No. of isolates[1]	Secondary metabolites
Penicillium commune	59	Cyclopiazonic acid
P. crustosum	15	Penitrem A; roquefortine C; occasionally terrestric acid
P. echinulatum	6	Viridicatin, cyclopenin, penechins
P. expansum	7	Citrinin; roquefortine C; chaetoglobosin X
P. roqueforti var. *roqueforti*	11	Roquefortine C; PR-toxin
P. roqueforti var. *carneum*	1	Roquefortine C, mycophenolic acid
P. solitum	40	Occasionally cyclopenin
P. spinulosum	1	Two unknown metabolites
Aspergillus versicolor	9	Sterigmatocystin
A. sydowii	6	Two unknown metabolites

[1] Number of isolates examined directly on CREAD, most of them from air samples from bread factories

Confirmation of identification based on secondary metabolites is not illustrated here, but in Tables 3 and 4 are listed well known metabolites which may be used. Most species produce several unidentified metabolites in addition to those mentioned in the Tables. *Eurotium* species produce a characteristic pattern of yellow, red, blue and orange metabo lites derived from the ascomata. *A. niger* produces two or three characteristic unknown metabolites. Two species, *P. corylophilum* and *Monascus ruber*, did not produce metabolites which could be detected with this method. However these species are of minor importance.

DISCUSSION

In comparison with other isolation media such as DRBC, colony sizes on DRYES and CREAD are large and sporulation of some species is reduced. But this may make differentiation of species easier to untrained eyes and enhances the production of secondary metabolites. To compensate for the larger colony diameters, the maximum acceptable number of colonies per plate is lower than the normal 50-100, and this must be compensated for by using additional dilutions.

More detailed investigations should be performed if unexpected results occur at confirmation, or if the mycoflora changes, i.e. if the relationship between morphology and metabolite profile differs significantly from the established patterns.

Species of other genera such as *Alternaria*, *Botrytis*, *Paecilomyces* and *Epicoccum* have also been detected using this procedure. The media used in this investigation may not be the optimal choice for growth, grouping and identification of these and perhaps for other genera as well, so we are continually checking and if necessary formulating other media for use in the procedures described here. The confirmation of some Penicillia not tested in this investigation may also require different substrates, as some species may have very specific substrate demands for the production of metabolites. But the solution to difficulties in detecting metabolites from certain species may also lie in optimisation of the TLC procedure.

Normally it is not possible to verify identifications on bread directly. This is due to the low water activity of bread, which inhibits production of secondary metabolites. One exception is *Eurotium* isolates, which are able to produce ascomata on bread. These isolates can be verified by direct TLC analysis of the infected bread, and the metabolite profile is practically identical to the profile on DRYES. All other isolates from bread have to be inoculated onto DRYES and CREAD to enable final identification.

Another possible approach to the final identification step is to combine the use of secondary metabolite profiles with micromorphological characteristics. Many species have very distinct micromorphological characters, which can sometimes be observed even with the stereomicroscope. As some of these characters can be seen on isolation media like DRYES (Frisvad *et al.*, 1992), it may be possible to use the microscope and/or the TLC method to confirm the identity of a species on the same substrate. The choice of method will depend on which is the easiest, sufficiently specific technique for practical purposes.

IMPLEMENTATION OF MYCOLOGICAL CONTROL IN THE FOOD INDUSTRY

We are in the process of implementing the identification procedures described here for several food companies. First, the important spoilage and toxigenic mycoflora of specific products have been characterised by us. We have then described the physiology, ecology and toxicology of the individual species, and their importance in the products. Then the mycological methods to be used, i.e. sampling schemes, substrates, incubation conditions, etc., have then been chosen. The company staff has then been trained to use the identification procedure specifically designed for their demands, and we have supplied photographs and standard metabolites for TLC of the important cultures.

In this way the companies should be able to perform an increasing part of their mycological control without assistance. They are able to find the infection routes of the species which are important in their particular production, and they have information on the important attributes of these species. This optimises opportunities to prevent infection and growth of important species in the critical process steps. Furthermore, knowledge of the composition of the normal spoilage mycoflora in any product provides information on which mycotoxins are relevant for mycotoxin control of the product.

Our experience with the implementation of these procedures has been limited, but very positive. Sometime in the future faster and more simple identification methods will become available, including immunoassays and DNA/RNA probe techniques. These methods should be introduced to detect specific, well known spoilage and toxigenic species where possible. They should be used along with the procedures presented in this paper, as it is continually necessary to observe any changes in the normal mycoflora, which may indicate a special problem and perhaps a need for detection and identification of further species.

REFERENCES

ENGEL, G. and TEUBER, M. 1978. Simple aid for identification of *Penicillium roqueforti* Thom. Growth in acetic acid. *Eur. J. Appl. Microbiol. Biotechnol.* **6**:107-111.

FILTENBORG, O. and FRISVAD, J.C. 1980. A simple screening-method for toxigenic moulds in pure cultures. *Lebensm.-Wissensch. Technol.* **13**: 128-130.

FILTENBORG, O., FRISVAD, J.C. and SVENDSEN, J.A. 1983. Simple screening method for molds producing intracellular mycotoxins in pure cultures. *Appl. Environ. Microbiol.* **45**: 581-585.

FILTENBORG, O. and FRISVAD, J.C. 1990. Identification of *Penicillium* and *Aspergillus* species in mixed cultures in Petri dishes using secondary metabolite profiles. *In* Modern Concepts in *Penicillium* and *Aspergillus* Classification, eds R.A. Samson and J.I. Pitt. pp. 27-36. New York: Plenum Press.

FLANNIGAN, B., HEALY, R.E. and APTA, R. 1986. Biodeterioration of dried barley malt. *In* Biodeterioration 6, eds S. Barry, D. Houghton, G.C. Llewellyn and C. O'Rear. pp. 300-305. Farnham Royal: Commonwealth Agricultural Bureaux.

FRISVAD, J.C. 1983. A selective and indicative medium for groups of *Penicillium viridicatum* producing different mycotoxins in cereals. *J. Appl. Bacteriol.* **54**: 409-416.

FRISVAD, J.C. and FILTENBORG, O. 1983. Classification of terverticillate Penicillia based on profiles of mycotoxins and other secondary metabolites. *Appl. Environ. Microbiol.* **46**: 1301-1310.

FRISVAD, J.C. and FILTENBORG, O. 1988. Specific *Penicillium* and *Aspergillus* mycoflora of different foods. *Proc. Japan. Assoc. Mycotoxicol.* Suppl. 1: 163-166.

FRISVAD, J.C., FILTENBORG, O., LUND, F. and THRANE, U. 1992. New selective media for the detection of toxinogenic mycoflora of cereal products, meat and cheese. *In* Modern Methods in Food Mycology, eds R.A. Samson, A.D. Hocking, J.I. Pitt and A.D. King. pp. 275-283. Amsterdam: Elsevier.

ICMSF. 1988. Microorganisms in Foods. Book 4. HACCP in Microbiological Safety and Quality. Oxford: Blackwell Scientific.

MOSSEL, D.A.A. 1986. Principles in media evaluation. *In* Methods for the Mycological Examination of Food, eds A.D. King, J.I. Pitt, L.R. Beuchat and J.E.L. Corry. pp. 65-66. New York: Plenum Press.

PITT, J.I., HOCKING, A.D. and GLENN, D.R. 1983. An improved medium for the detection of *Aspergillus flavus* and *A. parasiticus*. *J. Appl. Bacteriol.* **54**: 109-114.

PITT, J.I. and SAMSON, R.A. 1990. Systematics of *Penicillium* and *Aspergillus* - past, present and future. *In* Modern Concepts in *Penicillium* and *Aspergillus* Classification, eds R.A. Samson and J.I. Pitt. pp. 3-13. New York: Plenum Press.

THRANE. U, FILTENBORG, O., FRISVAD, J.C. and LUND, F. 1992. Improved methods for detection and identification of toxigenic *Fusarium* species. *In* Modern Methods in Food Mycology, eds R.A. Samson, A.D. Hocking, J.I. Pitt and A.D. King. pp. 285-291. Amsterdam: Elsevier.

NEW SELECTIVE MEDIA FOR THE DETECTION OF TOXIGENIC FUNGI IN CEREAL PRODUCTS, MEAT AND CHEESE

J. C. Frisvad, O. Filtenborg, F. Lund and U. Thrane

Department of Biotechnology
Technical University of Denmark
DK-2800 Lyngby, Denmark

SUMMARY

Four new media have been proposed for selective enumeration and direct identification of fungi associated with cereal products, bread, meat and cheese. Acetic acid Dichloran Yeast extract Sucrose agar (ADYS) is useful for selective isolation and diagnosis of fungi spoiling acid preserved bread, including *Penicillium roqueforti*, *Paecilomyces variotii* and *Monascus ruber*. Dichloran Creatine agar (CREAD) is an effective screening medium for *Penicillium* and *Aspergillus* species associated with meat, cheese, nuts and other lipid containing or proteinaceous foods. All species growing well on CREAD, except *P. solitum*, produce large amounts of specific secondary metabolites, including mycotoxins, enabling direct identification by thin layer chromatography. CREAD has also been used for monitoring the flora of air in bread and cheese factories. A third medium, Dichloran Rose bengal Yeast Extract Sucrose agar (DRYES), developed specifically for detection of *P. verrucosum*, can be effectively used for the qualitative detection of toxigenic *Penicillium* and *Aspergillus* species in foods. Most food-borne species can be identified directly using profiles of secondary metabolites. For the detection of *P. verrucosum* in cereals, DRYES can be improved by replacing rose bengal with 18% glycerol, to produce Dichloran Yeast extract Sucrose 18% Glycerol agar (DYSG). Diagnostic secondary metabolites also are produced in large amounts on this medium allowing direct confirmation of identity and actual mycotoxin production.

INTRODUCTION

To obtain a good picture of the fungi in foods, both dilution plating and direct plating on two or more selective media may be necessary. The combination of Dichloran Rose Bengal Chloramphenicol agar (DRBC; King *et al.*, 1979) and Dichloran 18% Glycerol agar (DG18; Hocking and Pitt, 1980) is effective for the general examination of filamentous fungi from foods (King *et al.*, 1986), but identification of many *Penicillium* and *Aspergillus* species requires subculturing onto identification media.

There is a genuine need for media selective for important species of *Penicillium* and *Aspergillus*, for example, detecting mycotoxins and other specific secondary metabolites directly without subculturing (Filtenborg *et al.*, 1992). Isolates of the same species should look alike, but be distinguishable from those of other species (Filtenborg and Frisvad, 1990). Many food-borne species of *Penicillium* look alike on media such as DRBC, DG18 and malt agars and some are also difficult to differentiate on morphological characters.

Experience with creatine sucrose agar (Frisvad, 1985) and Yeast Extract Sucrose agar (YES; Filtenborg and Frisvad, 1990) was used to develop new selective media for important groups of toxigenic filamentous fungi in foods.

For our work, selective media should restrict the growth of spreading fungi, be discriminating, selective and enhance sporulation; they should also promote production of secondary metabolites to permit species verification or direct identification. For the present investigation, data on the mycoflora associated with many foods (Frisvad and Filtenborg, 1988) assisted us in developing appropriate media, and restricted the number of species of interest. We also emphasised known toxigenic mould species, further reducing the number of species to be identified to species level. A series of well known media (King *et al.*, 1986) were also tested qualitatively for selectivity, support of secondary metabolite production and control of rapidly growing fungi.

We wished to develop a medium for the selective isolation of fungi growing on acid preserved (rye) bread, which also permitted mycotoxin production to allow confirmation of identity and direct determination of important mycotoxins. We also wished to find out whether other filamentous fungi could grow on a medium containing 0.5% acetic acid and if these fungi were natural contaminants of bread. The medium Acetic Dichloran Yeast extract Sucrose agar (ADYS) was developed from observations by Engel and Teuber (1978) who showed that of several species of *Penicillium* tested only *P. roqueforti* could grow in liquid Czapek Dox medium with 0.5% acetic acid added. Frisvad (1981) confirmed that a solid medium, glucose yeast extract agar plus 0.5% acetic acid could also be used to differentiate *P. roqueforti* from other Penicillia. ADYS was a logical development of these needs and observations.

YES was originally developed as a semisynthetic liquid medium for production of aflatoxins (Davis *et al.*, 1966), but was later shown to be an excellent general medium for production of several other mycotoxins (Scott *et al.*, 1970). YES agar was later used for diagnostic purposes and mycotoxin production (Frisvad and Filtenborg, 1989). As the brand of yeast extract in YES agar has a marked influence on production of secondary metabolites, Filtenborg *et al.* (1990) amended YES agar with the addition of magnesium sulphate and overcame this problem.

Engel and Teuber (1978) also tested liquid media containing creatine and nitrite as sole N-sources for *P. roqueforti* and found that all tested strains of this species could grow on such media. However, their results with other species of *Penicillium* were less promising. After testing with many *Penicillium* isolates, Frisvad (1981, 1985) showed that a solid medium with creatine as sole N-source and bromocresol purple as a pH indicator satisfactorily differentiated some species of *Penicillium* subgenus *Penicillium*. Good growth on this medium was correlated with association with foods containing proteins and lipids such as cheese and meat (Frisvad, 1988; Frisvad and Filtenborg, 1988). Creatine agar appears to be a useful selective basal medium for Penicillia and other fungi growing on such substrates. We report here on a qualitative evaluation of these four media.

MATERIALS AND METHODS

Media

The formulae of the media described here are as follows:
Acetic Dichloran Yeast extract Sucrose agar (ADYS) contains yeast extract, 20 g; sucrose, 150 g; $MgSO_4.7H_2O$, 0.5 g; dichloran 0.002 g (0.2% in ethanol, 1 mL); agar, 20 g; and water to 1 litre. After autoclaving, 0.5% glacial acetic acid is added.
Dichloran Creatine sucrose bromocresol agar (CREAD) contains creatine.H_2O, 3 g; sucrose, 30 g; KH_2PO_4, 1 g; $MgSO_4.7H_2O$ 0.5 g; KCl, 0.5 g; $FeSO_4.7H_2O$, 0.01 g; $ZnSO_4.7H_2O$, 0.01 g; $CuSO_4.5H_2O$, 0.005 g; bromocresol purple, 0.05 g; dichloran,

0.002 g (0.2% in ethanol, 1 mL); chloramphenicol, 0.1 g; agar, 20 g; water, 1 litre. The pH is adjusted to 4.8 before autoclaving (modified from Frisvad, 1985).

Dichloran Rose bengal Yeast Extract Sucrose agar (DRYES) contains yeast extract, 20 g; sucrose, 150 g; $MgSO_4.7H_2O$, 0.5 g; rose bengal, 0.025 g; dichloran, 0.002 g (2% in ethanol, 1 mL); chloramphenicol, 0.05 g; $ZnSO_4.7H_2O$, 0.01 g; $CuSO_4.5H_2O$, 0.005 g; agar, 20 g and distilled water to 1 litre. Chlortetracycline (0.05 g) was added after autoclaving (modified from Frisvad, 1983). If tap water is used Zn and Cu need not be added.

Dichloran Yeast extract Sucrose 18% Glycerol agar (DYSG) contains yeast extract, 20 g; sucrose, 150 g; KH_2PO_4, 1 g; $MgSO_4.7H_2O$, 0.5 g; dichloran, 0.002 g (0.2% in ethanol, 1 mL); chloramphenicol, 0.05 g; $ZnSO_4.7H_2O$, 0.01 g; $CuSO_4.5H_2O$, 0.005 g; agar, 20 g; glycerol (A.R.), 220 g and distilled water, 1 litre. Chlortetracycline (0.05 g) is added after autoclaving.

Growth and identification

The media were tested and compared with reference media, DRBC, DG18 and antibiotic amended Plate Count Agar (PCACC), using mass inocula and dilution plating of pure cultures from the IBT fungal culture collection at the Department of Biotechnology, Technical University of Denmark, Lyngby, Denmark. Dilution plating of 30 food and feed samples, and air sampling were also used.

Where possible, fungi were identified to genus and species level without subculturing. The following characters were used: conidial colour, reverse colour, degree of conidiogenesis, presence and colour of ascomata or sclerotia, conidial head structure as seen under a stereomicroscope, structure and ornamentation of microscopical structures as seen under the compound microscope and the profile of secondary metabolites as determined by the agar plug thin-layer chromatography (TLC) method (Filtenborg *et al.*, 1983)

Some of the fungi were isolated and inoculated on Czapek Yeast Extract agar (CYA) and Malt Extract Agar (MEA) (Pitt, 1979) in pure culture to verify their identity. Cultures on both CYA and MEA were also tested for production of secondary metabolites by the agar plug TLC method (Filtenborg *et al.*, 1983).

Statistical evaluation

Because different species of fungi react differently to selective agents, selective media are difficult to compare by the usual statistical methods. Species differences may cancel out when comparing general purpose media, but the differences can be very large when selective principles such as water activity, temperature, gas atmosphere, light and fungicides are employed. Ideally, all fungal species of interest should be compared on both new and reference media. With a desired relative precision of p% of the estimate of the fungal count and F as the assumed average colony number per plate after incubation the following number (r) of parallel plates should be used (Niemela, 1983):

$$r = 10000/p^2F \ldots\ldots (1)$$

For 5% precision and F=25 (a reasonable number for fungi), r=16, i.e. 16 parallel plates should be used for each sample and each medium. As a rule of thumb, 4 isolates of each species to be tested will just give an indication of the possibility of generalisation, 8-12 will give a fair picture of the species on the test media and 24 would be ideal (P.H.A. Sneath, personal communication), given that the isolates are from different geographical and ecological niches. Thus good precision and predictability in fungal colony counts and precise and general medium comparisons require a very large number of tests. In this paper, colony counts have been assumed to be Poisson distributed. A good approximation of the 95% confidence limits of one actual colony count is (Box *et al.*, 1978):

$$i + 2 - 2\sqrt{(i + 1)} < m < i + 2 + 2\sqrt{(i + 1)} \quad \ldots\ldots (2)$$

where m is the estimated mean and i the observed colony count on one plate. For 25 colonies, the 95% lower and upper confidence limits are 16 and 35 and the 99% limits are 14 and 41. For 50 colonies the limits are 37 and 66 and 34 and 71 for 95% and 99% confidence respectively (Box *et al.*, 1978, p. 644). Formula (2) was used for evaluating whether the media suggested here were significantly better or worse for important species than established media such as DG18 and DRBC.

RESULTS AND DISCUSSION

General performance of the media tested

Based on 100 food samples, a qualitative evaluation of the merits of the different media for isolation of filamentous fungi from foods are summarised in Table 1. In general, media with low carbohydrate and high nitrogen levels (PCACC, DCPA and AFPA) were poor for sporulation and production of secondary metabolites, but the latter two media are of value for special purposes (Pitt *et al.*, 1983; Andrews and Pitt, 1986). CREAD and ADYS were effective for the species selected for, as characteristic production of secondary metabolites by each species could be directly tested on the media. The remaining four media were of more general use: DRBC and DRYES could be used for most toxigenic fungi and DG18 and DYSG were suited for xerophilic fungi and most toxigenic Penicillia and Aspergilli including *P. verrucosum* and *A. flavus*. DRBC and DG18 supported sporulation and production of some secondary metabolites, while DRYES and DYSG had advantages like good differentiation of some *Penicillium* species and very good mycotoxin production.

Counts of filamentous fungi

Numbers of xerophilic fungi recovered on the low water activity media, DYSG and DG18, were not significantly different on 12 samples of spices. However counts of xerophilic fungi were significantly lower on DRYES than DYSG or DG18 (95% confidence limit). This was much more pronounced for *Aspergillus* Section *Restricti* and *Wallemia sebi* than for species of *Eurotium*. However species differences were found in the recovery of *Eurotium* on DRYES compared with DYSG and DG18. Further tests, both in food samples and pure cultures, are needed. Counts for important toxigenic species in food samples were not significantly different on DRYES, DYSG and DG18. Counts were obtained for *P. citrinum* (10 samples), *A. flavus* (5 samples), *P. aurantiogriseum* (35 samples) and *P. verrucosum* (5 samples). Again more samples need to be tested to provide general conclusions.

A series of food-borne *Penicillium* species (Frisvad, 1985) and *Aspergillus* species were tested for growth on CREAD. Among common Aspergilli, only *A. versicolor*, *A. sydowii*, *A. ustus* and *A. clavatus* were not inhibited.

Counts of *P. commune* from 15 samples, *P. crustosum* (7 samples), *P. solitum* (4 samples), *P. echinulatum* (1 sample), *P. expansum* (10 samples) and *A. versicolor* (13

Table 1. Observations on selectivity of reference and newly developed media for food-borne fungi

Medium[1]	Features
DRBC	Good control of Zygomycotina, *Fusarium*, *Botrytis*, *Trichoderma* and other fast growing fungi; good sporulation and recovery of field fungi; poor differentiation of *Penicillium* species; poor recovery of xerophiles; poor production of secondary metabolites.
DG18	Good control of fast growing fungi; poor sporulation and recovery of field fungi; poor differentiation of *Penicillium* species; good recovery of xerophiles, but occasionally poor sporulation; poor production of secondary metabolites in most cases.
PCACC	Poor control of fast growing fungi but excellent recovery of different species of Zygomycotina; poor sporulation of field fungi; poor differentiation of *Penicillium* species (and poor sporulation and conidium colour development); poor recovery of xerophiles; very poor production of secondary metabolites.
DRYES	Diagnosis of *Penicillium verrucosum*; reasonable control of fast growing fungi, occasionally heavy growth of *Rhizopus* and *Mucor* species; good recovery but poor sporulation of field fungi; poor recovery of *Wallemia* and *Aspergillus* section *Restricti*; reasonable recovery and excellent differentiation of *Eurotium* species; good recovery and differentiation of *Penicillium* species; excellent production of secondary metabolites.
DYSG	Diagnosis of *Penicillium verrucosum*; good control of fast growing fungi; poor recovery of field fungi; excellent recovery of xerophiles; growth rate of *Eurotium* occasionally too high; good recovery and differentiation of *Penicillium* species; production of secondary metabolites inhibited in *Penicillium* species other than *P. verrucosum*.
CREAD	Excellent recovery of *P. roqueforti*, *P. crustosum*, *P. solitum*, *P. commune*, *P. echinulatum*, *P. expansum*, *P. spinulosum*, *A. versicolor* and *A. sydowii*; poor growth of other food-borne fungi; good control of fast growing fungi; poor recovery of field fungi and xerophiles; excellent production of secondary metabolites by fungi listed.
ADYS	Excellent recovery of *P. roqueforti*, *Paecilomyces variotii* and *Monascus ruber*. No other filamentous fungi have been found to grow on this medium; good production of secondary metabolites by these species.

[1] DRBC, DG18, DRYS see Appendix; PCACC, Plate Count Agar with Chlortetracycline and Chloramphenicol; DYSG, Dichloran Yeast extract Sucrose 18% Glycerol agar; CREAD, acidified Dichloran Creatine agar; ADYS, Acetic Dichloran Yeast extract Sucrose agar.

samples) were not significantly different on CREAD, DRBC and DG18. However on crowded plates with more than 75 colonies, colonies of *P. commune* and *A. versicolor* were readily counted on CREAD, even in the presence of 10 to 40 times as many creatine negative strains. With cheese samples, colonies of *P. nalgiovense* were in some cases predominant on DG18 and DRBC, making reliable counting of *P. commune* impossible. However on CREAD, *P. nalgiovense* was strongly inhibited and *P. commune* and *A. versicolor* could be easily counted. Similarly, *A. versicolor* from spice samples was only counted with difficulty on DRBC and DG18 because of strong growth of other Aspergilli such as *A. niger*, *A. tamarii* and *A. flavus*. However, as most Aspergilli are strongly inhibited by creatine, *A. versicolor* could be easily counted on CREAD. Counts on ADYS were not significantly different from those on DG18, DRBC and CREAD but germination of three pure cultures of *P. roqueforti* var. *roqueforti* was retarded for up to 5 days. Ten air samples from factories showed no statistical differences between the number of *P. roqueforti* colonies on ADYS and DG18. Pure cultures of *Paecilomyces variotii* and *Monascus ruber* were not inhibited by 0.5% acetic acid, however, few colonies were observed in this study.

A summary (Table 1) lists observed advantages and disadvantages of commonly used dilution plating media for filamentous fungi, based on examination of 100 samples of cereals, spices, herbs, cheeses, meat and nuts.

Wallemia sebi, Aspergillus restrictus and *A. penicillioides* developed significantly better on DYSG than on DG18. On DG18 some colonies of *W. sebi* did not develop the characteristic brown conidia or reverse colour and the sporulation of many isolates of *Aspergillus* Section *Restricti* was poor. However these xerophiles did not develop at all on CREAD and DRBC, and only about one tenth of those recovered on DYSG were counted on DRYES. For this reason total counts are not reported here as they would only reflect the different recovery of these species. As mentioned earlier, the main purpose of these media is recovery of individual species. A very large number of food samples would be needed to test optimal recovery of all important toxigenic species. Total counts can mask interactions between species and media.

Direct identification to genus level

Direct identification of all fungi to genus level was only possible by using more than one medium. Again DYSG was good for the recovery and direct identification of xerophilic fungi, i.e. *Eurotium*, *Wallemia*, *Aspergillus* Section *Restricti*, *Penicillium* and *Chrysosporium* by observation under a stereomicroscope or the compound microscope. Field fungi such as *Botrytis*, *Alternaria*, *Drechslera*, *Cladosporium*, *Fusarium* and *Epicoccum* were best recovered on DRBC, but sporulation was often poor. This was most pronounced for *Epicoccum*, *Fusarium* and *Alternaria*, genera requiring light for good sporulation. The use of light to induce sporulation is not possible with media containing rose bengal, such as DRBC and DRYES, because rose bengal may be give rise to production of toxic singlet oxygen (Pady *et al.*, 1960; Madelin, 1987; Gams and van Laar, 1982). Rose bengal may be inhibitory for some fungi even without exposure to light (Madelin, 1987).

Direct identification to species level

Direct identification of all fungi to species level was not possible, but many isolates could be identified using the characteristics listed in Methods. The readily recognisable species included *Aspergillus niger*, *A. candidus*, *A. terreus* and *A. fumigatus*. The latter species may be mistaken for a *Penicillium* by superficial inspection. The use of a stereomicroscope assisted in confirming the identity. Common species of toxigenic filamentous fungi occurring on cereals, meat, cheese, bread and nuts are listed in Table 2, with characters that can be used directly for identification on isolation plates.

Identification of storage fungi to species level in stored cereals

Counts and appearance of *Wallemia* and *Aspergillus* Section *Restricti* on DYSG were significantly better than on DG18. Occasionally DG18 gave higher counts of *Eurotium* than other media, but the colonies had rather undefined margins and differentiation was poor. Spreading of *Rhizopus* and *Mucor* was best controlled on DYSG and DG18. Differences between media were less pronounced for other species, e.g. *Eurotium repens, E. chevalieri, P. aurantiogriseum, P. verrucosum, Cladosporium* species, *A. flavus, A. niger* and *A. versicolor.* DRYES and DYSG differentiated more *Eurotium* and *Penicillium*

Table 2. Direct identification of common food-borne filamentous storage fungi on DRYES agar

Fungus	Criterion[1]
Aspergillus candidus	White smooth conidia
A. flavus	Yellow green conidia, finely roughened
A. fumigatus	Blue green rough conidia, in columns
A. niger	Black or dark brown, rough conidia
A. ochraceus	Small yellow brown, smooth conidia
A. parasiticus	Dark green rough conidia
A. sydowii[2]	Blue green rough conidia
A. tamarii	Brown rough conidia, very thick walled
A. terreus	Cinnamon smooth ellipsoidal conidia, in columns
A. versicolor[2]	Slowly growing colonies, echinulate conidia
A. wentii	Few conidial heads, brown conidia
Emericella nidulans	Hülle cells
Eurotium species	Yellow cleistothecia
Monascus ruber	Stalked brown cleistothecia
Paecilomyces variotii	Light brown smooth ellipsoidal conidia, in long divergent chains
Penicillium aurantiogriseum var. *aurantiogriseum*	Blue green conidia, yellow reverse, poor sporulation
P. aurantiogriseum var. *polonicum*	Blue green conidia, yellow reverse, good sporulation
P. aurantiogriseum var. *melanoconidium*	Green conidia, good sporululation
P. viridicatum	Green conidia, yellow reverse, poor sporulation
P. brevicompactum	Rough conidia, good sporulation
P. chrysogenum	Smooth stipes, good sporulation
P. commune[2]	Poor sporulation
P. crustosum[2]	Good sporulation
P. echinulatum[2]	Dark green rough conidia
P. expansum[2]	Ellipsoidal conidia
P. griseofulvum	Small phialides, divergent rami
P. roqueforti var. *roqueforti*	Green reverse, smooth globose conidia, rough stipes
P. roqueforti var. *carneum*[2]	Low colonies, smooth globose conidia, rough stipes, pale reverse
P. solitum[2]	Dark green conidia, orange reverse
P. verrucosum	Violet brown reverse

[1] The plates are first examined under the stereomicroscope to identify the isolates to genus level, then mounted and examined under the compound microscope.
[2] These species grow well on CREAD.

species than DG18 and DRBC, an important factor when a large number of colonies are found in each plate. Production of specific secondary metabolites was best on DRYES, but a certain number of secondary metabolites were also produced on DRBC, DYSG and DG18 (Filtenborg *et al.*, 1992).

A detailed comparison with other media (Table 1) showed that DYSG was an even better selective and diagnostic medium for *P. verrucosum* than DRYES. Production of the three specific secondary metabolites, ochratoxin A, citrinin and verrucolone was higher on DYSG than on DRYES and much higher than on other media tested. The pronounced violet brown reverse colour of *P. verrucosum* and the absence of rose bengal made counting on DYSG very easy. The major drawback of DYSG is a rather high growth rate of *Eurotium* species, and that of DRYES is the poor control of *Rhizopus* growth in some cases. Nevertheless, the combination of DRYES and DYSG appeared to be of value for directly identifying important toxigenic species of filamentous storage fungi in cereals. The two latter media should be compared with DG18 using a large number of cereal samples containing ochratoxin A in order to detect whether the numbers of *P. verrucosum* recovered are statistically different.

Identification of *Penicillium* and *Aspergillus* species from bread, meat and cheese

The mycoflora of meat, cheese and nuts is significantly different from that of cereals (Frisvad and Filtenborg, 1988), because of large differences in food composition. Important toxigenic species on these high protein foods include *Aspergillus versicolor, Penicillium crustosum, P. commune, P. solitum, P. echinulatum, P. roqueforti* and *P. expansum.* These species grow well on creatine sucrose agar (Frisvad, 1981, 1985). Unlike the formulation of Frisvad (1985), CREAD used here was of low pH (4.8), to enable observation of the characteristic base production by *P. crustosum, P. echinulatum, P. commune* and *P. expansum.* Three other fungi grew well: *P. spinulosum, A. sydowii* and *A. versicolor.* The latter has been shown to be a very important contaminant of cheese (Northolt and Soentoro, 1988). Counts of *A. versicolor* were usually higher on CREAD than on other media, probably because this slow growing fungus is overgrown by *Eurotium* and other fungi on low a_w media (Figure 1). *Eurotium* species grow very poorly on CREAD.

On CREAD *P. solitum* has dark green conidia and a late production of base, while *P. roqueforti* var. *roqueforti* has a dark green reverse, no production of base, a high growth rate and a low colony. *P. aurantiogriseum* var. *polonicum*, which occasionally occurs on meat and nuts, grew quite well on CREAD, but the margin of the colonies was arachnoid and the fungus was clearly inhibited. *P. echinulatum* could be recognised by its rough, dark green conidia. *P. expansum* was the only creatine positive terverticillate *Penicillium* species with smooth stipes. Differentiation between *P. commune* and *P. crustosum* was difficult.

In addition, characteristic secondary metabolites are produced strongly on CREAD (Filtenborg *et al.*, 1992), so species can be identified using TLC. Other fungi found on cheese, meat and nuts, especially drier products, include *P. brevicompactum, P. chrysogenum* and *P. corylophilum*, all creatine negative species. These species can be recognised on DRYES or DYSG, although production of secondary metabolites by *P. chrysogenum* on DRYES and other low a_w media is weak. *P. chrysogenum* can be identified by its smooth stipes with quite divergent rami and smooth conidia or by meleagrin and roquefortine C production on media based on CYA. *P. brevicompactum* is easily recognised on DRYES by its characteristic *Aspergillus* like conidial heads under the stereomicroscope and by its micromorphology. Thus for mycological examination of these food types a combination of CREAD and DRYES, DRBC or DG18 can be used.

Direct identification of species occuring on acid preserved bread

From a large number of air samples from bread factories and direct plating of mouldy bread on ADYS, only *P. roqueforti* var. *roqueforti, P. roqueforti* var. *carneum, Paecilomyces variotii* and *Monascus ruber* were found. These taxa could be recognised by their

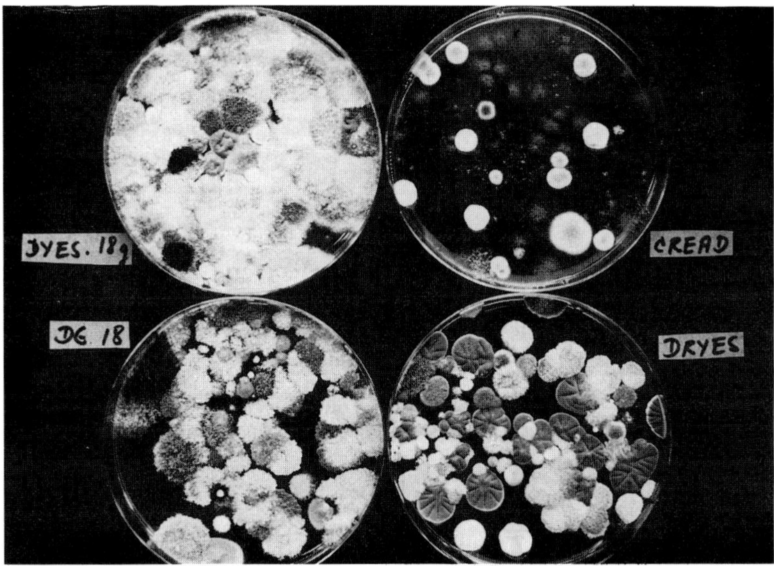

Figure 1. A sample of white pepper dilution plated on DYSG, CREAD, DG18 and DRYES. Note the strong growth of *A. niger*, *Eurotium* spp., *A. tamarii* and *Penicillium* spp, masking the small colonies of *A. versicolor*. On CREAD, *A. versicolor* can easily be counted.

conidial or mycelial colours and morphology. *P. roqueforti* has green, globose conidia, *Paecilomyces variotii* light brown, ellipsoidal conidia and *M. ruber* reddish brown mycelium and abundant ascomata. *P. roqueforti* var. *roqueforti* produced PR toxin on ADYS while *Paecilomyces variotii* produced viriditoxin (chemotype I). Chemotype II of *P. variotii*, producing patulin and looking more like a species of *Penicillium*, was not found on any of the samples tested. No significant difference was found in dilution plates of two isolates of *P. roqueforti* var. *roqueforti*. ADYS plates should be counted after seven days because some conidia germinate more slowly on this acetic acid medium than on other media.

REFERENCES

ANDREWS, S. and PITT, J.I. 1986. Selective medium for isolation of *Fusarium* species and dematiaceous hyphomycetes from cereals. *Appl. Environ. Microbiol.* **51**: 1235-1238.

BOX, G.E.P., W.G. HUNTER and J.S. HUNTER. 1978. Statistics for experimenters. New York: John Wiley and Sons.

DAVIS, N.D., DIENER, U.L. and ELDRIDGE, D.W. 1966. Production of aflatoxins B_1 and G_1 by *Aspergillus flavus* in a semisynthetic medium. *Appl. Microbiol.* **14**: 378-380.

DOMSCH, K.H., GAMS, W. and ANDERSON, T.-H. 1980. Compendium of Soil Fungi. New York: Academic Press.

ENGEL, G. and TEUBER, M. 1978. Simple aid for the identification of *Penicillium roqueforti* Thom. Growth in acetic acid. *Eur. J. Appl. Microbiol. Biotechnol.* **6**: 107-111.

FILTENBORG, O. and FRISVAD, J.C. 1990. Identification of *Penicillium* and *Aspergillus* species in mixed cultures in Petri dishes using secondary metabolite profiles. *In* Modern Concepts in *Penicillium* and *Aspergillus* Classification, eds R.A. Samson and J.I. Pitt. pp. 27-36. New York: Plenum Press.

FILTENBORG, O., FRISVAD, J.C. and SVENDSEN, J.A. 1983. Simple screening method for molds producing intracellular mycotoxins in pure cultures. *Appl. Environ. Microbiol.* **45**: 581- 585.

FILTENBORG, O., FRISVAD, J.C. and THRANE, U. 1990. The significance of yeast extract composition on metabolite production in *Penicillium*. *In* Modern Concepts in *Penicillium* and *Aspergillus* Classification, eds R.A. Samson and J.I. Pitt. pp. 433-440. New York: Plenum Press.

FILTENBORG, O., FRISVAD, J.C., LUND, F. and THRANE, U. 1992. Simple identification procedure for spoilage and toxigenic mycoflora of foods. *In* Modern Methods in Food Mycology, eds R.A. Samson, A.D. Hocking, J.I. Pitt and A.D. King. pp. 263-273. Amsterdam: Elsevier.

FRISVAD, J.C. 1981. Physiological criteria and mycotoxin production as aids in identification of common asymmetric Penicillia. *Appl. Environ. Microbiol.* **41**: 568-579.

FRISVAD, J.C. 1983. A selective and indicative medium for groups of *Penicillium viridicatum* producing different mycotoxins in cereals. *J. Appl. Bacteriol.* **54**: 409-416.

FRISVAD, J.C. 1985. Creatine-sucrose agar, a differential medium for mycotoxin producing terverticillate *Penicillium* species. *Lett. Appl. Microbiol.* **1**: 109-113.

FRISVAD, J.C. 1988. Fungal species and their specific production of mycotoxins. *In* Introduction to Food-borne Fungi, eds R.A. Samson and E.S. van Reenen Hoekstra. pp. 239-249. Baarn: Centraalbureau voor Schimmelcultures.

FRISVAD, J.C. and FILTENBORG, O. 1988. Specific mycotoxin producing *Penicillium* and *Aspergillus* mycoflora of different foods. *Proc. Jap. Assoc. Mycotoxicol.* **Suppl. 1**: 163-166.

FRISVAD, J.C. and FILTENBORG, O. 1989. Terverticillate Penicillia: chemotaxonomy and mycotoxin production. *Mycologia* **81**: 837-861.

GAMS, W. and VAN LAAR, W. 1982. The use of solacol (validamycin) as a growth reterdant in the isolation of soil fungi. *Netherland J. Plant Pathol.* **88**: 39-45.

HOCKING, A.D. and PITT, J.I. 1980. Dichloran-glycerol medium for enumeration of xerophilic fungi from low moisture foods. *Appl. Environ. Microbiol.* **39**: 488-492.

KING, A.D., HOCKING, A.D. and PITT, J.I. 1979. Dichloran-rose bengal medium for enumeration and isolation of molds from foods. *Appl. Environ. Microbiol.* **37**: 959-964.

KING, A.D., PITT, J.I., BEUCHAT, L.R. and CORRY, J.E.L. eds. 1986. Methods for the Mycological Examination of Food. New York: Plenum Press.

MADELIN, T.M. 1987. The effect of a surfactant in media for the enumeration, growth and identification of airborne fungi. *J. Appl. Bacteriol.* **63**: 47-52.

NIEMELA, S. 1983. Statistical evaluation of results from quantitative microbiological examinations. Nordic Committee on Food Analysis, NMKL Report 1, 2nd edn. Uppsala.

NORTHOLT, M.D. and SOENTORO, P.S.S. 1988. Fungal growth on foodstuffs related to mycotoxin contamination. *In* Introduction to Food-borne Fungi, eds R.A. Samson and E.S. van Reenen Hoekstra. pp. 231-238. Baarn: Centraalbureau voor Schimmelcultures.

PADY, S.M., KRAMER, C.L. and PATHAK, V.K. 1960. Suppression of fungi by light on media containing rose bengal. *Mycologia* **52**: 347-350.

PITT, J.I. 1979. The Genus *Penicillium* and its Teleomorphic States *Eupenicillium* and *Talaromyces*. London: Academic Press.

PITT, J.I., HOCKING, A.D. and GLENN, D.R. 1983. An improved medium for the detection of *Aspergillus flavus* and *A. parasiticus*. *J. Appl. Bacteriol.* **54**: 109-114.

SCOTT, P.M., LAWRENCE, J.W. and VAN WALBEEK, W. 1970. Detection of mycotoxins by thin-layer chromatography: application to screening of fungal extracts. *Appl. Microbiol.* **20**: 839-842.

IMPROVED METHODS FOR DETECTION AND IDENTIFICATION OF TOXIGENIC *FUSARIUM* SPECIES

U. Thrane, O. Filtenborg, J. C. Frisvad and F. Lund

Department of Biotechnology
Technical University of Denmark
DK-2800 Lyngby, Denmark

SUMMARY

A simple procedure for detection and direct identification of *Fusarium* species has been designed. The primary isolation medium, potato dextrose agar with iprodione and dichloran, is selective for *Fusarium* species. *Fusarium* isolates are identified directly on this medium by cultural characteristics and simple morphological characters. For confirmation thin layer chromatographic analysis of the cultures on the isolation substrate may be performed. Most routine laboratories, where only a limited number of *Fusarium* species will be of interest, can recognise *Fusarium* species by this method. Application of the method to the cereal industry is given here, as well as a simple synoptic key for identification of the five most common species in malting barley, *F. avenaceum*, *F. culmorum*, *F. equiseti*, *F. poae* and *F. tricinctum*.

INTRODUCTION

Taking into consideration that individual food and feed samples usually contain a limited number of *Fusarium* species (Nirenberg, 1989) and that routine laboratories examine a large number of similar samples, it is possible to design simple procedures for rapid detection and identification of food or feed associated *Fusarium* species. Routine laboratories can be expected to be able to recognise the few commonly encountered species (3-6). Less common species will occur from time to time, but identification of 70-80% of the *Fusarium* flora will usually cover the important species. Our examination of barley and malt during recent years indicates that the most commonly occurring *Fusarium* species are *F. avenaceum*, *F. culmorum*, *F. equiseti*, *F. poae* and *F. tricinctum*.

The selective substrate for *Fusarium* species Czapek Iprodione Dichloran agar (CZID) was developed for use in the cereal industry (Abildgren *et al.*, 1987), primarily to enhance the existing methods for detecting *Fusarium* in malting barley.

Since 1988 CZID has been tested in industrial laboratories in Finland, the Netherlands and Denmark for detection of *Fusarium* species in malting barley. The laboratories have included CZID with routine *Fusarium* counts on Potato Dextrose Agar (PDA) modified by addition of pentachloronitrobenzene (PCNB), Malt Salt Agar (MSA) and direct plating on filter paper as recommended by the European Brewing Convention (EBC) (Gyllang *et al.*, 1981). As a result of the initial testing programmes, the EBC Barley and Malt Microflora Subgroup conducted a collaborative examination in three laboratories of 10 barley samples on CZID, the modified PDA and filter paper. A minimum of 100 surface disinfected kernels per sample were direct plated on each substrate. In 24 of the 30 tests *Fusarium* was detected when plated on CZID (mean: 15 % *Fusarium* infection), while *Fusarium* was only detected in 15 cases (mean: 2 % infection) when plated on the modified PDA and in 16 cases (mean: 12 % infection) when using filter paper (P. Sigsgaard, unpubl.). Thus, *Fusarium* was detected much more frequently on CZID than on modified PDA and filter paper. CZID has now become the standard substrate for *Fusarium* counts in one brewery in Denmark, two malt houses in Finland, and at the Technical Research Centre of Finland, Espoo. Recently CZID has been recommended by EBC as the standard substrate for detection of *Fusarium* species in barley and malt (P. Sigsgaard, pers. comm., 1990).

CZID is difficult to use for rapid identification of *Fusarium* isolates to species level, so subculturing is often required. A pictorial album in colour of 17 common *Fusarium* species on CZID, PDA and oat meal agar modified by addition of PCNB and Triton X-100 has been published (Anonymous, 1989), but even that requires pure cultures.

CZID has been used regularly at this laboratory for isolating *Fusarium* by direct plating, but the isolates have always been subcultured for identification. Attempts to simplify identification by adding analysis by thin layer chromatography (TLC) for production of secondary metabolites on CZID have been unsuccessful as no metabolites useful for confirmation of identity have been detected so far. However, the use of both PDA and Yeast Extract Sucrose agar (YES), ensures production of the important metabolites in detectable amounts (Thrane, 1986).

This study was undertaken to develop a simple screening procedure for *Fusarium* species by incorporating iprodione, the selective principle from CZID, in Potato Dextrose Agar and Yeast Extract Sucrose Agar, followed by direct identification (Thrane, 1989; Filtenborg and Frisvad, 1990). Direct identification on isolation substrates requires a combined use of two or more characters. The characters proposed are: a) cultural appearance (mycelium, pigmentation), b) morphology of conidia and conidiophores, and c) metabolite profile on the isolation substrate (agar plug TLC).

MATERIALS AND METHODS

Direct plating

Malting barley and other grains were examined for *Fusarium* by direct plating of surface disinfected samples onto the media described below. Kernels were surface disinfected by continuously shaking with 1 % sodium hypochlorite (freshly prepared) for 2 min. Kernels were set into the medium before agar solidified to prevent germination. Plates were incubated at 25°C for 7 days under alternating black light and daylight fluorescent tubes (12 hr cycle).

Dilution plating

Conidial suspensions were prepared from cultures grown on Spezieller Nährstoffarmer Agar (SNA; Nirenberg, 1976), diluted in 0.9 % NaCl in water. Suspensions were inoculated on the media described below. Plates were incubated at 25°C for 7 days under alternating black light and daylight fluorescent tubes (12 hr cycle).

Colony diameters

A small experiment was conducted where colonies arising from dilution of conidial suspensions of pure cultures were plated onto CZID, DCPA, DRBC, DRYES and PDID. Colonies were counted and growth rates or colony diameters measured.

Media

Potato Dextrose Iprodione Dichloran agar (PDID) consisted of potato dextrose broth (Difco), 24 g; $ZnSO_4.7H_2O$, 0.01 g; $CuSO_4.5H_2O$, 0.005 g; chloramphenicol, 0.05 g; dichloran, 0.002 g (0.2% in ethanol, 1 mL); agar, 20 g; water, 1 litre. pH was adjusted to 6.5-7.0 before autoclaving. After autoclaving and cooling to 60°C, chlortetracycline hydrochloride, 0.05 g (0.5% sterile aqueous solution, 10 mL) and iprodione, 0.003 g [0.6 % Rovral 50WP (50% iprodione) sterile aqueous suspension, 1 mL] were added before pouring.

Yeast Extract Sucrose Iprodione Dichloran agar (YESID) consisted of yeast extract (Difco), 20 g; sucrose, 150 g; $MgSO_4.7H_2O$, 0.5 g; $ZnSO_4.7H_2O$, 0.01 g; $CuSO_4$ $5H_2O$, 0.005 g; chloramphenicol, 0.05 g; dichloran, 0.002 g (0.2% in ethanol, 1 mL); agar, 20 g; water, 1 litre. After autoclaving and cooling to 60°C, chlortetracycline hydrochloride, 0.05 g (0.5% sterile aqueous solution, 10 mL) and iprodione, 0.003 g [0.6 % Rovral 50WP (50% iprodione) sterile aqueous suspension, 1 mL] were added before pouring.

Formulations for CZID are given by Abildgren *et al.* (1987), Dichloran Chloramphenicol Peptone Agar (DCPA) by Andrews and Pitt (1986), Dichloran Rose Bengal Chloramphenicol agar (DRBC) by King *et al.* (1979) and Dichloran Rose bengal Yeast Extract Sucrose agar (DRYES) by Frisvad (1983).

Secondary metabolite production

For determination of secondary metabolites, the agar plug TLC method of Filtenborg *et al.* (1983) was used. The developing system used was toluene/ethyl acetate/90 % formic acid, 5/4/1 v/v/v (TEF). The retardation factor, R_{fg}, relative to the retardation factor of an external griseofulvin standard (R_{fg} 1.0), was calculated (Thrane, 1986).

RESULTS AND DISCUSSION

Initial studies indicated that PDID was just as selective for *Fusarium* as CZID, and gave slightly higher *Alternaria* counts (unpublished data). PDID also has the advantage that mycologists working with *Fusarium* will have had experience with PDA (or potato sucrose agar, PSA), so the morphological and cultural characteristics already published in most *Fusarium* monographs and manuals can be used directly on PDID cultures. The full profile of secondary metabolites was not produced on PDID but still the key metabolites were produced, equivalent to the metabolite production in PSA cultures (Thrane, 1986).

On YESID the effect of the fungicides was strongly reduced. No selection for *Fusarium* was observed, as several *Penicillium*, *Alternaria* and *Rhizopus* species showed good growth on YESID together with *Fusarium*. Increasing the amount of fungicides was unsuccessful as *Fusarium* species were strongly inhibited. The mixture of iprodione and dichloran has to be adjusted carefully. Furthermore, the morphological structures observed on YESID were too poor to be of any value.

Colony growth rates were much higher on DCPA and DRYES than on CZID, PDID and DRBC. Counting a high number of colonies was therefore more difficult (Table 1). Up to 100 colonies were easily counted on CZID, PDID and DRBC but such counts were impossible on DCPA and DRYES due to a 5-10 fold higher growth rate.

CZID and PDID were suitable for dilution plating as no selective inhibition was observed. DRBC is less suitable for *Fusarium*, as incubation under alternating light, recommended for sporulation, may trigger the transformation of rose bengal into inhibitory compounds. The red colour of the substrate is also a disadvantage.

Table 1. Counts (means of duplicates) and colony diameters (mm) of six *Fusarium* species on five different media

	CZID		PDAID		DCPA		DRBC		DRYES	
	Count	Diam.	Count	Diam.	Count	Diam.	Count	Diam.	Count	Diam.
F. avenaceum	92	3	122	2-3	NC[1]	30	107	10	139	15
F. equiseti	86	10	133	10	NC	>40	146	10	NC	>40
F. culmorum	61	10	80	10	NC	>30	75	10-18	NC	>40
F. verticillioides	>200	4	>200	4	NC	>40	>200	4	NC	>40
F. poae	>200		>200		NC		>200		NC	
F. pallidoroseum	>200		>200		NC		>200		NC	

[1] NC, not countable because growth rates too high

Secondary metabolite production varied widely on the substrates used. The *Fusarium* species examined produced detectable amounts of secondary metabolites on YESID and PDID. However, metabolite production on CZID, DCPA and DRBC was too poor to be of diagnostic value in routine identifications. An acceptable compromise between morphology, cultural appearance and secondary metabolites was put together using PDID as isolation substrate.

Direct identification of *Fusarium* species in the cereal industry

The following procedure, applied to barley and malt, is being implemented by Danish laboratories:

1. Surface disinfected kernels are surface plated onto PDID before agar solidifies.
2. Plates are incubated for 7 days at 25°C under alternating light.
3. Cultural appearance and pigmentation are then used in an attempt to identify cultures by comparison with colour slides of reference cultures.
4. Cultural characters are observed by stereomicroscope and microscopic morphology on a slide preparation in water plus 0.25 % Triton-X.
5. Secondary metabolites are examined by agar plug TLC on PDID cultures.
6. Cultures can then be identified from the synoptic key below.

Data from an industrial laboratory implementing the application are very promising. During an 18 month period, 41 barley and malt samples were plated onto PDID and DRYES in addition to routine counts on Czapek Dox agar (CZ in Raper and Thom, 1949), Dichloran 18% Glycerol agar (DG18; Hocking and Pitt, 1980) and MSA. A total of 2050 kernels were plated on each substrate.

The majority of the samples were infected by *Alternaria* species and the infection rate was quite high, but caused no interference with *Fusarium* identification (Table 2). The *Fusarium* infection was low: only 27 samples were positive, with a mean of 8 % infection when plated on PDID (Table 2). Other substrates gave a lower *Fusarium* count.

Isolates from PDID cultures were checked for identity by the senior author. Seventy of the 107 isolates (65%) were correctly identified by the industrial laboratory (Table 3).

Table 2. Fungi on 41 barley and malt samples as examined by an industrial laboratory using direct plating on five media

	PDID		DRYES		DG18		CZ		MSA	
	Samples[1]	Rate[2]	Samples	Rate	Samples	Rate	Samples	Rate	Samples	Rate
Fusarium spp.	66	8	7	3	7	2	37	6	NC[3]	
Alternaria spp.	NC		80	74	83	77	NC		NC	
Eurotium and *Aspergillus* spp.	NC		24	5	34	16	NC		27	12
Penicillium spp.	NC		27	12	17	6	NC		NC	

[1] Percent of samples infected
[2] Mean infection rate of infected samples
[3] NC, not counted

Table 3. Identification of 107 isolates of fungi from barley and malt by an industrial laboratory using techniques described in this paper

	No. of isolates	Isolates identified correctly	Percentage
Fusarium poae	51	38	75
F. tricinctum	28	19	68
F. avenaceum	11	6	55
F. equiseti	11	4	37
F. culmorum	4	3	75
Microdochium nivale	2	-	-
Total	107	70	65

F. poae was often identified by colony texture and the odour, whereas *F. avenaceum* and *F. tricinctum* were recognised by the red colour in the substrate and the production of antibiotic Y. Pigmentation in patches in *F. tricinctum* cultures, and orange sporodochia of *F. avenaceum* were often used for discrimination of these species. The low accuracy in identification of *F. equiseti* was due to poor sporulation and weak TLC patterns.

Of other genera occurring on malt and barley *Epicoccum nigrum* needs to be taken into consideration, as this red- and yellow pigmented species is also able to grow on PDID. Some cultures of *E. nigrum* may be mistaken for a *Fusarium* species on the basis of cultural appearance and pigmentation, but microscopic differences are readily recognisable. Using TLC analysis, *E. nigrum* is characterised by a bright yellow spot at R_{fg} 1.68 and three blue spots at R_{fg} 1.30, 1.02, and 0.73, respectively, when the TLC plate is inspected under UV-light (365 nm).

Synoptic key for common *Fusarium* species on malt and barley

1. Reverse colour on PDID
   Rose to red: *F. avenaceum*, *F. culmorum*, *F. poae*, *F. tricinctum*
   Pale to yellow: *F. avenaceum*, *F. equiseti*, *F. poae*, *F. tricinctum*
   Brownish: *F. avenaceum*, *F. equiseti*

2. Morphology
   One type of conidium: *F. equiseti*, *F. poae*
   Two or more types of conidia: *F. avenaceum*, *F. poae*, *F. tricinctum*
   Pyriform to globose conidia: *F. poae*, *F. tricinctum*
   Sporodochia present: *F. avenaceum*, *F. culmorum*, *F. equiseti*, *F. tricinctum*

3. Secondary metabolites detected by TLC
   Antibiotic Y [yellow fluorescence (365 nm), R_{fg} 0.91]: *F. avenaceum*, *F. tricinctum*
   Fusarin C (faint yellow in daylight, R_{fg} 0.58): *F. culmorum*, *F. poae*
   Fusarochromanones [blue fluorescence (365 nm), R_{fg} 0.1 + 0.5]: *F. equiseti*
   Zearalenones [blue fluorescence (254 nm), R_{fg} 1.35 + 1.74]: *F. culmorum*, *F. equiseti*

Characteristics of common *Fusarium* species in barley and malt

F. avenaceum

PDID: light floccose mycelium, orange to brownish sporodochia.
Reverse colour: pale to rose (red), occasionally brownish tinged.
Morphology: two types of conidia, 1-3 celled 'cigar-shaped' conidia from aerial mycelium and long and slender ('needle shaped') conidia from substrate mycelium (sporodochia).
Secondary metabolites on TLC: antibiotic Y.

F. culmorum

PDID: fast growing mycelium, often with yellow tips, red-brown (orange) sporodochia.
Reverse colour: rose to vinaceous red.
Morphology: one type of conidium, 4-7 celled, with notched ends.
Secondary metabolites on TLC: fusarin C, zearalenones.

F. equiseti

PDID: light (yellowish-brownish) floccose mycelium.
Reverse colour: pale, salmon to light brownish.
Morphology: one type of conidium, 4-7 celled, with tapering apical cell and distinct foot-shaped basal cell. Chlamydospores often present.
Secondary metabolites on TLC: fusarochromanones, zearalenones.

F. poae

PDID: fast growing white to yellowish floccose mycelium. May have sweet, fruity odour.
Reverse colour: pale, yellowish to rose.
Morphology: two types of conidia, globose to pyriform 1-2 celled conidia commonly borne in the aerial mycelium, and falcate to kidney-shaped 3-5 celled conidia produced in low numbers.
Secondary metabolites on TLC: fusarin C.

F. tricinctum

PDID: pale sparse mycelium often with annular rings from the illumination periods.
Reverse colour: red to vinaceous red, as pigmentated patches, or occasionally yellow pigmented.
Morphology: two types of conidia, pyriform and fusoid/cylindrical 1-2 celled conidia from the aerial mycelium and falcate 3-5 celled conidia from the substrate mycelium.
Secondary metabolites on TLC: antibiotic Y.

REFERENCES

ABILDGREN, M.P., LUND, F., THRANE, U. and ELMHOLT, S. 1987. Czapek-Dox agar containing iprodione and dicloran as a selective medium for the isolation of *Fusarium* species. *Lett. Appl. Microbiol.* **5**: 83-86.

ANDREWS, S. and PITT, J.I. 1986. Selective medium for isolation of *Fusarium* species and dematiaceous hyphomycetes from cereals. *Appl. Environ. Microbiol.* **51**: 1235-1238.

ANONYMOUS. 1989. *Fusarium* Picture Album. Espoo, Finland: Technical Research Centre of Finland.

FILTENBORG, O. and FRISVAD, J.C. 1990. Identification of *Penicillium* and *Aspergillus* species in mixed cultures in Petri dishes using secondary metabolite profiles. *In* Modern Concepts in *Penicillium* and *Aspergillus* Classification, eds R.A. Samson and J.I. Pitt, pp. 27-37. New York: Plenum Press.

FILTENBORG, O., FRISVAD, J.C. and SVENDSEN, J.A. 1983. Simple screening method for molds producing intracellular mycotoxins in pure cultures. *Appl. Environ. Microbiol.* **45**: 581-585.

FRISVAD, J.C. 1983. A selective and indicative medium for groups of *Penicillium viridicatum* producing different mycotoxins in cereals. *J. Appl. Bacteriol.* **54**: 409-416.

GYLLANG, H., KJELLÉN, K., HAIKARA, A. and SIGSGAARD, P. 1981. Evaluation of fungal contamination on barley and malt. *J. Inst. Brewing* **87**: 248-251.

HOCKING, A.D. and PITT, J.I. 1980. Dichloran-glycerol medium for enumeration of xerophilic fungi from low-moisture foods. *Appl. Environ. Microbiol.* **39**: 488-492.

KING, A.D., HOCKING, A.D. and PITT, J.I. 1979. Dichloran-rose bengal medium for enumeration and isolation of molds from foods. *Appl. Environ. Microbiol.* **37**: 959-964.

NIRENBERG, H.I. 1976. Untersuchungen über die morphologische und biologische Differenzierung in der *Fusarium*-Sektion *Liseola*. *Mitteilung aus der Biologische Bundesanstalt für Land- und Forstwirschaft*. Berlin-Dahlem. **169**: 1-117.

NIRENBERG, H.I. 1989. Identification of Fusaria occurring in Europe on cereals and potatoes. *In Fusarium*: Mycotoxins, Taxonomy and Pathogenicity, ed. J. Chelkowski, pp. 179-193. Amsterdam: Elsevier.

RAPER, K.B. and THOM, C. 1949. A Manual of the Penicillia. Baltimore: Williams and Wilkins.

THRANE, U. 1986. Detection of toxigenic *Fusarium* isolates by thin layer chromatography. *Lett. Appl. Microbiol.* **3**: 93-96.

THRANE, U. 1989. *Fusarium* species and their specific profiles of secondary metabolites. *In Fusarium*: Mycotoxins, Taxonomy and Pathogenicity, ed. J. Chelkowski, pp. 199-225. Amsterdam: Elsevier.

COMPARISON OF SEVERAL MEDIA AND METHODS FOR DETECTION AND ENUMERATION OF TOXIGENIC *FUSARIUM* SPECIES

L. B. Bullerman[1] and Diane I. West[2]

[1]*Department of Food Science and Technology*
University of Nebraska
Lincoln, NE 68583-0919, U.S.A.

[2]*ConAgra Frozen Foods,*
Six ConAgra Drive
Omaha, NE 68102-5006, U.S.A.

SUMMARY

Several media were tested for the ability to detect and enumerate *Fusarium* species. These included Nash Snyder medium (NS), Modified Czapek-Dox medium (MCz), Modified Czapek-Dox medium + Pentachloronitrobenzene (MCz + PCNB), Tomato Juice Agar (Difco), V-8 Juice Agar, Czapek-Dox Agar (Difco) and Potato Dextrose agar (Difco) + tetracycline. Stock cultures of *Fusarium*, *Aspergillus*, *Penicillium*, *Alternaria*, *Cladosporium* and *Rhizopus* species were grown on these media and the nature and appearance of growth determined. Media were also tested by direct plating of surface sterilised and unsterilised whole kernel corn samples, and by plate counts. NS was the most selective medium for *Fusarium*, by both direct plating and plate count techniques as only *Fusarium* species grew on this medium. The disadvantages of NS medium were that *Fusarium* species required long incubation times (7 days) for colony formation and showed poor colour development. The next most selective media for *Fusarium* were MCz and MCz+PCNB. Both were fairly selective for *Fusarium* but permitted growth of other mould genera. However, *Fusarium* species were readily distinguishable from other moulds, with colour development more characteristic of *Fusarium* than on NS. Incubation times were also shorter (5 days). All of the other media lacked selectivity and were not suitable for isolation and enumeration of *Fusarium*.

INTRODUCTION

The genus *Fusarium* includes numerous mycotoxin producing species which are commonly found on cereal grains, cereal foods and feeds (Marasas *et al.*, 1984b). Most *Fusarium* species are plant pathogens and invade plant tissues and developing seeds (grain) in the field (Mills, 1989). Some are able to persist in harvested and stored grain, and grow in storage if moisture contents become favourable (Mills, 1989).

Two species of considerable interest are *Fusarium graminearum* Schwabe and *Fusarium moniliforme* Sheldon. *F. graminearum* can invade both corn and wheat in the field, causing corn ear rot and wheat scab (head blight) (Mills, 1989). *F. graminearum* produces deoxynivalenol (DON), zearalenone, nivalenol and fusarenon X (Marasas *et al.*, 1984a). *F. moniliforme* invades corn, sorghum and rice. It is seed and soil borne, and very common in corn plants, ultimately invading kernels in the field and persisting into storage (Hesseltine and Bothast, 1976; Marasas *et al.*, 1979). *F. moniliforme* produces a number of mycotoxins including moniliformin and fumonisins (Gelderblom *et al.*, 1988; Mills, 1989).

Other potentially mycotoxigenic *Fusarium* species of importance in cereals include *F. sporotrichioides*, *F. poae*, *F. semitectum*, *F. equiseti*, *F. culmorum*, *F. subglutinans* and *F. oxysporum* (Mills, 1989). These species produce a range of mycotoxins including zearalenone, deoxynivalenol, T-2 toxin, diacetoxyscirpenol, neosolaniol, butenolide, and nivalenol (Mills, 1989). A critical need therefore exists for development of selective media for the detection and enumeration of *Fusarium* species in foods and feeds.

Fusarium species are often associated with necrotic plant tissue, which may be extensively colonised by other fungi and bacteria, so selective media have long been used to isolate *Fusarium* (Nelson *et al.*, 1983). Two such media are Nash and Snyder medium (Nash and Snyder, 1962) and modified Czapek-Dox medium (Nelson *et al.*, 1983). This paper compares these media with several others commonly used to culture Fusaria, for detecting *Fusarium* species in corn and for plate counts of *Fusarium* in processed foods.

MATERIALS AND METHODS

Culture Media

Nash and Snyder (NS) medium was prepared according to Nelson *et al.* (1983) and consisted of peptone (Difco), 15.0 g; KH_2PO_4, 1.0 g; $MgS0_4 \cdot 7H_20$, 0.5 g; agar (Difco), 20.0 g; pentachloronitrobenzene, 1.0 g; and distilled water, 1.0 L. The medium was autoclaved at 121°C for 15 min. Streptomycin sulphate stock solution, 20.0 mL and neomycin sulphate stock solution, 12.0 mL, were added to each litre of medium just prior to pouring. Stock solutions contained streptomycin sulphate 5.0 g/100 mL and neomycin sulphate 1.0 g/100 mL in separate distilled water solutions. Stock solutions were sterilised by filtration through 0.45 μm membrane filters.

Modified Czapek Dox (MCz) medium, prepared according to Nelson *et al.* (1983) consisted of dextrose, 20.0 g; KH_2PO_4, 0.5 g; $NaNO_3$, 2.0 g; $MgSO_4 \cdot 7H_20$, 0.5 g; yeast extract, 1.0 g; 1% ferrous sulphate solution, 1.0 mL; agar (Difco), 20.0 g and distilled water, 1.0 L. It was sterilised by autoclaving at 121°C for 15 min. Dichloran (1.0 mL of stock solution), streptomycin sulphate (20 mL of stock solution) and tetracycline (20 mL of stock solution) were added just before pouring. Stock solutions contained streptomycin sulphate (5.0 g/100 mL distilled water), tetracycline (0.2 g/100 mL distilled water) and dichloran (0.2 g/100 mL ethanol) and were sterilised by filtration through a 0.45 μm membrane filter.

Modified Czapek Dox medium plus pentachloronitrobenzene (MCz + PCNB) was MCz with the addition of 1.0 g/L of pentachloronitrobenzene. Tomato Juice agar (TJ), Czapek-Dox agar (Cz) and Potato Dextrose Agar (PDA) were all standard Difco agars (Difco Laboratories, Detroit, MI). Potato Dextrose Agar plus tetracycline (PDAT) was prepared by adding 40 μg/L of tetracycline to PDA after autoclaving. V-8 juice agar was prepared by neutralising and clarifying V-8 juice with calcium carbonate ($CaCO_3$) and centrifuging at 3000 rpm. Supernatant (200 mL) was made to 1 L with distilled water, and 15 g of agar added (Romero and Gallegly, 1963).

Fungal Cultures.

Pure cultures of 23 different fungal species were used. *Fusarium* species with numbers of each in brackets were: *F. acuminatum* (1), *F. avenaceum* (1), *F. crookwellense* (1), *F. culmorum* (1), *F. equiseti* (1), *F. graminearum* (5), *F. moniliforme* (6), *F. oxysporum* (1), *F. poae* (3), *F. proliferatum* (3), *F. sporotrichioides* (3), *F. subglutinans* (3) and *F. tricinctum* (2). Other fungi tested included *Aspergillus flavus*, *A. parasiticus*, *A. ochraceus*, *A. oryzae*, *Penicillium cyclopium*, *P. martensii* and *P. viridicatum*, and one species each of *Rhizopus*, *Cladosporium* and *Alternaria*.

Inoculation and incubation

Pure cultures were inoculated onto agar plates of each of the different media in three point inoculation patterns. Cultures were incubated for 5-7 days and growth rates observed.

Food Samples

Corn samples (6) were obtained from the Lincoln Grain Inspection Service, Lincoln, NE. Tortillas, oatmeal and wheat germ obtained from retail sources were also examined.

Corn samples were surface sterilised by soaking for 1 min in 5.25% sodium hypochlorite, rinsed three times in sterile distilled water and dried on sterilised paper towels. Both surface sterilised and unsterilised corn kernels were direct plated, 10 kernels per plate, onto each medium except TJ and Cz. Tortillas were cut into c.a. 1.0 cm squares and direct plated. NS and MCz + PCNB were also compared with MCz, V-8 and PDAT by direct plating of surface sterilised kernels of corn in advanced stages of deterioration. Plates were incubated for 5 to 10 days and observed for mould growth.

Corn samples were also ground in a laboratory mill (Glen Mills, Maywood, NJ, USA), diluted with phosphate buffer, mixed in a Colworth Stomacher, serially diluted in phosphate buffer and spread plated onto all media except TJ and Cz. The rolled oats and wheat germ were diluted and plated similarily. Plates were incubated for 5 to 7 days and counted using a Quebec colony counter.

RESULTS AND DISCUSSION

Pure culture studies

Fusarium species grew well on NS and MCz, while species of other genera were suppressed. NS was more suppressive than MCz, and also resulted in slower growth of *Fusarium* species. The best growth, colony characteristics and colour development of *Fusarium* species was observed on MCz, V-8 and PDAT. TJ agar was a dark colour which interfered with visual observations of colony colours. Color development on Cz was less than on MCz. *Fusarium* species grew readily on MCz. Growth of other genera was suppressed for a time, but they eventually grew and formed recognisable colonies. Based on these results TJ and Cz agars were eliminated from further study.

Direct plating

Results of direct plating studies are summarised in Table 1. Unsterilised corn kernels were heavily contaminated with spreading, Mucoraceous fungi which quickly overgrew the V-8, MCz and PDAT plates. However, NS restricted the growth of spreading moulds but allowed the growth of *Fusarium* species and species of other genera. Surface sterilised corn kernels showed a high percentage internal infection with *Fusarium* which was readily detectable on NS, MCz and PDAT, while V-8 allowed growth of spreading moulds in some samples. NS and MCz media appeared to give the best combination of *Fusarium* growth along with suppression of growth of other moulds.

Table 1. Percentage mould infection of corn kernels as determined on Nash and Snyder (NS), modified Czapek's (MCz), modified Czapek's + PCNB (MCz + PCNB) and Potato Dextrose + Tetracycline (PDAT) agars

Corn Sample	NS	MCz	MCz + PCNB	PDAT
1	1.0	4.0	4.0	1.0
2	ND[1]	2.0	3.0	1.0
3	ND	5.0	1.0	ND
4	4.0	3.0	4.0	2.0
5	ND	2.0	2.0	2.0
6	ND	3.0	2.0	2.0

[1]ND - None detected

Table 2. Total mould counts of several foods as determined on Nash and Snyder (NS), Modified Czapek's (MCz), Modified Czapek's + PCNB (MCz + PCNB), and Potato Dextrose + Tetracycline (PDAT) agars

Food	NS	MCz	MCz + PCNB	PDAT
Wheat germ	ND[1]	$1.4x10^3$	$3.5x10^2$	$1.1x10^3$
Yellow corn	$3.1x10^4$	$4.1x10^4$	$3.2x10^4$	$5.5x10^4$
White corn	$3.8x10^3$	$1.4x10^4$	$1.1x10^4$	$1.0x10^4$

[1]ND - None detected

In comparisons using badly deteriorated corn kernels, PDAT and V-8 were readily overgrown with spreading moulds even from surface sterilised kernels. They were judged to be unsuitable for isolation and enumeration of *Fusarium*. MCz allowed growth of *Fusarium* species as well as most other genera. Spreading moulds quickly became a problem when the plates were incubated for longer times. *Fusarium* species were recognisable on MCz but other moulds were not sufficiently suppressed to permit easy isolation. MCz + PCNB gave results similar to those obtained with MCz but with slightly longer suppression of other moulds.

Overall, NS medium gave the best results in terms of suppression of competing moulds and growth of *Fusarium* species. The main disadvantage to NS was that growth of *Fusarium* species was slower, requiring 7 to 10 days to form recognisable colonies, and with less colour development. MCz and MCz + PCNB permitted more rapid growth of *Fusarium* species, allowing earlier enumeration and identification, but competing moulds grew if incubation was too long. NS was the best medium for detection, enumeration and identification of *Fusarium* species, but MCz and MCz + PCNB were also very good and had the advantage of supporting more rapid growth of *Fusarium* species, and better colour development.

Dilution plating

Counts of foods on NS were slightly lower than on MCz, MCz + PCNB and PDAT (Table 2). No moulds were detected in wheat germ on NS and counts in ground corn were slightly lower on NS than the other media. This was due to the fact that only *Fusarium* moulds were detected on NS, whereas *Aspergillus, Penicillium* and other moulds grew on the other media. PDAT permitted growth of spreading moulds which made counting difficult. No moulds were detected in the oatmeal on any of the media.

CONCLUSIONS

Overall, NS was most selective for *Fusarium* species, both for direct plated corn kernels and for plate counts, while competing moulds were suppressed. MCz was somewhat less selective than NS for *Fusarium*, as other mould genera were able to grow on prolonged incubation. MCz + PCNB suppressed growth of competing moulds only slightly longer than MCz. NS had the disadvantage of requiring longer incubation than MCz, and permitted less colour development by *Fusarium* colonies. MCz did not suppress the growth of other moulds as well as NS. Of the media tested, NS exhibited the best qualities for the overall detection, enumeration and isolation of *Fusarium* species. Purified *Fusarium* cultures should be transferred from NS to other media such as V-8, PDAT and Cz for maintenance and identification studies as soon as possible.

REFERENCES

GELDERBLOM, W.C.A., JASKIEWICZ, K., MARASAS, W.F.O, THIEL, R.G., HORAK, M. J., VLEGGAAR, R. and KRIEK, N.P.J. 1988. Fumonisins - novel mycotoxins with cancer-promoting activity produced by *Fusarium moniliforme*. *Appl. Environ. Microbiol.* **54**: 1806-1811.

HESSELTINE, C.W. and BOTHAST, R.J. 1976. Mold development in ears of corn from tasseling to harvest. *Mycologia* **69**: 328-340.

MARASAS, W.F.O., KRIEK, N.P.J., WIGGINS, V.M., STEYN, P.S., TOWERS, D.K. and HASTIE, T. J. 1979. Incidence, geographic distribution, and toxicity of *Fusarium* species in South African corn. *Phytopathology* **69**: 1181-1185.

MARASAS, W.F.O., KRIEK, N.P.J., FINCHAM, J.E. and VAN RENSBURG, S.J. 1984a. Primary liver cancer and oesophageal basal cell hyperplasia in rats caused by *Fusarium moniliforme*. *Int. J. Cancer* **34**: 383-387.

MARASAS, W.F.O., NELSON, P.E., and TOUSSOUN, T.A. 1984b. Toxigenic *Fusarium* Species. Identity and Mycotoxicology. University Park, PA: Pennsylvania State University Press.

MILLS, J.T. 1989. Ecology of mycotoxigenic *Fusarium* species on cereal seeds. *J. Food Prot.* **52**: 737-742.

NASH, S.M. and SNYDER, W.C. 1962. Quantitative estimations by plate counts of propagules of the bean root rot *Fusarium* in field soils. *Phytopathology* **52**: 567-572.

NELSON, P.E., TOUSSOUN, T.A. and MARASAS, W.F.O. 1983. *Fusarium* species: an Illustrated Manual for Identification. University Park, PA: Pennsylvania State University Press.

ROMERO, S. and GALLEGLY, M.E. 1963. Oogonium germination in *Phytophthora infestans*. *Phytopathology* **53**: 899-903.

EVALUATION OF METHODS FOR SELECTIVE ENUMERATION OF *FUSARIUM* SPECIES IN FEEDSTUFFS

D. E. Conner

Department of Poultry Sciences
Auburn University,
Auburn, AL 36849-5416, USA.

SUMMARY

Dichloran Chloramphenicol Peptone Agar (DCPA) containing crystal violet (CV) was evaluated for its ability to support growth of *Fusarium* spp. while inhibiting growth of *Aspergillus* spp. and *Penicillium* spp. A concentration of 1.0 μg/mL CV was inhibitory to all test cultures, while 0.5 μg/mL CV allowed selective growth of *Fusarium* spp. In subsequent studies, DCPA containing 0.5 μg/mL CV (DCPCVA), DCPA and Dichloran Rose Bengal Chloramphenicol agar (DRBC) were evaluated for efficacy in selectively enumerating *Fusarium* spp. from grain based feedstuffs. Additionally, the effect of holding feedstuff samples at -20°C for 24 hr prior to plating was determined. Samples of untreated and frozen feedstuffs were stomached for 2 min in 0.1% peptone water (PW), serially diluted with PW, surface inoculated onto test media and incubated upright at 25°C for 5 days. Both DCPA and DCPCVA were effective in suppressing *Aspergillus* and *Penicillium* growth, and populations of *Fusarium* spp. were more easily distinguished on these media than on DRBC. Holding samples at -20°C for 24 hr had no effect on fungal populations enumerated.

INTRODUCTION

Because of the relation of many *Fusarium* spp. to disease manifestations in poultry, there is a need to monitor poultry feeds for populations of these fungi (Mirocha, 1984). Pitt and Hocking (1985) recommended Dichloran Chloramphenicol Peptone Agar (DCPA) for the enumeration and isolation of *Fusarium* spp. from soil, grains and animal feeds. In commercial feeds, crystal violet (CV) is often used as an antifungal additive; however, according to some marketing information, CV is less effective against *Fusarium* spp. than against other fungi. This paper reports an evaluation of the addition of low levels of CV to DCPA to further suppress growth of fungi other than Fusaria while allowing full development of Fusaria. In subsequent studies, DCPA containing 0.5 μg/mL CV (DCPCVA), DCPA and Dichloran Rose Bengal Chlorampenicol agar (DRBC) were evaluated for efficacy in selectively enumerating *Fusarium* spp. from feedstuffs based on grain. Additionally, the effect of treating feedstuff samples by freezing on recovery of *Fusarium* species was determined.

MATERIALS AND METHODS

Evaluation of crystal violet

DCPA, prepared according to directions given by Pitt and Hocking (1985), was supplemented with 0, 0.5, 1.0, 2.0, 4.0 or 8.0 μg/mL CV (Sigma Chemical Co., St. Louis,

MO, USA), surface inoculated with conidial suspensions of *Fusarium roseum* R6136, *F. equiseti* R4238, *Aspergillus flavus* (feed isolate) or *Penicillium chrysogenum* (provided by G. Morgan-Jones, Auburn University) and incubated upright at 25°C for 5 days. Colony size and development of the cultures on DCPA containing the various levels of CV were determined by visual examination.

Evaluation of enumeration methods

Feedstuffs analysed. Six different poultry feeds or feed ingredients were analysed (Table 1). All samples were obtained from the research feed mill, Department of Poultry Science, Auburn University.

Table 1. Feedstuffs analysed

Corn meal	Broiler feed
Rice meal	Layer feed
Chick feed	Breeder feed

Treatment by freezing. Prior to plating and enumeration, samples (*ca.* 100 g) of each feedstuff were held at -20°C for 24 hr.

Media. Three media were evaluated for their efficacy in selectively enumerating *Fusarium* spp. in feedstuffs, both untreated and treated by freezing. These were DCPA prepared according to Pitt and Hocking (1985); Crystal Violet supplemented DCPA (DCPCVA), DCPA supplemented with CV to give a final concentration of 0.5 μg/m; and DRBC, prepared according to King et al. (1986). All media were cooled to 45-50°C before pouring, and plates were allowed to surface dry for 24 hr at room temperature before inoculation.

Enumeration procedure

Samples (25 g) of feedstuffs both untreated and after treatment by freezing were mixed with 225 mL of 0.1% aqueous peptone (PW) and stomached for 2 min. Serially diluted (PW) samples were inoculated (0.1 mL) onto duplicate plates of the three media within 30 sec. of stomaching. Samples were surface plated, and plates incubated upright at 25°C for 5 days. Plates on which 15-150 typical *Fusarium* colonies (as determined by colony morphology and microscopic examination) had formed were used for enumeration purposes. Subjective evaluations regarding ease of counting and growth of other fungi were also made.

RESULTS AND DISCUSSION

Evaluation of crystal violet

Unsupplemented DCPA allowed excellent growth of both *F. roseum* and *F. equiseti*, and allowed some growth of *A. flavus* and *P. chrysogenum* (Table 2). At a concentration of 0.5 μg/mL of CV, growth of *F. roseum* and *F. equiseti* was not impaired; however, growth of *A. flavus* and *P. chrysogenum* was sparse. At 1.0 μg/mL of CV, the Fusaria were inhibited and *A. flavus* and *P. chrysogenum* did not grow. None of the test fungi grew on DCPA containing $\geq$ 2.0 μg/mL of CV.

Table 2. Fungal growth on dichloran chloramphenicol peptone agar containing various levels of crystal violet

CV concentration (µg/mL)	Colony development			
	F. roseum	*F. equiseti*	*P. chrysogenum*	*A. flavus*
0	++++	++++	+++	+++
0.5	++++	++++	+	+
1.0	+	+	-	-
2.0	-	-	-	-
4.0	-	-	-	-
8.0	-	-	-	-

Evaluation of enumeration methods

Populations of *Fusarium* spp. recovered from untreated and treated feedstuffs on the three media are given in Table 3.

Table 3. Recovered populations ($\log_{10}$ cfu/g) of *Fusarium* spp.

Sample	DCPA		DCPCVA		DRBC	
	Untreated	Frozen[1]	Untreated	Frozen	Untreated	Frozen
Corn meal	4.71	4.70	4.79	3.34	4.96	3.60
Rice meal	4.95	4.53	4.60	3.58	4.18	4.57
Chick feed	4.45	4.59	4.20	4.48	3.41	3.28
Broiler feed	3.97	3.81	3.69	2.97	3.98	3.54
Layer feed	4.48	4.23	4.48	4.48	3.41	3.11
Breeder feed	4.81	4.41	4.79	4.30	4.69	3.89

[1] Samples held at -20°C for 24 hr prior to plating.

There were only small differences in populations of *Fusarium* spp. recovered on the three media evaluated. Treatment by freezing had little effect on populations of *Fusarium* spp. and was not effective in suppressing recovery of other fungi on DRBC. Within particular samples, more *Aspergillus* and *Penicillium* species were recovered on DRBC than on DCPA and DCPCVA, indicating that DCPA and DCPCVA were effective in suppressing growth of *Aspergillus* and *Penicillium* species while allowing good development of *Fusarium* spp. This allowed counting and identifying of *Fusarium* colonies to be carried out more readily.

Results indicate that 0.5 µg/mL of CV in DCPA did not adversely affect the growth of *F. roseum* and *F. equiseti*, but was effective in inhibiting *A. flavus* and *P. chrysogenum* when tested under laboratory conditions. In analysis of feed samples, incorporation of 0.5 μg/mL of CV into DCPA offered no obvious advantage over unsupplemented DCPA in selectively enumerating *Fusarium* spp. However, supplementation of DCPA with CV may offer some advantages in enumerating Fusaria from other types of products.

REFERENCES

KING, A.D., PITT, J.I., BEUCHAT, L.R. and CORRY, J.E.L., eds. 1986. Methods for the Mycological Examination of Food. p. 297. New York: Plenum Press.

MIROCHA, C.J. 1984. Mycotoxicoses associated with *Fusarium*. *In* The Applied Mycology of *Fusarium*, eds M.O. Moss and J.E. Smith. pp. 141-156. Cambridge: Cambridge University Press.

PITT, J.I. and HOCKING, A.D. 1985. Fungi and Food Spoilage. p. 39. Sydney: Academic Press.

COLLABORATIVE STUDY ON MEDIA FOR DETECTION AND DIFFERENTIATION OF *ASPERGILLUS FLAVUS* AND *A. PARASITICUS*, AND THE DETECTION OF AFLATOXIN PRODUCTION

J. I. Pitt

CSIRO Division of Food Processing
P.O. Box 52
North Ryde, N.S.W. 2113, Australia

With collaboration from: K. Akerstrand, N. Braendlin, J.C. Frisvad, D. Heperkan, K.A. Wheeler and P. Nielsen.

SUMMARY

This collaborative study had three aims: (i) to compare the efficacy of Aspergillus Flavus and Parasiticus Agar (AFPA) and Aspergillus Differential Medium (AMD) with and without antibiotics against Dichloran Rose Bengal Chloramphenicol Agar (DRBC) as a standard medium; (ii) to test the efficacy of DRBC for differentiating *A. flavus* from *A. parasiticus*, in comparison with microscopic examination as a standard method; and (iii) to test the ability of Coconut Extract Agar (CEA) to detect aflatoxin production by *A. flavus* and *A. parasiticus*.

Results showed that ADM was ineffective in the absence of antibiotics because of bacterial overgrowth. Counts on ADM with antibiotics and AFPA were comparable with DRBC. However, enumeration on the two differential media was much easier than on DRBC, while AFPA was much more effective than ADM with antibiotics in the presence of spreading fungi such as *Mucor* and *Rhizopus*.

All participants agreed that DRBC was an effective medium for distinguishing *A. flavus* from *A. parasiticus*. Comparison with standard cultures of each species was strongly advised.

CEA was reasonably effective at detecting aflatoxin production. Some isolates were reported to produce a yellow fluorescence as well as the blue one characteristic of aflatoxins. Again the use of a standard aflatoxigenic isolate as a control was considered essential.

INTRODUCTION

Aspergillus Differential Medium (ADM) was described by Bothast and Fennell (1974) for the rapid detection of *Aspergillus flavus* and *A. parasiticus*, the main producers of the carcinogenic aflatoxins. The essential ingredient of ADM was reported to be ferric citrate, which induces formation of an intense orange yellow colour in the reverse of colonies of these two species. This colour, accurately described as Cadmium Orange (Kornerup and Wanscher, 1978), was rarely observed under colonies of other fungi.

ADM as originally formulated lacked agents to control bacteria and spreading fungi. Pitt *et al.* (1983) modified ADM by adding dichloran and chloramphenicol. Adjustment of medium composition resulted in an optimal formulation, named Aspergillus Flavus and Parasiticus Agar (AFPA). Enumeration could be carried out after only 2 days incubation at 30°C.

Beuchat (1984) modified ADM by adding 0.01% chloramphenicol after sterilising to produce ADM plus antibiotic (ADMA). He compared ADMA with AFPA, and reported that both media were of value, were usually more effective at 30°C than at 37°C, and that the orange yellow reverse colour was produced more quickly on AFPA and was of higher intensity.

Differentiation of *A. flavus* from *A. parasiticus* is important because the latter species makes G as well as B aflatoxins, and frequently at higher concentrations. Also, higher percentages of *A. parasiticus* isolates are toxigenic (Klich and Pitt, 1988). It has been found that colonies of these two species show consistently different conidial colours on DRBC. *A. flavus* colonies are yellow green, and those of *A. parasiticus* dark green (Hocking and Pitt, in preparation).

This study was primarily designed to collaboratively test the efficacy of ADM, ADMA and AFPA for detection and enumeration of *A. flavus* and *A. parasiticus*. A second aim was to collaboratively test the efficacy of DRBC for differentiating these species.

A third aim was to examine the effectiveness and ease of use of a coconut agar medium for detecting aflatoxin production by *A. flavus* and *A. parasiticus*. Such a medium would be of great utility in laboratories which lack the facilities to carry out thin layer chromatography, or where large numbers of isolates need to be screened for the ability to produce aflatoxins. The medium studied was that of Davis *et al.* (1987).

MATERIALS AND METHODS

Samples

The protocol for this collaborative study requested the use of both dilution plating and direct plating techniques, with a minimum of 3-4 samples to be examined by each method. Samples for dilution plating were to be suitable for blending or Stomaching, i.e. crushed nuts, corn flour, etc. Samples were to be stomached or blended with 0.1% peptone water in the ratio 1:10, diluted 1:10 to 10^{-2} or 10^{-3}, then 0.1 ml portions surface plated.

Samples for direct plating were to be particulate foods, i.e. peanuts, corn kernels or cut pieces of larger nuts. Samples (50 pieces) were to be surface disinfected (2 min) in household or laboratory chlorine bleach diluted 1:10 with water. After chlorine treatment, samples were to be rinsed once with sterile water then plated directly onto media, at the rate of 7-12 pieces per plate.

Media

The following media were to be used: Aspergillus Differential medium (ADM; Bothast and Fennell, 1974), ADM with chloramphenicol (ADMC; Beuchat, 1984); Aspergillus Flavus and Parasiticus Agar (AFPA; Pitt *et al.*, 1983), Dichloran Rose Bengal Chloramphenicol agar (King *et al.*, 1986) and Coconut Extract Agar (CEA; Davis *et al.*, 1987).

ADM contains tryptone, 1.5%, yeast extract, 1.0%, ferric citrate, 0.05% and agar, 1.5%. ADMA is ADM with pH adjusted to 6.5 and with chloramphenicol, 0.01%, added before autoclaving. AFPA contains bacteriological peptone, 1.0%; yeast extract, 2.0%; ferric ammonium citrate, 0.05%; chloramphenicol, 0.01%; dichloran, 0.1% of a 0.2% stock solution in ethanol; and agar, 1.5%. CEA, slightly modified from Davis *et al.*

(1987) was to be made as follows: take 100 g desiccated or shredded coconut, homogenise (5 min) with hot water (500 mL) and filter through 4 layers of cheese cloth. Add agar (1.5%), heat to dissolve and autoclave.

Protocol

Experiment 1. Evaluation of media for enumerating *A. flavus* and *A. parasiticus*.

Carry out sample preparation as above. Spread plate appropriate dilutions and direct plate surface sterilised particles onto ADM, ADMA, AFPA and DRBC. Incubate all plates except DRBC at 28-30°C for 2-3 days, then count colonies with bright orange-yellow reverse colours only. Do not continue incubation beyond 4 days.

Incubate DRBC plates at 25°C for 5 days. Count *A. flavus* and *A. parasiticus* by inspection under a stereomicroscope, using yellow green *Aspergillus* heads as the criterion.

Report counts for each dilution plating sample and medium, as log count/g, and for direct plating as percentage invasion of particles by *A. flavus* plus *A. parasiticus*.

Experiment 2. Evaluation of a method for distinguishing *A. flavus* from *A. parasiticus*.

Inoculate colonies identified as *A. flavus* and *A. parasiticus* in Experiment 1 onto DRBC plates, preferably as one colony each of four isolates per plate. Also inoculate colonies of standard *A. flavus* and *A. parasiticus* isolates (supplied). Incubate plates at 25°C for 5 days or more. Attempt to differentiate between *A. flavus* and *A. parasiticus* on the basis of conidial colour on the DRBC plates, scoring yellow green colonies as *A. flavus* and dark green colonies as *A. parasiticus*, using standard colonies as a reference. Subsequently, make wet mounts and examine each colony under the compound microscope at high (400X) magnification. Distinguish *A. flavus* from *A. parasiticus* on the basis of conidial wall texture, shape and size (Klich and Pitt, 1988).

Report results as numbers of *A. flavus* and *A. parasiticus*, comparing the two techniques.

Experiment 3. Evaluation of CEA for assessing aflatoxin production by *A. flavus* and *A. parasiticus*.

Plate colonies from Experiment 1 onto CEA, by a technique similar to that in Experiment 2. Incubate at 25-30°C for 2-5 days. Examine inverted plates under long wave length ultraviolet light. Aflatoxin production is indicated by a characteristic blue fluorescence at colony centres, or more widespread. Yellow-green fluorescence should be ignored. If plastic Petri dishes are used, pale fluorescence around the edges of the plates (from the plastic itself) should also be ignored.

Separately report percentage of *A. flavus* and *A. parasiticus* colonies producing aflatoxin.

RESULTS

Experiment 1. Evaluation of media for enumerating *A. flavus* and *A. parasiticus*

Collaborators all reported examining a high percentage of negative samples. They also reported that ADM (without antibiotics) was ineffective due to bacterial overgrowth, and few results were obtained. Pooled positive results for the other three media from all investigators are given in Table 1.

Table 1. Log counts of *Aspergillus flavus* plus *A. parasiticus* dilution plated on three media[a]

Sample	ADMC	AFPA	DRBC
Nutmeg	2.32	2.40	2.38
Corn flour	3.46	3.48	3.46
Ground hazelnut	2.58	2.65	2.62
Soy flour	1.78	1.90	1.90
Soy flour	2.00	2.00	2.00
Pistachios	2.30	2.30	2.00
Buckwheat flour	2.00	2.70	2.00
Barley	2.00	2.78	-
Corn meal	2.54	2.48	2.00
Spice mix	3.36	3.32	3.33
White pepper	1.30	2.54	-
Oat meal	2.65	2.60	3.04
Corn	4.66	4.76	-
Corn	2.88	2.78	-
Corn	3.11	2.93	4.48
Corn	4.76	4.82	5.04
Corn	4.00	4.26	4.92
Corn	2.60	2.65	-
Average[b]	2.83	2.94	3.08

[a] ADMC, Aspergillus differential medium with chloramphenicol; AFPA, Aspergillus flavus and parasiticus agar; DRBC, Dichloran rose bengal chloramphenicol agar.
[b] Average of the 10 samples where data were available for all three media.

Table 2. Percentage infection of samples by *Aspergillus flavus* plus *A. parasiticus* as determined by direct plating on three media[a]

Sample	ADMC	AFPA	DRBC
Hazelnuts	4	2	4
Hazelnuts	0	0	6
Hazelnuts	20	0	20
Pistachios	32	50	30
Pistachios	8	10	10
Pistachios	20	20	20
Black pepper	0	10	10
Black pepper	20	50	10
Green coffee	2	2	4
Peanuts	14	12	14
Corn	72	84	44
Corn	84	94	54
Corn	92	96	54
Corn	20	16	10
Average corn	67	73	41
Average others	10	13	11

[a] ADMC, Aspergillus differential medium with chloramphenicol; AFPA, Aspergillus flavus and parasiticus agar; DRBC, Dichloran rose bengal chloramphenicol agar.

From the small amount of data available, no significant differences were apparent between the results for the three media. No statistical differences were found. For samples with data for all three media, average counts were ranked DRBC > AFPA > ADMC, but differences were small.

Results for direct plated samples are given in Table 2. Again the small number of samples prevents the drawing of any general conclusions. However, DRBC was significantly less effective that the other two media. Averages, both for corn, which was highly infected, and other commodities, where infection was lower, was in the order AFPA > ADMC > DRBC.

Collaborators reported that enumeration using ADMA or AFPA was simpler and more precise than using DRBC, as the diagnositc reverse colour was readily observed. When spreading fungi such as *Mucor* and *Rhizopus* were present on the plates, AFPA was a much more effective enumeration medium than ADMA.

Experiment 2. Evaluation of a method for distinguishing *A. flavus* from *A. parasiticus*

All collaborators reported that differentiation of *A. flavus* from *A. parasiticus* was effectively accomplished on DRBC. Table 3 provides a summary of the data obtained. Operators varied in their ability to distinguish the two species on DRBC, but the maximum error, as judged by microscopical identification, was 4/66 identifications; the least was zero. Considering that none of the operators had any prior training in the use of this technique, the results are highly satisfactory.

Experiment 3. Evaluation of CEA for assessing aflatoxin production by *A. flavus* and *A. parasiticus*

Data obtained from three collaborators in the test of CEA as a detection medium for aflatoxin production are summarised in Table 4. Collaborator comment was that attempting to observe more than one colony fluorescing on each CEA plate was very

Table 3. Comparison between differentiation of *A. flavus* and *A. parasiticus* on DRBC and by standard microscopic techniques

Collaborator	No. of isolates	Based on conidial colour on DRBC		Based on microscopic morphology	
		A. flavus	*A. parasiticus*	*A. flavus*	*A. parasiticus*
1	100	83	17	83	17
2	25	25	0	25	0
3	27	25	2	24 [a]	2
4	66	47	19	44 [b]	18 [b]
5[c]	100	74	26	75	25

[a] One doubtful

[b] Three presumptive *A. flavus* with rough conidia; one presumptive *A. parasiticus* smooth, i.e. 4 incorrect from 66 determinations on DRBC.

[c] Number of colonies examined not recorded; figures are percentages.

Table 4. Detection of aflatoxin production on coconut extract agar

	A. flavus			*A. parasiticus*		
Collab-orator	No. tested	Aflatoxin on CEA	Aflatoxin by TLC	No. tested	Aflatoxin on CEA	Aflatoxin by TLC
1	83	34	nt	17	17	17
2	24	1	nt	1	1	1
3	22	7 [a]	11	2	2	2
4[a] Run 1	100	52	nt	100	100	nt
2	100	44	nt	100	96	nt

[a] Number of colonies examined not recorded; figures are percentages.

difficult; after initial trials, the usual procedure was to inoculate only one isolate at a single point on each Petri dish. Collaborators also commented that observing the blue fluorescence was sometimes difficult because of interfering yellow fluorescent compounds.

Two collaborators obtained satisfactory results from this technique. It is recommended for use in the routine laboratory, but with the reservation that operators require training, and comparisons should be made with positive and negative controls. Control by checking against aflatoxin assays using TLC from time to time also seems a necessary precaution to prevent loss of consistency in using the method.

REFERENCES

BEUCHAT, L.R. 1984. Comparison of Aspergillus Differential Medium and Aspergillus flavus/parasiticus Agar for enumerating total yeasts and molds and potentially aflatoxigenic Aspergilli in peanuts, corn meal and cowpeas. *J. Food Prot.* **47**: 512-519.

BOTHAST, R.J. and FENNELL, D.F. 1974. A medium for rapid identification and enumeration of *Aspergillus flavus* and related organisms. *Mycologia* **66**: 365-369.

DAVIS, N.D., IYER, S.K. and DIENER, U.L. 1987. Improved method of screening for aflatoxin with a coconut agar medium. *Appl. Environ. Microbiol.* **53**: 1593-1595.

KING, A.D., PITT, J.I., BEUCHAT, L.R. and CORRY, J.E.L., eds. 1986. Methods for the Mycological Examination of Food. New York: Plenum Press.

KLICH, M.A. and PITT, J.I. 1988. Differentiation of *Aspergillus flavus* from *A. parasiticus* and other closely related species. *Trans. Br. Mycol. Soc.* **91**: 99-108.

KORNERUP, A. and WANSCHER, J.H. 1978. Methuen Handbook of Colour. 3rd ed. London: Eyre Methuen.

PITT, J.I., HOCKING, A.D. and GLENN, D.R. 1983. An improved medium for the detection of *Aspergillus flavus* and *A. parasiticus*. *J. Appl. Bacteriol.* **54**: 109-114.

7

NEW AND ALTERNATIVE TECHNIQUES FOR DETECTING FUNGI IN FOODS

RAPID DETECTION OF HEAT RESISTANT FUNGI IN FRUIT JUICES BY AN IMPEDIMETRIC METHOD

Per V. Nielsen

Department for Biotechnology
Technical University of Denmark
DK-2800 Lyngby, Denmark

SUMMARY

An automated impedance test was developed for detection of heat resistant moulds in fruit juice products. Growth in apple juice resulted in significant change in capacitance, and reliable detection, but changes in impedance and conductance were insufficient for this purpose. A comparative study of fruit juice supplemented with various media formulations showed that liquid media are preferable to solid media. Addition of yeast extract (0.75% w/w), potassium dihydrogen phosphate (0.6% w/w) and ammonium sulphate (0.1% w/w) to the fruit juices markedly improved curve quality resulting in earlier and more reproducible detection. The detection limit of the method is one spore per millilitre, which can be detected within 100 hours.

INTRODUCTION

Heat resistant moulds are an important spoilage problem for the food processing industry. The most frequently isolated species belong to *Byssochlamys* (*B. fulva* and *B. nivea*), *Talaromyces* (*T. macrosporus* - formerly *T. flavus* var. *macrosporus* - and *T. bacillosporus*) and *Neosartorya* (*N. fischeri* var. *glabra* and *N. fischeri* var. *spinosa*) (Pitt and Hocking, 1985; Nielsen and Samson, 1992). *Eupenicillium* and *Hamigera* species have also been isolated occasionally from spoiled fruit products (Samson *et al.*, 1992). Heat resistant ascospores of these species are ubiquitous in soil and cause problems in heat processed fruit products containing soil contamination. Besides deteriorating the product, most of these fungi also produce mycotoxins (Roland *et al.*, 1984; Frisvad *et al.*, 1990; Nielsen and Samson, 1992)

Analysing these products for contamination by heat resistant ascospores is quite laborious and time consuming, requiring incubation for at least 7 days and ideally up to 30 days (Hocking and Pitt, 1984). It is therefore of great interest to develop a rapid method for detection of heat resistant fungi.

Impedance monitoring has been established as a useful tool for rapid detection of bacterial contamination of a wide range of foods, and now this technique is emerging into the area of food mycology. Besides the fundamental work by Williams and Wood (1986), most emphasis has been directed towards detection of yeasts in various products: fruit mix (Fleischer *et al.*, 1984), orange juice (Zindulis, 1984) and wine (Henschke and Thomas, 1988). Recently a detailed study of impedance change induced by food spoilage moulds has resulted the in development of a medium for general impedimetric examination of food products (Watson-Craik *et al.*, 1990).

The basis of impedance microbiology is the monitoring of changes in chemical composition of the growth medium as a consequence of microbial metabolism. Impedance (Z) is the resistance towards flow of an electrical current in the substrate. It consists of two components, conductance (G) which is associated with the mobility and number of ions in the substrate, and capacitance (C) which is associated with changes in close proximity to the electrode. A detailed explanation of these parameters is given by Firstenberg-Eden and Eden (1984).

One of the drawbacks of using impedimetry in mycology is that quantification usually is not possible when monitoring samples with mixed flora on nonselective media. Instead, the method monitors the overall metabolic activity of the fungi in the test substrate, which may be juice or a medium derived from it. If the medium used for detection is similar to the actual product the method may assist in shelf life prediction, a more informative test than a total plate count (Williams, 1989). Impedance techniques are especially useful for sterility testing, as the limit of detection is one propagule per test well, which will typically hold 0.1 to 1.0 mL of the sample.

A general overview of rapid methods in food mycology is presented by Williams (1989) and Firstenberg-Eden and Eden (1984). This study was designed to develop a sensitive impedimetric method for detection of heat resistant moulds in fruit juices.

Sequential experiments performed were: impedimetric detection of fungal growth in fruit juice; the direct impedimetric detection of fungal growth in pasteurised fruit juice packs; the use of supplementary media for optimising detection; and testing limit of fungal detection in the presence of the optimal supplementary medium.

MATERIALS AND METHODS

Fungi

The following isolates were used: *Neosartorya fischeri* var. *fischeri* (IBT 3023), *N. fischeri* var. *glabra* (IMI 102173), *N. fischeri* var. *spinosa* (IBT 3001), *Talaromyces macrosporus* (IBT 4821) and *Byssochlamys nivea* (IBT 10071). IBT numbers are from the collection at the Department of Biochemistry, The Technical University of Denmark. Ascospores were harvested from Oat meal Agar (OA) or Malt Extract Agar (MEA) plates incubated for 3 to 4 weeks at 30°C. Asci, cleistothecia and gymnothecia were disrupted by twice treating suspensions in peptone water (0.5% (w/v)) with Tween 80 (0.1% (w/v)) for 5 min in an ice cooled water bath sonicator. After settling for 2 min the ascospore suspensions were filtered through sterile glass wool and stored at -30°C until use.

Substrates

Fruit juices, tested either with or without addition of various substrates, were: apple juice (pasteurised, Rynkeby, Denmark), nine-fruit juice (frozen fruit juice concentrate containing a mixture of pineapple, orange, passionfruit, mandarin, lemon, lime, mango, papaya, and banana, Gat Food Canneries, Israel); and blackcurrant juice (pasteurised, Rynkeby, Denmark).

Media supplements were tested in combination for their efficacy in improving detection on the Bactometer: Difco yeast extract, (0.0, 0.25, 0.5, 0.75%); KH_2PO_4 (0.0, 0.33, 0.5, 0.66, 1.0, 1.67%); $(NH_4)_2PO_4$ (0.0, 0.1, 0.2%); $(NH_3)H_2PO_4$ (0.0, 0.33%), and agar (0, 1.0%, poured into the juice, 1.5% prepoured). Not all combinations were analysed as the trial was performed using fractional factorial designs as described by Box *et al.* (1978). The optimum concentrations of supplements are listed in Table 1.

Table 1. Supplements for improved capacitance detection of heat resistant fungi in fruit juice

Yeast extract	2.25%
KH_2PO_4	1.8%
$(NH_4)_2SO_4$	0.3%

Impedimetry

The basic procedure for testing fruit juices for the presence of heat resistant moulds is outlined in Table 2. The experiments were performed on a Bactometer B123-2 (Vitek systems, UK) capable of simultaneously monitoring 256 samples distributed in 16 disposable modules of 16 wells. The instrument was capable of monitoring change in impedance and its two components conductance and capacitance over time. Incubation temperature was 30°C. Uninoculated juice or juice spiked with a suspension of ascospores were pasteurised at 80°C for 15 or 30 min. Aliquots (1.0 mL) were dispensed into wells prefilled with 0.5 or 1.0 mL of supplementary media. Media containing agar were either prepoured or dispensed after the juice had been distributed into the wells.

Experiments were also performed in 4 oz bottles (approx. 115 mL). The bottles were filled with 100 mL fruit juice leaving in average 15 mL of headspace inoculated with one mL of various spore suspensions, mounted with stainless steel electrodes, sealed and pasteurised for 15, 30 or 60 min at 80°C.

Table 2. Procedure for testing fruit juices for contamination with heat resistant moulds

Prepour 0.5 mL supplements (Table 1) into modules
Heat 5 mL juice in a small test tube for 15 min at 80°C or 30 min at 75°C
Dispense 1.0 mL aliquots in 4 wells
Monitor changes in capacitance at 30°C for 100 hours

Fractional factorial designs

Impedance detection times recorded by the Bactometer and the quality of the impedance curves was judged on acceleration, i.e. the rate of change in the signal during growth and maximal total change. These results were compared by Yates's algorithm and the reverse Yates's algorithm (Box *et al.*, 1978). By these methods the isolated effect of each media component (main effect) and the effects of the interactions between media components were first estimated. The significant effects are then used to build a model describing the combined effect of the tested media components. Based on the model an optimal combination of media components was determined.

RESULTS AND DISCUSSION

A preliminary experiment in which *Neosartorya fischeri* var. *fischeri* was inoculated into apple juice adjusted to pH 6.0 with 1.0 M NaOH resulted in consistent and rapid change in capacitance, but poor change in conductance and impedance (Figure 1). Consequently growth was detected only from the capacitance readings. Zindulis (1984), Williams and Wood (1986) and Watson-Craik *et al.* (1989) also found that the capacitance signal was

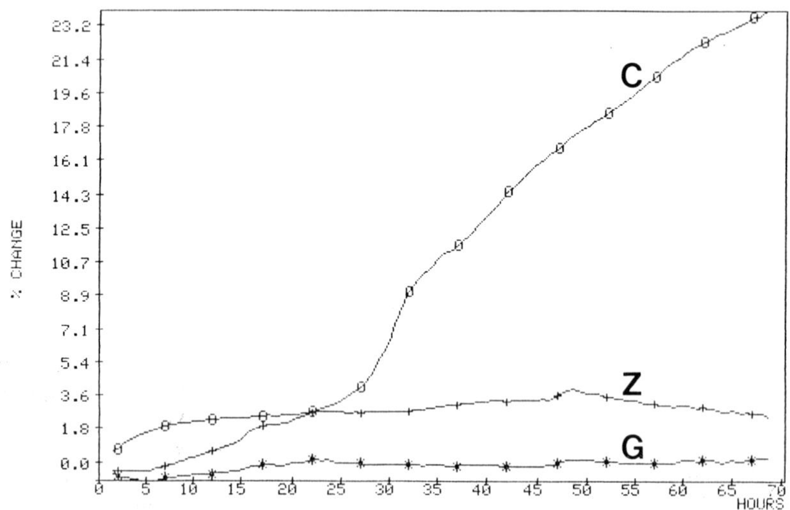

Figure 1. Changes in impedance (Z), conductance (G) and capacitance (C) during growth of *Neosartorya fischeri* var. *fischeri* in apple juice, pH 6.0 at 25°C.

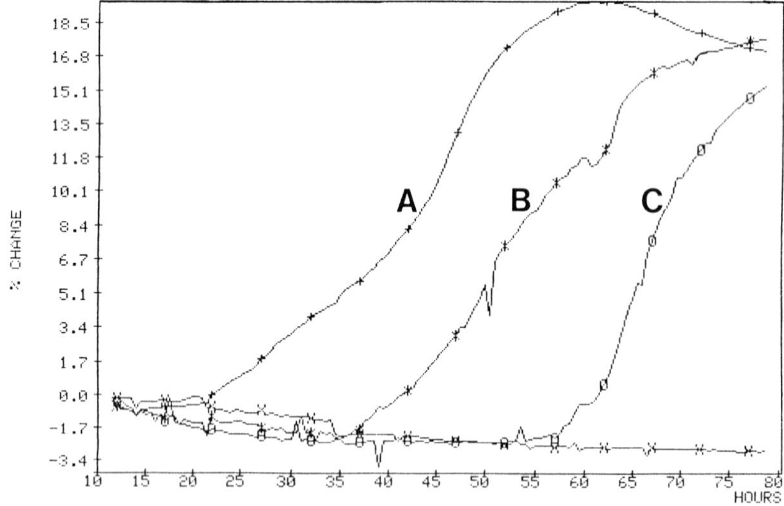

Figure 2. Capacitance change during growth of *N. fischeri* var. *spinosa* in bottles filled with nine-fruit juice. A, 10^4; B, 10^2; C, 10^0 ascospores per mL; and D, control. Bottles were pasteurised at 80°C for 30 min and incubated at 30°C. Detection times as determined by the Bactometer were 22.6, 37.5, and 54.0 hours in bottle A, B, and C respectively.

superior to conductance when monitoring mould growth. Media promoting detection of fungal growth by monitoring the conductance signal have been described (Connolly *et al.*, 1988) but the basic composition of these media is very different from fruit juices and they are sensitive to medium supplementation especially for nitrogeneous food products (Watson-Craik *et al.*, 1989). Based on these results only the capacitance signal was used in the following optimisation of supplementary media studies.

Figure 2 shows the change in capacitance during incubation at 30°C in nine-fruit juice spiked with various concentration of *N. fischeri* var. *spinosa*. A clear increase in detection time, i.e. the time when the curve starts to accelerate, occurs for each hundred fold decrease in numbers of ascospores. Detection of the lowest number of ascospores, ca. 1 cfu per mL, was remarkably rapid considering the limited headspace. However, the headspace and thus the available amount of oxygen may not be important, as the impedimetric detection occurred long before visible colonies were evident. As the large bottles were inconvenient to use, the same juice was also tested in bactometer modules with and without various supplementary media constituents (Figure 3). The capacitance curve and detection time were significantly affected. A total of 64 combinations of media constituents was tested. The most important findings, leading to the construction of an effective supplementary medium, are highlighted below.

Increasing concentrations of KH_2PO_4 enhanced the capacitance change during growth (Figure 4). This may have been due to the buffering effect of phosphate, as the capacitance signal is strongly affected by pH (Zindulis, 1984). Watson-Craik *et al.* (1990) found some similarities between titration curves of media with various phosphate concentrations and capacitance change during growth in the same media. Increasing buffer concentration reduces the capacitance change due to hydrogen ions produced during the early growth of the mould and increases time to detection. However, the detection delay means a larger biomass, hence hydrogen ion production is now greater, resulting in a more rapid capacitance change.

Figure 5 compares wells prefilled with liquid and solid media to which heat treated fruit juice was subsequently added. The better curve quality obtained in liquid media may be due to quicker diffusion and hence more ready access to medium components. Solidifying the heat treated fruit juice by adding hot agar did not in general improve the curve quality. Depending on the fungus and fruit juice, the solidified media resulted in everything from somewhat sharper curves with similar or slightly delayed detection times (*N. fischeri* var. *glabra* and *B. nivea*) to curves with weak acceleration and no detection (*T. macrosporus*). Liquid supplements are therefore preferable when monitoring heat resistant moulds in fruit juices, and are easier to use than solid media.

Schaertel *et al.* (1987) and Watson-Craik *et al.* (1990) found that agar media gave sharper curves and shorter detection times, but in the work reported here, this was not always the case. The reason is unclear: perhaps microaerophilic conditions occur when 1.0 mL juice is mixed with 0.5 mL agar medium as compared to the adequate gas exchange occurring when 0.1 mL is added to the top of a solid medium.

Yeast extract facilitated impedimetric detection of fungal growth in fruit juices by decreasing detection time and producing sharper curves. The highest concentration used in this study (0.75% of the total volume) gave the best results (data not shown).

The statistical analysis of the data resulted in the optimal supplements listed in Table 1. However, this was determined on the basis of a restricted number of components and concentration ranges and may not be the definitive optimum. In subsequent tests the supplements promoted rapid detection of a number of heat resistant moulds (Figure 6).

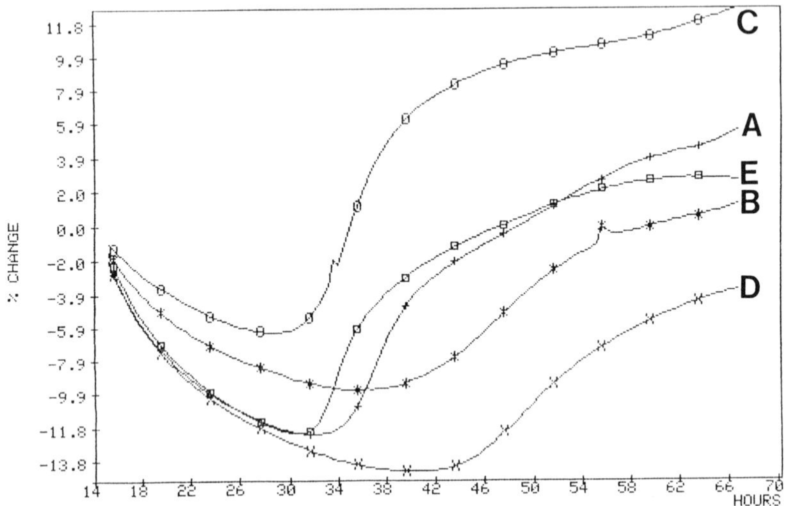

Figure 3. Capacitance changes during growth of *N. fischeri* var. *spinosa* in modules containing 1.0 mL of nine-fruit juice (A). This was supplemented with: B, 1.0 mL water; C, 1.0 mL water containing 1.0% yeast extract and 2.0% KH_2PO_4; D, 1.0 mL water containing 2.0% agar; and E, 1.0 mL water containing 1.0% yeast extract, 2.0% KH_2PO_4 and 2.0% agar. Initial number of viable ascospores was 10^2; incubation temperature 30°C.

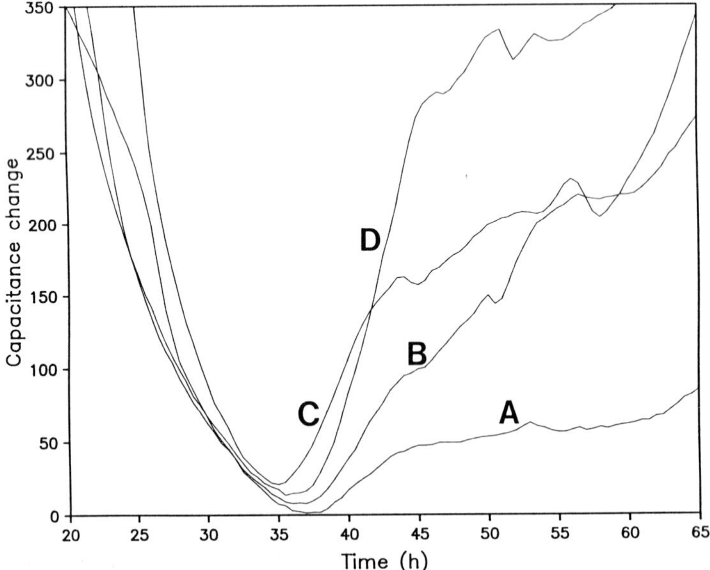

Figure 4. The effect of various concentrations of KH_2PO_4 on capacitance change during growth of *N. fischeri* var. *glabra* in nine-fruit juice. A, 0.00%; B, 0.33%; C, 1.00 %; and D, 1.67%.

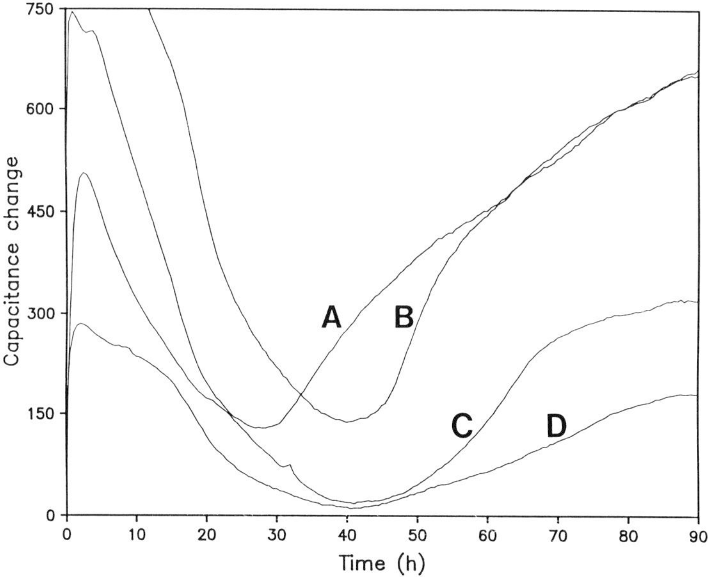

Figure 5. The effect of agar on capacitance changes during growth of *N. fischeri* var. *glabra* in nine-fruit juice. A and B, 0.0%; C and D, 1.0%. The initial number of viable ascospores was 2.3×10^3 in A and C and 2.3×10^1 in B and D.

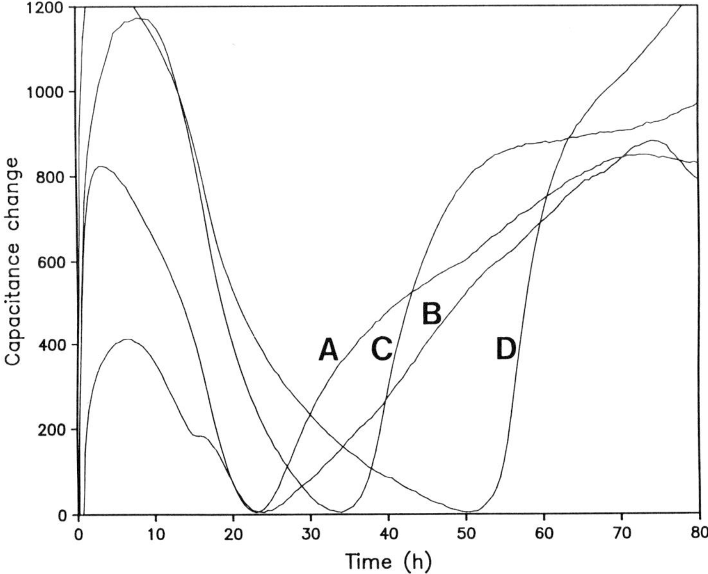

Figure 6. Capacitance change during growth of several heat resistant moulds in pasteurised nine-fruit juice supplemented as in Table 1. A. *Neosartorya fischeri* var. *glabra* (IMI 102173), 4.8×10^3/mL, B, *N. fischeri* var. *spinosa*, 4.6×10^3; C, *Talaromyces flavus*, 3×10^1; D, *Byssochlamys nivea*, 2.1×10^2.

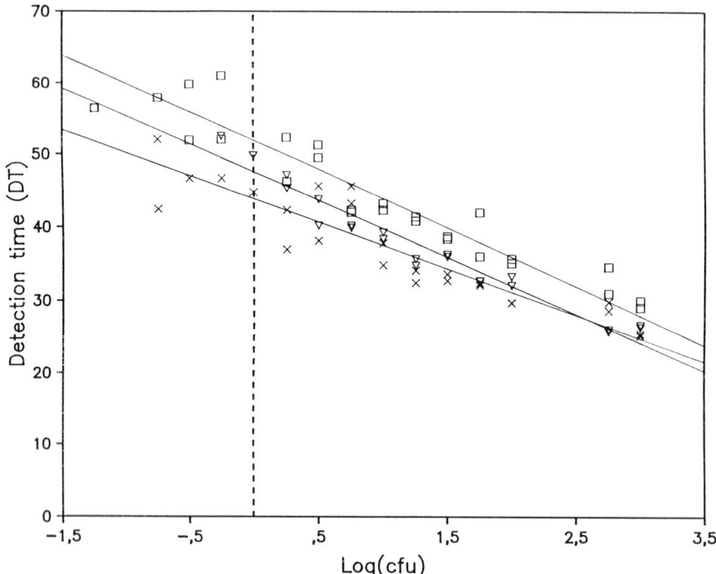

Figure 7. Correlation between initial number of viable ascospores and detection time in blackcurrant juice (□), apple juice (×) and nine-fruit juice (▽).

A reasonably good correlation was observed between initial numbers of viable ascospores and the detection time for *N. fischeri* var. *glabra*, added to various fruit juices (Figure 7). Differences between juices were not significant. The correlation was best at viable ascospore numbers higher than 10 per mL. At lower numbers, statistical and biological variations resulted in lower correlations.

The general problem of monitoring low numbers of fungi is a typical one in analysing heat treated products (Hocking and Pitt, 1984). Calculation of spore or colony numbers from detection times is not feasible, as growth rates depend on the fungus, the degree of sublethal stress and the juice. The test for heat resistant fungi in a fruit juice will therefore in most cases be a sterility test, as even one viable ascospore per mL may be unacceptable. As the distribution of spores in samples taken out for analysis follows a Poisson distribution, four times one mL must be analysed and found free of viable ascospores, i.e. a detection time of more than 100 hours, to conclude with 95% confidence that the product contains less than one viable ascospore per mL. From Figure 7, it is clear that one ascospore of *N. fischeri* var. *spinosa* is detected after approximately 60 hours. Similar or slightly longer detection times were found for the other fungi tested.

Tests with undiluted fruit juice concentrates gave very poor capacitance changes even when the fungi were able to grow in the concentrate. Diluting concentrates before analysis is therefore recommended.

Contamination with spore forming bacteria did not appear to be a problem in these experiments. If such problems did occur, the antibiotics generally used in mycological media, e.g. 50 ppm chloramphenicol and 50 ppm chlortetracycline, may be added. This has no or little effect on curve quality and detection time (Williams and Wood, 1986; Watson-Craik *et al.*, 1990).

REFERENCES

BOX, G.E.P., HUNTER, W.G. and HUNTER, J.S. 1978. Statistics for Experimenters: an Introduction to Design, Data Analysis, and Model Building. New York: John Wiley and Sons.

CONNOLLY, P., LEWIS, S.J. and CORRY, J.E.L. 1988. A medium for the detection of yeasts using a conductimetric method. *Int. J. Food Microbiol.* **7**: 31-40.

FIRSTENBERG-EDEN, R. and EDEN, G. 1984. Impedance Microbiology. New York: John Wiley and Sons.

FLEISCHER, M., SHAPTON, N. and COOPER, P.J. 1984. Estimation of yeast numbers in fruit mix for yogurt. *J. Soc. Dairy Technol.* **37**: 63-65.

FRISVAD, J.C., FILTENBORG, O., SAMSON, R.A. and STOLK, A.C. 1990. Chemotaxonomy of the genus *Talaromyces*. *Antonie van Leeuwenhoek* **57**: 179-189.

HENSCHKE, P.A. and THOMAS, D.S. 1988. Detection of wine-spoiling yeasts by electronic methods. *J. Appl. Bacteriol.* **44**: 123-133.

HOCKING, A.D. and PITT, J.I. 1984. Food spoilage fungi. II. Heat resistant fungi. *CSIRO Food Res. Q.* **44**: 73-82

NIELSEN, P.V. and SAMSON, R.A. 1992. Differentiation of food-borne taxa of *Neosartorya*. *In* Modern Methods in Food Mycology, eds R.A. Samson, A.D. Hocking, J.I. Pitt and A.D. King. pp. 159-168. Amsterdam: Elsevier.

PITT, J.I. and HOCKING, A.D. 1985. Fungi and Food Spoilage. Sydney: Academic Press.

ROLAND, J.O., BEUCHAT, L.R., WORTHINGTON, R.E. and HITCHCOCK, H.L. 1984. Effect of sorbate, benzoate, sulfur dioxide and temperature on growth and patulin production by *Byssochlamys nivea* in grape juice. *J. Food Prot.* **47**: 237-241.

SAMSON, R.A., VAN REENEN-HOEKSTRA, E.S. and HARTOG, B.J. 1992. Influence of pre-treatment of raspberry pulp on the detection of heat resistant moulds. *In* Modern Methods in Food Mycology, eds R.A. Samson, A.D. Hocking, J.I. Pitt and A.D. King. pp. 155-158. Amsterdam: Elsevier.

SCHAERTEL, B.J., TSANG, N. and FIRSTENBERG-EDEN, R. 1987. Impedimetric detection of yeast and mold. *Food Microbiol.* **4**: 155-163.

WATSON-CRAIK, I.A., AIDOO, K.E. and ANDERSON, J.G. 1989. Introduction of conductance and capacitance changes by food-borne fungi. *Food Microbiol.* **6**: 231-244.

WATSON-CRAIK, I.A., AIDOO, K.E. and ANDERSON, J.G. 1990. Development and evaluation of a medium for the monitoring of food-borne moulds by capacitance changes. *Food Microbiol.* **7**: 129-145.

WILLIAMS, A.P. 1989. Fungi in foods - rapid detection methods. *In* Rapid Methods in Food Microbiology, eds M.R. Adams and C.F.A. Hope. pp. 255-272. Amsterdam: Elsevier.

WILLIAMS, A.P. and WOOD, J.M. 1986. Impedimetric estimation of moulds. *In* Methods for the Mycological Examination of Food, eds A.D. King, J.I. Pitt, L.R. Beuchat, and J.E.L. Corry, pp. 230-238. New York: Plenum Press.

ZINDULIS, J. 1984. A medium for the impedimetric detection of yeast in foods. *Food Microbiol.* **1**: 159-167.

USE OF IMPEDIMETRY TO STUDY THE EFFECT OF ANTIFUNGAL AGENTS ON FUNGAL GROWTH

M. O. Moss

School of Biological Sciences
University of Surrey
Guildford, Surrey GU2 5XH, U.K.

SUMMARY

The measurement of impedance, capacitance or conductance changes due to their biochemical activities, is an effective means for detecting and quantifying microorganisms, particularly bacteria and yeasts, in the food industry. Direct relationships between time to signal appearance and initial viable counts can be used to determine parameters such as generation time and exponential growth constants by experimentally manipulating the inoculum size. The impedimeter can then be used to study the influence of environmental factors such as temperature on these parameters, and also the effect of sublethal concentrations of biocides.

The Bactometer 32 (Bactomatic Inc., Princeton, NJ, USA) has been used to detect trichothecene mycotoxins by their effect on yeast growth. The same experimental framework permits study of the interaction of temperature and antifungal agents on the growth and metabolic activity of filamentous fungi. However, the filamentous growth form and the possibility of morphological differentiation may make the interpretation of results more difficult.

INTRODUCTION

The measurement of impedance, capacitance or conductance changes in a medium, arising from the biochemical activity of metabolically active microorganisms, has become an established means of detecting and quantifying microorganisms in the food industry. The technique is especially effective for bacteria and yeasts, for which it is possible to obtain a direct relationship between the time taken for the appearance of a signal and the initial viable count (Cady *et al.*, 1978; Zindulis, 1984). Methods have been devised for the selective detection of, for example, *Salmonella* (Bullock and Frodsham, 1989) and Lancefield Group D streptococci (Neaves *et al.*, 1988) in foods. Although developments in instrumentation have primarily been for the enumeration and detection of microorganisms, especially for the quality assurance of foods and raw materials used in the food industry, obsolete instruments such as the Bactometer 32 (Bactomatic Inc., Princeton, NJ, USA) can be used to determine parameters such as generation time and exponential growth constants by experimentally manipulating the inoculum size. The impedimeter can then be used to study the influence of environmental factors such as temperature and biocidal chemicals on these parameters. Because the trichothecene mycotoxins have demonstrable antifungal activity, impedimetry methods have been evaluated (Adak *et al.*, 1987a; b). *Hyphopichia burtonii* and *Hansenula fabianii* were shown to be especially sensitive to these toxins, the no effect level for T-2 toxin being as low as 0.018 μg/mL and 0.012 μg/mL respectively, for these two yeasts.

The Bactometer 32 has been used at this University as a teaching tool, to demonstrate the interaction of temperature and chloramphenicol concentration on the growth of *Escherichia coli* in brain heart infusion broth. The same experimental framework permits study of the interactions of temperature with chemicals such as antifungal agents on the growth and metabolic activity of filamentous fungi. However, the filamentous growth form and the possibility of morphological differentiation may make the interpretation of results more difficult.

Appropriate media for the rapid impedimetric detection of filamentous fungi have been developed (Williams and Wood, 1986; Watson-Craik *et al.*, 1989; 1990). The results of preliminary studies utilising Malt Extract Glucose medium (MEG; Williams and Wood, 1986) to determine the influence of subinhibitory concentrations of chloramphenicol on the growth of *Aspergillus parasiticus* are presented here.

MATERIALS AND METHODS

Equipment

A Bactometer 32 instrument was used in conjunction with modules taking 8 samples and their appropriate controls.

Medium and fungus

The medium used, MEG, contained malt extract, 2%; glucose, 2%, pH 5.5, sterilised by autoclaving at 121°C for 10 min (Williams and Wood, 1986). Chloramphenicol solutions were prepared in the same medium to provide a range of concentrations double those finally required. The fungus used was *Aspergillus parasiticus* NRRL 3145, maintained on slopes of Czapek Dox agar.

Spore suspensions

A stock conidial suspension in 0.2% Tween 80, containing approximately 10^7/mL, was prepared from a 10 day old slope culture. One mL was transferred to 100 mL of MEG and the suspension shaken on a rotary shaker for 30 min. Four ten fold dilutions were prepared by transferring 1 mL spore suspension to 9 mL MEG with vigorous shaking at each dilution. Microscopic examination demonstrated that most conidia had separated but some chains were still present.

Impedimetry

Duplicate wells were inoculated with 0.5 mL of spore suspension, or sterile MEG in controls, plus 0.5 mL of the appropriate chloramphenicol solution, and modules incubated in the Bactometer at 29.8°C. Detection times, maximum rate of activity, dF/dt(max) indicated by counting flips per hour, and total activity, F(total), estimated by counting the number of flips in the signal, were determined directly from the chart recorder output. Inoculum levels were determined using triplicate spread plates on malt extract agar.

RESULTS AND DISCUSSION

The results of two experiments are summarised in Tables 1 and 2, and compared with similar Bactometer data for *Escherichia coli* (Table 3).

A striking difference between the results with *A. parasiticus* and *E. coli* is the effect of inoculum size on the shape of the signal. In the case of *E. coli* the detection time is the only parameter influenced by inoculum size, over the range of 10 to 10^5 cfu/mL. Impedimetry is therefore a very useful method for enumeration and for determining generation times, and hence exponential growth constants. Increasing concentrations of chloramphenicol caused an increase in generation time, a decrease in dF/dt, but apparently did not influence total activity (Table 3). An increase in temperature from 34°C to 37° and 43°C also caused an increase in sensitivity of *E. coli* to chloramphenicol.

However, inoculum size clearly influenced the shape of the signal, as well as the detection time, for *A. parasiticus*, making it difficult to calculate a parameter equivalent to generation time. Bearing in mind that germ tubes are polarised structures and their subsequent development involves branching, it is to be expected that the behaviour of moulds will be qualitatively different from that of single celled microorganisms. Although the biomass of a filamentous fungus may increase exponentially during the early phase of growth, the elongation of an individual hypha is often linear over significant periods of time and the exponential increase in biomass, and total hyphal length, is mainly due to branching of the hyphae (e.g. Kotov and Reshetnikov, 1990).

In studying fungal growth by impedimetry, Watson-Craik *et al.* (1990) used chloramphenicol and streptomycin (both at 75 μg/mL). Williams and Wood (1986) demonstrated that chloramphenicol and oxytetracycline (both at 100 μg/mL) caused a significant increase

Table 1. The effect of inoculum size and chloramphenicol in the range 0-50 mg/L on growth of *Aspergillus parasiticus* in malt extract glucose at 29.8°C as measured in the Bactometer

Chloram. (μg/mL)	Parameter	Inoculum (cfu/mL) 1.9x10^4	1.9x10^3	1.9x10^2	1.9x10^1
0	DT (h)[1]	14.3	18.7	26.2	32.5
	τ (h)[2]	-	1.3	2.3	1.9
	dF/dt[3]	4.0	2.7	2.8	3.1
	F (tot)[4]	49	43	38	34
12.5	DT (h)	13.9	18.4	24.4	30.0
	τ (h)	-	1.4	1.8	1.7
	dF/dt	4.3	2.3	2.8	3.1
	F (tot)	45	41	38	nd[5]
25	DT (h)	14.7	21.8	26.8	34.5
	τ (h)	-	2.1	1.5	2.3
	dF/dt	4.2	2.0	2.7	2.8
	F (tot)	44	36	35	nd
50	DT (h)	14.7	18.9	26.8	34.7
	τ (h)	-	1.3	2.4	2.4
	dF/dt	4.1	2.5	3.1	2.7
	F (tot)	43	39	35	nd

[1] DT: Detection time
[2] τ: Doubling time
[3] dF/dt: Maximum number of flips per hour
[4] F (tot): total number of flips in the signal
[5] nd: not determined

Table 2. The effect of inoculum size and chloramphenicol in the range 0-200 mg/L on growth of *Aspergillus parasiticus* in malt extract glucose at 29.8°C as measured in the Bactometer

Chloram. (μg/mL)	Parameter	Inoculum (cfu/mL) 3.7x10⁴	3.7x10³	3.7x10²	3.7x10¹
0	DT (h)[1]	14.1	22.7	40.8	50.1
	τ (h)[2]	-	2.6	2.4	2.9
	dF/dt[3]	4.1	2.6	2.6	3.2
	F (tot)[4]	46	41	37	35
50	DT (h)	15.7	20.5	34.8	49.0
	τ (h)	-	1.4	4.3	4.3
	dF/dt	3.6	2.2	2.2	3.0
	F (tot)	41	39	35	35
100	DT (h)	15.7	22.8	43.8	49.4
	τ (h)	-	2.1	6.3	1.7
	dF/dt	3.7	2.6	3.2	3.0
	F (tot)	41	38	36	33
200	DT (h)	17.9	39.7	52.2	63.3
	τ (h)	-	6.6	3.8	3.3
	dF/dt	3.0	2.2	2.3	2.0
	F (tot)	36	31	28	nd[5]

[1, 2, 3, 4, 5] See Table 1 for footnotes

Table 3. The interaction of temperature and chloramphenicol concentration on parameters measured with a Bactometer using *Escherichia coli* in brain heart infusion broth

Temp. °C	Parameter	Concentration of chloramphenicol (μg/mL) 0	0.63	1.25	1.87
34	τ[1]	19.0	23.3	24.6	34.9
	dF/dt[2]	9	6	5	3
	F (tot)[3]	21	18	22	20
37	τ	18.0	23.4	32.4	48.8
	dF/dt	9	6	4	2
	F(tot)	19	19	18	16
43	τ	29.0	75.9	nd[4]	nd
	dF/dt	7	6	nd	nd
	F(tot)	19	19	nd	nd

[1] τ: doubling time in minutes
[2] dF/dt: maximum number of flips per hour
[3] F(tot): total number of flips in the signal
[4] nd: not detected

in detection time and a reduction in the strength of the impedance signal with *A. ochraceus*. Tables 1 and 2 show that chloramphenicol alone has no reproducible effect on *A. parasiticus* at concentrations less than 100μg/mL.

These experiments confirm that impedimetric methods for the routine enumeration of filamentous fungi will not be straight forward, although detailed studies have demonstrated progress in the development of appropriate media (Watson-Craik *et al.*, 1990). However, impedimetric methodology allows a large number of similar experiments to be carried out over a relatively short period of time, so the Bactometer may provide a useful tool for studying the influence of various factors on the physiological activity of moulds.

REFERENCES

ADAK, G.K., CORRY, J.E.L. and MOSS, M.O. 1987a. Use of impedimetry to detect trichothecene mycotoxins. 1. Screen for susceptible microorganisms. *Int. J. Food Microbiol.* **5**: 1-13.

ADAK, G.K., CORRY, J.E.L. and MOSS, M.O. 1987b. Use of impedimetry to detect trichothecene mycotoxins. 2. Limits of sensitivity of four microorganisms to T-2 toxin and the effect of solvent and test medium. *Int. J. Food Microbiol.* **5**: 15-27.

BULLOCK, R.D. and FRODSHAM, D. 1989. Rapid impedance detection of salmonellas in confectionary using modified LICNR broth. *J. Appl. Bacteriol.* **66**: 385-391.

CADY, P., HARDY, D., MARTINS, S., DUFOUR, S.W. and KRAEGER, S.J. 1978. Automated impedance measurements for rapid screening of milk microbial content. *J. Food Prot.* **41**: 277-283.

KOTOV, V. and RESSHETNIKOV, S.V. 1990. A stochastic model for early mycelial growth. *Mycol. Res.* **94**: 577-586.

NEAVES, P., WADDELL, M.J. and PRENTICE, G.A. 1988. A medium for the detection of Lancefield Group D cocci in skimmed milk powder by measurement of conductivity changes. *J. Appl. Bacteriol.* **65**: 437-448.

WATSON-CRAIK, I.A., AIDOO, K.E. and ANDERSON, J.G. 1989. Induction of conductance and capacitance changes by food-borne fungi. *Food Microbiol.* **6**: 231-244.

WATSON-CRAIK, I.A., AIDOO, K.E. and ANDERSON, J.G. 1990. Development and evaluation of a medium for the monitoring of food-borne moulds by capacitance changes. *Food Microbiol* **7**: 129-145.

WILLIAMS, A.P. and WOOD, J.M. 1986. Impedimetric estimation of moulds. *In* Methods for the Mycological Examination of Food, eds A.D. King, J.I. Pitt, L.R. Beuchat and J.E.L. Corry. pp. 230-238. New York: Plenum Press

ZINDULIS, J. 1984. A medium for the impedimetric detection of yeasts in foods. *Food Microbiol.* **1**: 159-167.

DETERMINATION OF VOLATILE COMPOUNDS FOR THE DETECTION OF MOULDS

P. Adamek, B. Bergström, T. Börjesson and U. Stöllman

Swedish Institute for Food Research
P.O. Box 5401,
S-40229 Gothenberg, Sweden

SUMMARY

Analysis of volatile metabolites produced by fungi is an important new technique for the detection of fungal growth in cereals. This paper presents results from such studies. Volatiles in the head space of cultivation containers were trapped on polymer adsorbents by flushing with a continuous airstream. The volatiles were then desorbed thermally and concentrated in a cold trap (-196°C) before injection onto combined gas chromatography mass spectrometry instrumentation. Volatiles produced during the growth on wheat of pure cultures of *Aspergillus amstelodami*, *Aspergillus flavus*, *Penicillium cyclopium* and *Fusarium culmorum* were studied. Important metabolites were methylfuran, 2-methylpropanol and 3-methylbutanol. The formation of odourous metabolites was studied by cultivating four *Penicillium* species (*P. ochrochloron*, *P. digitatum*, *P. chrysogenum* and *P. cyclopium*) on Czapek agar and malt extract agar. These compounds were sensorially characterised by GC-sniffing analysis and chemically identified by mass spectrometry. A combination of different esters gave fruity aromas while mushroom aromas originated from 8-carbon compounds. Musty and earthy aromas came from volatiles with retention times above 20 minutes. The pattern of volatiles produced by *A. flavus* and *A. oryzae* on solid and liquid media was investigated. No specific differences between the species were noticed during the growth on solid media. However, after 10 days growth on liquid medium the pattern of volatiles differed significantly for the two species. These qualitatively different volatiles could be used to distinguish *A. flavus* from *A. oryzae*.

INTRODUCTION

Detection of mould contamination in commodities such as cereals by traditional methods is time consuming and not necessarily related to actual fungal activity or to the extent of deterioration. A number of indirect methods, such as seed germination, discolouration or estimation of free fatty acids may be used to estimate fungal activity. These methods are often simple and rapid, but are not directly related to the metabolic activity of the fungi. The odour of cereals, detected by sniffing, is more related to fungal metabolic activity. This procedure is also rapid, but has a number of drawbacks. It is subjective, and could be injurious to the analyst's health. Futhermore, only a fraction of the fungal volatiles with a characteristic odour can be detected. One improvement to this technique was the use of gas chromatography for the classification of corn odours (Dravnieks *et al.*, 1973).

The search for better methods is of great importance. One of the most promising new techniques is analysis of volatile metabolites produced during fungal growth. This can be developed into a simple and rapid method with many potential applications, with high sensitivity and a good chance of representative sampling of all fungi.

A combination of gas chromatography and mass spectrometry was used by Kaminski *et al.* (1974) to identify the volatile compounds produced by different fungi during growth on wheat meal. This method makes it possible to detect volatiles present in low concentrations in large volumes (Hyde *et al.*, 1983). The quality of cereals in long term storage has also been monitored by this method (Abramson *et al.*, 1983). A further development of the method is the use of a porous polymer to trap the volatiles (Kaminski *et al.*, 1985; Harris *et al.*, 1986).

In the study reported here, three separate experiments were carried out. The first was designed to screen and identify volatiles associated with fungal growth on wheat. The second was designed to identify odourous compounds produced by different *Penicillium* species and the third investigated differences in the pattern of volatile metabolites produced by *Aspergillus flavus* and *A. oryzae*.

METHODS

Fungi

The following species and strains of fungi were used: *Eurotium amstelodami*, SLU 10-904; *Aspergillus flavus*, CBS 242.73, CBS 118.62 and CBS 569.65; *A. oryzae*, CBS 570.65, CBS 574.65 and CBS 818.72; *Fusarium culmorum*, SLU 10-717; *Penicillium cyclopium*, SLU 10-878; *P. chrysogenum*, SIK 5.2.24; *P. digitatum*, CBS 658.68 and *P. ochrochloron*, CBS 338.59. *Fusarium culmorum* was maintained on oatmeal agar (oatmeal, 3%; agar, 2%). All other strains were maintained on 2% Malt extract Agar (MA).

Preparation of spore suspension and inoculation.

Each mould strain was grown on MA for production of conidia. After incubation for about 10 days at 20°C, 10 mL of a 0.85% NaCl solution was added to each agar plate and conidia were liberated using a glass rod. Conidial suspensions were filtered through a cotton cloth before being inoculated onto the growth substrate.

Cultivation media

Winter wheat with an original moisture content of 11.5% was used as cultivation medium in the first experiment. The moisture content was raised to 25% by adding small portions of distilled water. Equilibrium was reached by shaking on a laboratory shaker for 20 hours. After this treatment, 500 g of wheat was put in a cultivation container and autoclaved at 121°C for 15 min. In the second experiment MA and commercial Czapek agar (Cz, Difco) were used as cultivation media.

As well as Cz, a modified Czapek solution which contained 1.0 g K_2HPO_4, 10 mL Czapek concentrate (Pitt, 1979), 5.0 g powdered yeast extract (Difco), 20 g glucose, 0.1 g $ZnSO_4.7H_2O$, 0.05 g $CuSO_4.5H_2O$ and 1000 mL water, was used in the third experiment.

Equipment for cultivation and sampling

The equipment for the first experiment is illustrated in Figure 1. Fernbach culture flasks (2 L), provided with an inlet and an outlet for the transport of volatiles by continuous flushing with an airstream, were used for cultivation.

To maintain constant moisture content in the wheat, air was moistened in two water-filled flasks connected in series. To protect the container against contamination by volatiles from the air, the air stream was passed through a polymer adsorbent (Tenax GC, 60-80 mesh). Glass tubes filled with glass wool were attached to the inlet and outlet to protect the container and environment from microbial contamination. Volatile metabolites were collected in Chromosorb 102, mesh 100-120.

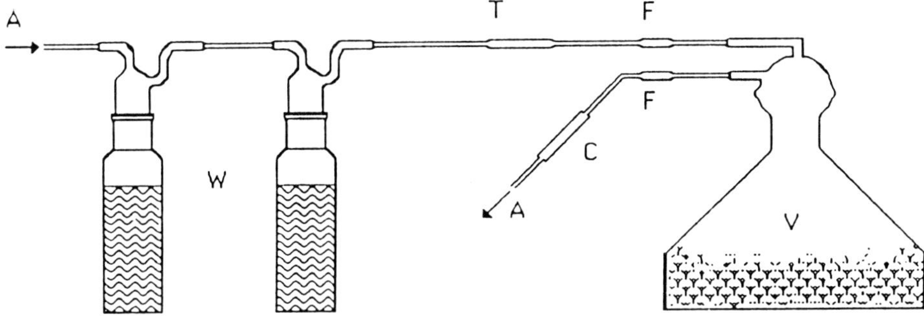

Figure 1. Equipment for cultivation of fungi and sampling of volatiles in the first experiment. A, airflow; W, water-filled flasks for moistening the air; T, Tenax GC adsorbent; F, spore filters; V, cultivation container; C, Chromosorb 102 adsorbent.

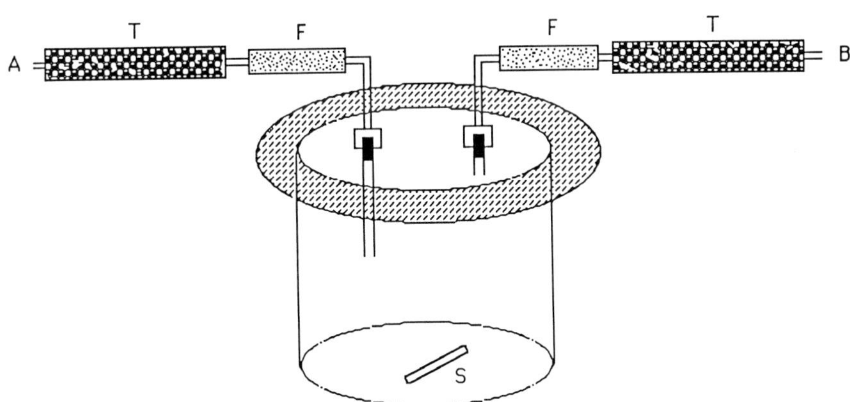

Figure 2. Equipment for growth of fungi and sampling of volatiles in the second and third experiment. A, air inlet; B, air outlet; F, spore filter; S, magnetic stirrer; T, Tenex GC adsorbent.

The equipment for the second and third experiment is illustrated in Figure 2. A cylindrical container of stainless steel, 72 mm in diameter and 50 mm high was fitted with a glass lid having an inlet and an outlet. To prevent contamination the containers were equipped with Tenax GC adsorbents and microbial filters. The volatile metabolites were collected in adsorbent Tenax GC. Slow magnetic stirring was used in the third experiment during mould cultivation in liquid medium.

Sampling of volatiles

Fungi were cultivated individually for 10 to 14 days and a continuous stream of air (10 mL/min; 14.4 L air/day) was passed through the cultivation containers. Sampling periods were 8 or 17 hours.

Analysis of volatile compounds

After sampling, the adsorbents were dried by a helium flow (20 mL/min) for 20 minutes at room temperature. The volatiles were then desorbed thermally at 110°C using a helium stream (20 mL/min) for 20 minutes. The released compounds were concentrated in a cold trap (-196°C) before injection onto a gas chromatograph column by heating the cold trap to 140°C. The gas chromatograph used was a Varian 3700 with a flame ionization detector and a fused silica capillary column (60 x 0.32 mm i.d.) containing chemically bonded methyl-polysiloxan, DB1, 1.0 μm. The temperature program used was 25 to 220°C with 2 min at the initial temperature and a heating rate of 4°C/min. The carrier gas was helium (4 mL/min). The data collection system was by Hewlett-Packard HP 3367. A Finnigan 9610-4023 gas chromatography-mass spectrometry system was used for identification of volatiles. The US National Bureau of Standards library of mass spectra was used.

Analysis of carbon dioxide

The metabolic activity of the fungi was followed by gas chromatographic measurement of the CO_2 content in the airstream from the cultivation containers. A Hewlett Packard gas chromatograph HP 5890 with TCD detector and packed column (Chromosorb 101, 60-80 mesh, 2 m x 1/8 in. o.d.) was used for isothermal analysis (25°C). The carrier gas was helium (16 mL/min). The data collection system was a Hewlett-Packard HP 3367.

RESULTS AND DISCUSSION

Experiment one: volatiles associated with fungal growth in wheat

As evidenced by CO_2 production growth of each fungus increased continuously throughout the experiment. The development of CO_2 is presented in Figure 3.

The concentration of some volatile compounds such as 2-methylfuran and 2-methylpropanol rose during the whole experiment. The production of other compounds such as 2-methyl-1-butanol and 3-methyl-1-butanol increased in the beginning of the experiment, but after 2-3 days of cultivation the concentration of these volatiles diminished (Table 1).

Experiment two: odours produced by *Penicillium* species

The headspace above freshly autoclaved MA contained a large amount of 3-methyl-1-butanal. Considerable quantities of ethanol, 2-propanone and hexanal were also present. On both media, 8-carbon compounds associated with a mushroom smell were produced. These compounds were also reported by Kaminski *et al.* (1974).

Table 1. Volatiles in headspace gases produced by different fungi during 6 days cultivation on wheat

Compound	*E. amstelodami*		*A. flavus*		*F. culmorum*		*P. cyclopium*	
	Max. amount	Day produced	Max. amount	Day produced	Max. amount	Day produced	Max. amount	Day produced
2-Methylfuran	58	6	119	6	-		100	6
2-Methyl-1-propanol	6	2-6	84	5	202	6	200	4
2-Methyl-1-butanol	81	2	-		-		-	
3-Methyl-1-butanol	-		86	2	-		35	2
3-Octen-2-ol	6	3-5	-		-		48	4
1-Octen-3-ol	10	3-5	-		-		45	4
2-Pentanon	28	3	-		-		-	
Monoterpenes	-		-		17	6	-	
Sesquiterpenes	-		-		36	5	7	5

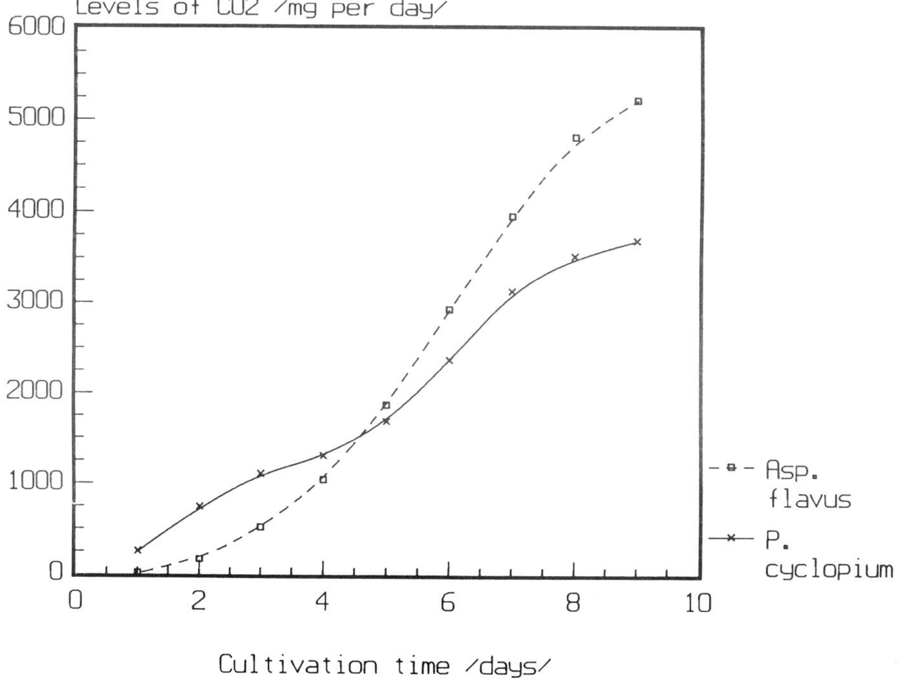

Figure 3. Production of CO_2 measured as mg of CO_2 per day during the cultivation of *A. flavus* and *P. cyclopium* on wheat in the first experiment

Table 2. Volatile flavour compounds produced by some *Penicillium* species during cultivation on 2% malt agar and Czapek agar

Volatile compound	Retention time	Description of flavour	Species	Flavour intensity
Ethyl 2-butenoate	17.9	Fruity	*P. ochrochloron*	2
Ethyl propionate	11.4	Fruity	*P. digitatum*	2
Ethyl propionate	11.9	Fruity	*P. ochrochloron*	2
4-Heptene-2-one	20.6	Banana	*P. chrysogenum*	2
Hexanol	19.9	Banana	*P. digitatum*	3
Isoamyl acetate	18.8	Pear	*P. ochrochloron*	3
Methylethyl butanoate	17.3	Fruity, apple	*P. ochrochloron*	2
3-Methyl-1-butyl acetate	18.4	Pear	*P. digitatum*	3
Methylethyl propionate	13.6	Pear, banana	*P. ochrochloron*	3
3-Octanol	24.8	Citrus	*P. chrysogenum*	3
Propyl acetate	12.2	Fruity, forest	*P. ochrochloron*	1
1-Octene-3-ol		Mushroom	*P. chrysogenum*	3
3-Octanone		Mushroom	*P. chrysogenum*	5

Table 3. Volatile flavour compounds produced by *Penicillium ochrochloron* on 2% malt agar and Czapek agar

Volatile compound	Medium		Description of flavour	Flavour intensity
	Cz	MA		
2,3-Butanedione	x	x	Butter	4
Dichloroethane	x		Sour, rancid	2
Ethanol	x	x	Alcoholic	1
Ethyl acetate		x	Acetone	4
Ethyl butanoate		x	Caramel	2
Ethyl 2-butenoate		x	Fruity	2
Ethyl pentenoate		x	Sweet	2
Ethyl propionate		x	Fruity	2
Isoamyl acetate		x	Pear	3
1,5-Hexadiene	x			0
Methylbutyl acetate		x	Pear	1
2-Methylbutanoic acid		x	Apple	2
3-Methyl-1-butanol		x	Acetone	2
2-Methyl-3-buten-1-ol	x			0
2-Methylethyl propionate		x	Pear	3
2-Methylfuran	x		Smoke	1
2-Methylpropanal		x	Burned	3
3-Methylbutanoic acid		x	Rancid	3
3-Methylhexane	x			0
Propyl acetate		x	Fruity	1
Terpene alcohol	x		Mouldy	3
2,2,4-Trimethylheptane	x		Cat	1

Table 4. Volatile flavour compounds produced by *Penicillium chrysogenum* on malt extract agar and Czapek agar

Volatile compound	Medium		Description of flavour	Flavour intensity
	Cz	MA		
2,3-Butanedione	x	x	Butter, cream caramel	
2-Butanol		x	pungent, rancid	3
2-Butanol	x			0
2-Butenone		x	Ethereal, cream caramel	3
Cyclooctene	x	x	Glue	1
1,2-Dichloroethane	x		Sour, rancid	1
Dimethylethyl propionate		x	Geranium	1
Ethanol	x		Alcoholic	1
Dimethylpropanoic acid	x		Grass	2
4-Heptene-2-one		x	Banana	2
3-Methyl-1-butanol		x	Sweet, burned	2
3-Methyl-3-butene-1-ol	x		Burned	2
2-Methyl-1-propenol		x	Sweet, burned	2
3-Octanol		x	Citrus	3
3-Octanol	x			2
3-Octanone	x	x	Mushroom, geranium	5
1-Octene-3-ol	x	x	Mushroom	3
7-Octene-4-ol		x	Geranium	
Pentanal		x	Grass	
2-Propanol	x	x		0
3-Pentanone	x		Sour, burned	
2-Pentene-1-ol		x	Acid, sourish	

Table 5. Volatile compounds produced by *P. digitatum* on 2% malt agar

Volatile compound	Description of flavour	Intensity
2,3-Butanedione	Butter, cream caramel	3
Ethanol	Alcoholic, aromatic	2
Ethyl acetate	Acetone	4
Ethyl propionate	Fruity, pear	2
Hexanol	Banana	3
Hexyl acetate	Decayed, old potatoes	1
2-Hydroxy-2-butanone	Caramel	2
Methyl acetate		0
2-Methyl-l-propenol	Decayed	2
2-Methylpropanone	Butter	1
2-Methylpropyl acetate	Acetone, pear, sourish	2
3-Methyl-l-butanol	Decayed	3
3-Methyl-l-butyl acetate	Pear, banana	3
Octanal	Medicinal, mint	3
Pentanal	Burned	1
2-Propanone		0
2-Propenyl butanoate	Sourish, fruity caramel	2
Propyl acetate	Sulphur, potatoes	1

Generally speaking, *Penicillium* species growing on MA produced more volatiles with strong aromas than when grown on Cz (Tables 2-6). A frequently reported metabolite, 3-methylbutanol (Kaminski *et al.*, 1974; Harris *et al.*, 1986; Börjesson *et al.*, 1989) was produced by all species on MA but not on Cz. During growth on Cz, several earthy, musty compounds were more prominent than during growth on MA.

P. ochrochloron growing on MA produced many flavour esters such as ethyl propionate, propyl acetate, ethyl acetate, 2-methylethyl propionate and 2-methyl butanoate (Table 3). The total smell was fruity on MA and musty and earthy on Cz.

P. chrysogenum produced a strong smell of mushrooms on both agars. The smell came mainly from two compounds, 3-octanone and 1-octen-3-ol (Table 4).

On MA, *P. digitatum* produced the largest number of flavour compounds detected, which mainly occurred in low concentrations. The identified volatiles are shown in Table 5. The total smell from this species is difficult to describe. Because of poor growth, no volatiles were reported from growth on Cz.

The odour of *P. cyclopium* was musty and earthy on both media. This odour came from compounds which occurred in very low concentrations and had retention times of more than 20 minutes. During growth on Cz, 3-octanone and 1-octene-3-ol were also found (Table 6).

Experiment three: volatile metabolites of *A. flavus* and *A. oryzae*

When *Aspergillus flavus* and *A. oryzae* were grown for 2 weeks on solid media, no species specific differences in the production of volatiles were detected. Growth, as detected by increasing CO_2 production, continued throughout the experiment. However, the development of CO_2 during growth in liquid media was quite different (Figure 4). The production of CO_2 reached a peak of 200-300 mg per day after 2-3 days cultivation.

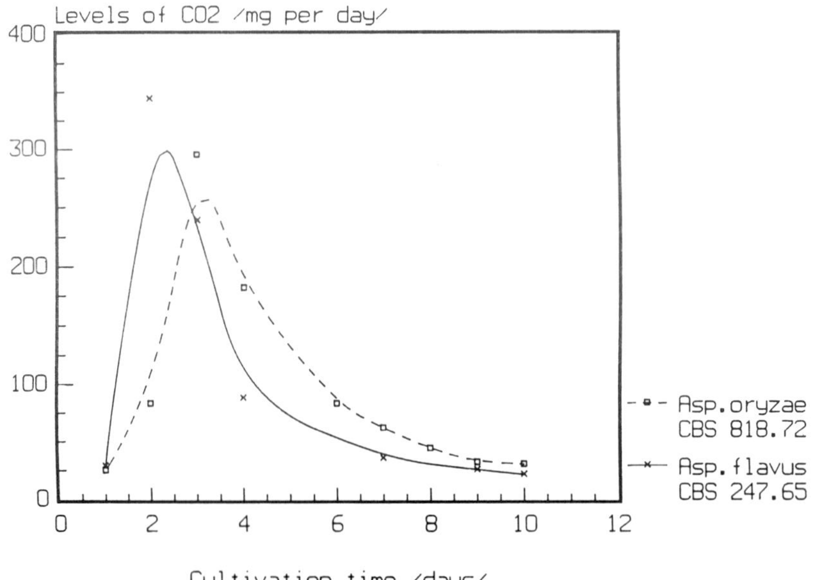

Figure 4. Production of CO_2 measured as mg of CO_2 per day during the cultivation of *A. flavus* and *A. oryzae* in liquid Czapek medium in the third experiment

Table 6. Volatile flavour compounds produced by *Penicillium cyclopium* on 2% malt agar and Czapek agar

Volatile compound	Medium		Description of flavour	Flavour intensity
	Cz	MA		
2,3-Butanedione	x	x	Butter, cream caramel	3
l-Butanol	x		Fruity, sour	1
2-Butanol		x		0
1,2-Dichloroethane		x	Sour, rancid	3
Ethanol	x	x	Alcoholic	1
2-Ethoxyethylbenzene		x	Woody	1
Ethylbenzene	x			0
Hexanal	x	x		0
Hexanone		x	Sweet	1
Hexene		x	Sourish, sweet	1
2-Methylbutane		x		0
2-Methyl-l-propanol		x	Burned	2
	x		Burned	1
3-Methyl-l-butanol		x	Burned	2
3-Methyl-2-butene-1-ol	x	.	Ethereal	1
Octanal		x	Fruity, sweet, flower	2
3-Octanol	x		Citrus	3
3-Octanone	x		Mushroom, geranium	1
l-Octene-3-ol	x		Mushroom	3
2,3-Pentanedione	x		Butter, cream caramel	1
l-Propanol	x		Sourish	2
		x		1
2-Propanol	x	x		0
2-Propanone		x	Decayed	1
l-Propen-2-ol-formate		x	Plastic	1

Table 7. Differences in production of two unidentified volatile components by *Aspergillus flavus* and *A. oryzae* growing in liquid medium.

Culture	Sample 1		Sample 2	
	Component A (ng/day)	Component B (ng/day)	Component A (ng/day)	Component B (ng/day)
A. flavus CBS 242.73	0	0	0	0
A. flavus CBS 118.62	0	0	0.24	0.46
A. flavus CBS 569.65	0	0	0	0
A. oryzae CBS 570.65	28.14	19.72		
A. oryzae CBS 574.65	5.14	6.09	1.12	1.55
A. oryzae CBS 818.72	2.82	6.39	4.25	10.48

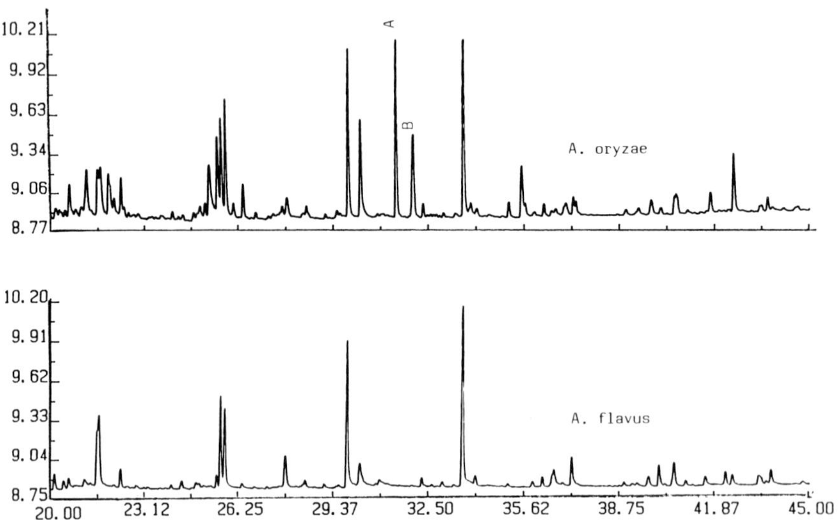

Figure 5. Comparison of the late fractions chromatograms showing the slowest eluting volatile compounds from the cultivation of *A. flavus* and *A. oryzae* respectively

A. flavus and *A. oryzae* gave similar patterns of volatiles during the first 5 days of growth in liquid medium, the predominant compounds being 2-methylpropanol, 3-methylbutanol and in most cases also ethanol. In the second phase of growth, days 6-10, the patterns of volatiles changed. The production of CO_2 and alcohols decreased strongly, but the production of other compounds increased. The patterns of these metabolites were found to differ between the two species. Two compounds, A and B, distinguished the strains of *A. oryzae* from the strains of *A. flavus* (Figure 5 and Table 7). Identification of these volatiles and study of production as a function of cultivation time are continuing.

REFERENCES

ABRAMSON, D., SINHA, R.N., and MILLS, J.T. 1983. Mycotoxins and odor formation in barley stored at 16 and 20% moisture in Manitoba. *Cereal Chem.* **60**: 350-355.

BÖRJESSON, T., STÖLLMAN, U., ADAMEK, P., and KASPERSSON, A. 1989. Analysis of volatile compounds for detection of molds in stored cereals. *Cereal Chem.* **66**: 300-304.

DRAVNIEKS, A., REILICH, H.G., and WHITFIELD, J. 1973. Classification of corn odor by statistical analysis of gas chromatographic patterns of headspace volatiles. *J. Food Sci.* **38**: 34-39.

HARRIS, N.D., KARAHADIAN, C. and LINDSAY, R.C. 1986. Musty aroma compounds produced by selected molds and actinomycetes on agar and whole wheat bread. *J. Food Prot.* **49**: 964-970.

HYDE, W., STAHR, H.M., SOGARD, D., and LERDAL, D. 1983. Volatiles analysis for mold detection in feeds and grain. Proc. 3rd Int. Symp. Vet. Lab. Diag. **41**: 323-331.

KAMINSKI, E., STAWICKI, S. and WASOWICZ, E. 1974. Volatile flavour compounds produced by molds of *Aspergillus*, *Penicillium* and fungi imperfecti. *Appl. Microbiol.* **27**: 1001-1004.

KAMINSKI, E., WASOWICZ, E., PRZYBYLSKI, R. and STAWICKI, S. 1985. Schnellmetoden zur Feststellung von Qualitätsveränderingen bei der Lagerung von Weizen. *Veroffentlichungen-Arbeitsgemeinschaft Getreideforschung* **198**: 41-51.

PITT, J.I. 1979. The Genus *Penicillium* and its Teleomorphic States *Eupenicillium* and *Talaromyces*. London: Academic Press.

EFFECT OF COPPER CONCENTRATION IN GROWTH MEDIA ON CONIDIATION IN *ASPERGILLUS* AND *PENICILLIUM*

A. D. King[1], N. J. Charley[2], M. Flores[1], A. D. Hocking[2] and C. Royer[1]

[1]*USDA ARS Western Regional Research Center*
800 Buchanan Street
Albany, CA 94710, USA

[2]*CSIRO Division of Food Processing*
P.O. Box 52
North Ryde, NSW 2113, Australia

SUMMARY

Components of microbiological media are becoming more highly purified to ensure uniformity from batch to batch. However, this purification process has led to mineral deficiencies in some media. In some fungal cultures, particularly *Penicillium* and *Aspergillus* species, mineral deficiencies can cause atypical and restricted conidiogenesis patterns. Addition of copper and zinc in correct concentrations can alleviate these problems.

INTRODUCTION

Czapek Yeast extract Agar (CYA) was developed as a standard medium for *Penicillium* taxonomy (Pitt, 1979). One of several functions of yeast extract was to act as a source of trace elements, and this function appeared to be fulfilled adequately. However, in recent years, poor conidial production or colours in some *Penicillium* species have been obtained in several laboratories. The possibility that a lack of trace elements was responsible prompted the series of experiments described in this paper.

METHODS

Media

Media initially tested were CYA (Pitt, 1979), CYA plus 5 ppm copper, CYA plus 10 ppm zinc and CYA plus 5 ppm copper and 10 ppm zinc. Subsequently, a range of copper and zinc levels was added to CYA to determine the concentrations resulting in maximum conidiogenesis. CYA plus 0, 0.5, 1, 2, 3, 4, and 5 ppm copper and 0, 1, 2, 4, 6, 8 and 10 ppm zinc were tested. In addition, copper (5 ppm) was tested in combination with various levels of zinc and similarly 10 ppm zinc was tested in combination with various levels of copper.

Fungi

Five fungi were used: *Aspergillus flavus* FRR 2755, *A. niger* FRR 3040, *Penicillium crustosum* FRR 943, *P. glabrum* FRR 2210, and *P. roqueforti* FRR 2437.

Inoculation and incubation

Petri dishes were inoculated at three points from conidial suspensions in semisolid agar (0.2%) plus surfactant (Tween 80, 0.05%; Pitt, 1979). Plates were incubated for 7 days at 25°C before evaluation. Colony diameters were also measured.

Evaluation of conidial production

The extent of conidiogenesis was measured by a four step visual rating and using a colour measuring instrument, the Minolta Chroma Meter (Minolta Corp., Meter Division, Ramsey, NJ 07446, USA). The Minolta Chroma Meter measured conidial colour by a standardised electronic flash. Colours were evaluated using the L* a* b* colour notation system and from those values a Whiteness Index (WI) was calculated.

$$WI = 100 - \sqrt{(100\text{-}L)^2 + a^2 + b^2}$$

WI values range from 0 (black) to 100 (white) so, in these experiments, a lower value indicated increased conidial formation.

RESULTS

Colony diameters on the four media were similar for the *Aspergillus* cultures. Although colony diameters for *Penicillium* cultures did not show a consistent pattern, colony growth rate did not appear to be influenced by the mineral supplements.

The WI values were lowest (highest conidial formation) for the copper and zinc supplemented media for *A. niger*, *P. crustosum* and *P. roqueforti* and the copper supplemented media for *A. niger* and *P. glabrum*. The visual rating of conidial formation ranked the copper supplemented CYA as highest for all five fungi and the copper and zinc supplemented next highest (Table 1). These data clearly showed that copper and zinc supplementation of the media enhanced the formation of conidia.

Table 1. Visual conidiogenesis rating[1] of mineral supplemented Czapek Yeast Agar (CYA)

Medium supplement (ppm)	*A. flavus*	*A. niger*	*P. crustosum*	*P. glabrum*	*P. roqueforti*
0	1	1	0	0	1
0.5 Cu	3	4	3	2	3
1.0 Cu	3	4	3	4	3
5.0 Cu	3	4	3	4	3
0	1	1	0	1	0
1 Zn	1	1	1	2	0
10 Zn	1	1	1	1	1
5 Cu + 2 Zn	2	4	3	3	3
5 Cu + 10 Zn	3	4	3	4	4
10 Zn + 1 Cu	2	1	3	4	4
10 Zn + 5 Cu	2	4	3	4	4

[1] Visual Rating: 0, No conidia; 4, Abundant conidia

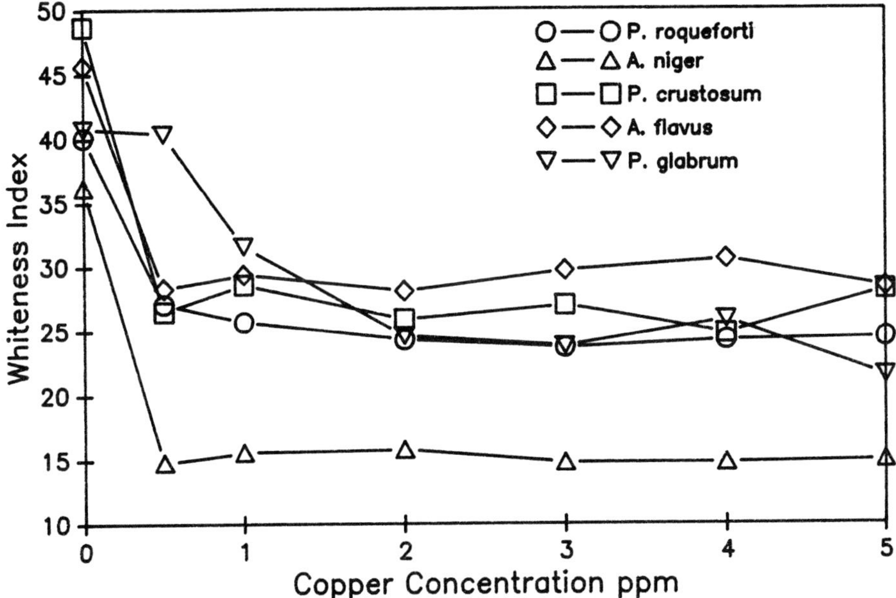

Figure 1. Effects of copper concentration on conidial formation by five fungi. Smaller Whiteness Index values are a measure of darker colour indicating greater conidial formation.

Figure 1 shows the level of copper plotted against WI for the five moulds. The values were calculated from the average of three readings of colony colour. With the exception of *P. glabrum*, all cultures showed an increase in conidial formation at copper concentrations greater than 0.5 ppm added to CYA compared with unsupplemented CYA. *P. glabrum* required 2 ppm added copper before conidial formation was uniform. All other treatments with added zinc or zinc and copper combinations gave similar results. Whether measured by colour meter or by visual inspection of the colonies, copper and zinc supplemented cultures gave the same reponse as copper alone. Zinc without copper did not enhance the formation of conidia. Colony diameters were not influenced by the mineral supplements.

Two similar experiments conducted in Sydney with the same strains produced different results. Sporulation of cultures grown on the supplemented media were not different from the control media. Presumably the water in Sydney, or the agar used for the media, contributed the necessary trace minerals.

The recommendation from this study is to supplement media that appear to be giving atypical colonies with at least 0.5 ppm copper. The formula for CYA in the *Penicillium* Laboratory Guide (Pitt, 1988) recommends adding copper and zinc to Czapek concentrate to provide final levels of 5 ppm and 10 ppm respectively.

REFERENCES

PITT, J.I. 1979. The Genus *Penicillium* and its Teleomorphic States *Eupenicillium* and *Talaromyces*. Sydney: Academic Press.

PITT, J.I. 1988. A Laboratory Guide to Common *Penicillium* Species. 2nd ed. North Ryde, NSW: CSIRO Division of Food Processing.

INFLUENCE OF MEDIUM ON GRAIN MYCOFLORA OBSERVED BY DIRECT PLATING

Kerstin Åkerstrand

Biology Division
National Food Administration
751 26 Uppsala, Sweden

SUMMARY

An evaluation of media used for direct plating of seed grains has shown that use of more than one medium is often necessary to obtain reliable results about the endogenous mycoflora. Dichloran 18% Glycerol agar (DG18) and Dichloran Rose bengal Yeast Extract Sucrose agar (DRYES) are the most effective media for examining grains, but one or two other media such as Dichloran Rose Bengal Chloramphenicol agar (DRBC) or Czapek Dox agar with iprodione (CZID) may also be valuable.

INTRODUCTION

When examining whole seed grains, enumeration of fungi by dilution plating is inadequate. Information on seed quality is best obtained by direct plating of surface disinfected seeds. The choice of media depends on the aim of the investigation. A single medium is seldom enough if a broad knowledge of the endogenous mycoflora is required.

In our laboratory we have tested many media described in the literature. The present study was designed to test the efficacy of several common media for the examination of seed grains.

METHODS

Grain samples. Eight grain samples were studied, three of wheat, three of rye, and one each of oats and buckwheat.

Media. Six media were examined: Czapek Dox agar with 50 mg/L each of chlortetracycline and chloramphenicol; Dichloran Rose Bengal Chloramphenicol agar (DRBC); Dichloran 18% Glycerol agar (DG18); Dichloran Chloramphenicol Peptone Agar (DCPA; Andrews and Pitt, 1986); Dichloran Rose bengal Yeast Extract Sucrose agar (DRYES) (Frisvad, 1983); and Czapek Dox agar with Iprodione (CZID; Abildgren *et al.*, 1987).

Plating. Whole grains were surface disinfected with sodium hypochlorite solution containing 3500 mg/L active chlorine (BDH sodium hypochlorite 3.5% solution, diluted 1:10) for two minutes (Andrews, 1986). Disinfected particles were rinsed twice in sterile water before plating 25 grains on each medium.

Incubation. Plates were incubated upright at 25°C for 7 days.

Table 1. Frequency of isolation of important mould genera on different media

Sample	Medium[2]	Frequency of different genera (%)[1]								
		Alt.	*Fus.*	*Clad.*	*Asp.*	*Eur.*	*Penicillium* *verr.*	*Penicillium* spp.	Muc.	Other
Wheat 1	Cz	66	-	-	-	-	-	2	-	32
	DRBC	84	-	-	2	-	-	-	-	18
	DG18	-	-	10	-	82	-	16	-	72
	DRYES	-	-	-	-	-	n.d.[3]	-	-	-
Wheat 2	Cz	72	4	16	18	-	-	16	-	24
	DRBC	48	-	18	36	-	-	14	4	38
	DG18	10	-	28	36	70	-	18	8	32
	DRYES	-	-	-	-	-	2	2	-	-
Wheat 3	CZID	34	8	-	10	-	-	2	-	48
	DRBC	22	-	6	20	-	-	18	-	46
	DG18	2	2	2	12	6	-	36	-	52
	DRYES	-	-	-	-	-	n.d.	-	-	-
Rye 1	Cz	90	-	2	-	-	-	-	-	2
	DRBC	32	-	6	4	2	-	6	-	50
	DG18	-	-	10	4	20	-	6	-	82
	DRYES	-	-	-	-	-	n.d.	-	-	-
Rye 2	Cz	22	-	4	18	2	4	16	6	58
	DRBC	34	-	6	20	4	10	14	2	42
	DG18	10	-	-	18	62	6	20	4	30
	DRYES	-	-	-	-	-	14	30	-	-
Rye 3	CZID	4	-	-	-	-	-	-	-	32
	DRBC	14	-	6	2	-	-	-	-	20
	DG18	-	4	6	-	6	-	-	-	42
	DRYES	-	-	-	-	-	n.d.	-	-	-
Oats	Cz	2	2	-	-	-	-	90	-	-
	DRBC	6	-	-	-	-	-	62	-	6
	DG18	8	-	-	-	6	-	44	-	-
	DRYES	-	-	-	-	-	6	-	-	-
Buckwheat	Cz	38	30	22	-	-	-	-	-	-
	DRBC	40	26	24	2	-	-	-	-	-
	DG18	32	14	24	-	-	-	-	-	-
	DCPA	22	48	8	-	-	-	-	-	-

[1] *Alt*, *Alternaria*; *Fus.*, *Fusarium*; *Clad.*, *Cladosporium*; *Asp.*, *Aspergillus*; *Eur.*, *Eurotium*; Muc., Mucoraceous fungi; *verr.*, *P. verrucosum*.
[2] Cz, Czapek Dox agar with antibiotics; DRBC, Dichloran Rose Bengal Chloramphenicol agar; DG18, Dichloran 18% Glycerol agar; DRYES, Dichloran Rose Bengal Yeast Extract Sucrose agar; CZID, Czapek Dox agar with Iprodione; DCPA, Dichloran Chloramphenicol Peptone Agar.
[3] n.d., not detected

RESULTS

The results of the experiments are shown in Table 1. Comments on the relative merits of each medium follow.

Czapek Dox agar supplemented with antibiotics is a general purpose medium which permits growth of many important species of field fungi as well as storage fungi. Sometimes, however, rapidly growing species are too vigorous, obscuring slowly growing species, which may escape detection.

DRBC is a good alternative to Czapek Dox agar with antibiotics, with more restricted growth of colonies and better limitation of spreading fungi.

On DG18, colonies are very distinct, and the most rapidly growing Zygomycetes are inhibited. *Eurotium* species are favoured, and counts of these important storage fungi are usually high. *Eurotium* species can, however, sometimes conceal other fungi. Sprouting is inhibited.

DCPA was developed to favour isolation of *Fusarium*. However, mycelial growth is very sparse, and no typical colours are developed. *Alternaria* is readily detected.

DRYES is important when *Penicillium verrucosum* is to be detected. An incubation temperature of 12°C is preferred, as this inhibits some interfering fungi.

CZID is effective for detecting *Fusarium* species, and is preferred for this genus. Other fungi are not completely inhibited, but only grow slowly. *Alternaria* produces conidia very well on this medium.

CONCLUSIONS

This investigation has shown that a single medium is insufficient for examination of the mycoflora of seed grains. DG18 and DRYES are adequate for storage fungi, but heavy growth of *Eurotium* or *Wallemia* may inhibit species of *Penicillium* or *Aspergillus*. In those cases, DRBC is recommended as a third medium. For the detection of field fungi including *Fusarium* and *Alternaria*, still another medium may be necessary.

REFERENCES

ABILDGREN, M.P., LUND, F., THRANE, U. and ELMHOLT. S. 1987. Czapek-Dox agar containing iprodione and dicloran as a selective medium for the isolation of *Fusarium* species. *Lett. Appl. Microbiol.* **5**: 83-86.

ANDREWS, S. 1986. Optimization of conditions for the surface disinfection of sorghum and sultanas using sodium hypochlorite solutions. *In* Methods for the Mycological Examination of Food, eds A.D. King, J.I. Pitt, L.R. Beuchat and J.E.L. Corry. pp. 28-32. New York: Plenum Press.

ANDREWS, S. and PITT, J.I. 1986. Selective medium for isolation of *Fusarium* species and dematicaeous Hyphomycetes from cereals. *Appl. Environ. Microbiol.* **51**: 1235-1238.

FRISVAD, J.C. 1983. A selective and indicative medium for groups of *Penicillium viridicatum* producing different mycotoxins in cereals. *J. Appl. Bacteriol.* **54**: 409-416.

METHODS FOR DETECTION AND IDENTIFICATION OF TECHNOLOGICALLY IMPORTANT FUNGI IN BARLEY AND MALT

Annette Lillie and M. Jakobsen

Alfred Jørgensen Laboratory Ltd
Frydendalsvej 30
1809 Frederiksberg, Denmark

SUMMARY

In the field, during storage and in the malting process, microflora and, in particular, mould contamination may dramatically affect the quality of barley and malt used in beer manufacture. For example, *Fusarium* and *Aspergillus* species may generate byproducts considered to be potential causes for excessive foam production ("gushing") when beer bottles are opened.

On the other hand, proteins such as hydrolytic enzymes and growth regulating factors secreted by microorganisms may have a positive effect on malt quality and hence on performance in the brewery. In this paper, methods for qualitative and quantitative evaluation of the mycoflora of barley and malt are reviewed. Our techniques for preliminary examination of barley samples, including pretreatment, selection of media and incubation conditions are presented. Taxonomic studies including thin layer chromatographic techniques are also reported.

The possibility of developing rapid screening methods for detection of technologically important fungi is discussed.

INTRODUCTION

Barley kernels may be contaminated with a variety of fungi. The composition of the mycoflora of barley and the level of contamination are determined by the growing conditions, primarily temperature and humidity. The most common field fungi isolated from barley are *Alternaria*, *Aureobasidium*, *Cladosporium*, *Epicoccum* and *Fusarium* species. The dominant storage fungi are *Aspergillus*, *Penicillium*, *Mucor* and *Rhizopus* species (Gyllang *et al.*, 1977; Flannigan *et al.*, 1982). *Stemphylium* and *Helminthosporium* species are also frequently isolated (Amaha *et al.*, 1973).

Some of the undesirable changes caused by growth of fungi and other microorganisms on barley are increase in respiration, darkening of the kernel, decrease in germination capacity and excessive foam production ("gushing") of beer produced from malted barley (Gjertsen *et al.*, 1963; Haikara, 1983). Growth of fungi in barley used in malting and brewing may also lead to an increase in α-amylase activity, a higher degree of modification of malt, increase of proteolytic activity with elevated wort nitrogen contents, and a higher degree of attenuation during fermentation (Prentice and Sloey, 1960; Haikara, 1983; Bol *et al.*, 1985), all changes apparently beneficial to the brewing process. Further, wort colour increases with increase of fungal activity during malting (Haikara, 1983; Bol *et al.*, 1985). Changes caused by the metabolic activity of fungi can take place in the field, during storage of barley and during malting.

The malting process is described briefly here. After cleaning and sorting, barley is steeped at 15-17°C, and takes up water, after 2 days to about 42-46% (w/w). The water is changed two to four times during the steeping period. After steeping the grain is allowed to germinate in layers approximately 1 metre deep. The grain is turned automatically 2-3 times every 24 hours, and is ventilated with a forced flow of air at 15°C and 100% R.H. Germination takes place over 6 to 8 days. During this process, called malting, the kernels produce cellulases, amylases, proteinases and peptidases responsible for the degradation of cellulose, starch and proteins. To stop germination and the enzymatic activity, the grains are then kiln dried at 80-115°C for 24 h to 2-4% (w/w) moisture. After drying the germ tubes are removed.

The problem of beer gushing became known as far back as 1913 when a Danish brewery experienced severe problems. The problem was significant in the 1950s (Gjertsen *et al.*, 1963), and has also occurred in recent years. It is of great concern to brewers.

In the early 1960s, the use of weathered barley for malting and subsequently the growth of micro-organisms in the field, during storage and malting were established as the primary causes of gushing (Prentice and Sloey, 1960; Gjertsen *et al.*, 1963). A direct relationship between growth of fungi during malting and gushing of beer produced has been reported (Gyllang and Martinson, 1976b).

Gushing is limited to certain geographical areas, especially Scandinavia and other parts of the world with a similar climate. The fungi most often reported as being responsible for gushing are *Fusarium* and *Aspergillus* spp. but also *Penicillium*, *Stemphylium* and *Rhizopus* spp. have been reported as active gushing inducers (Amaha *et al.*, 1973).

Haikara (1983) inoculated flowering barley in the field with different strains of *Fusarium avenaceum* and *F. culmorum* for two successive years and found that only *F. culmorum* caused gushing in beer. It thus appears that gushing is induced by specific species of fungi. Experimentally, *Aspergillus amstelodami* and *A. fumigatus* have been shown to cause gushing when added to steeping water (Gyllang and Martinson, 1976a).

METHODS

Detection of fungi in barley and malt

The standard method for detection of fungi, especially *Aspergillus* and *Fusarium*, in barley and malt is based on a surface sterilisation (1% hypochlorite) followed by direct plating on agar media.

Malt salt agar (European Brewing Convention, 1981) is used for the detection of *Aspergillus*. It contains malt extract, 20 g; NaCl, 75 g; agar 15 g; distilled water to 1 L, and is pH 4.9. For the detection of *Fusarium*, Czapek agar (Difco), pH 7.3 is used.

For detecting *Aspergillus*, 50 kernels are treated with 1% hypochlorite (1 min) followed by two rinses in sterile 7% (w/w) NaCl. The kernels are then direct plated on 10 Petri dishes with Malt salt agar and incubated at 30°C for 4 days.

For detecting *Fusarium*, 50 kernels are treated with 1% hypochlorite (1 min), but this time followed by two rinses in sterile water. After plating, the samples are incubated for 6 days at room temperature with a photoperiod of 12 h under UV light (360 nm) followed by 12 h dark.

After incubation, the percentages of kernels infected with *Aspergillus* and *Fusarium* spp. are calculated. In Denmark, these methods have been used to set standards for malting barley: kernel infection with these fungi should not exceed 30 and 10 percent, respectively.

RESULTS

Fungi detected in barley

Table 1 shows a summary of some results obtained with samples drawn at different times of the year from three malting houses in different localities in Denmark.

Results from 1979 and 1980 were very similar, with about one third of the samples having more than 30% of kernels infected with *Aspergillus* and about one tenth of the samples with more than 10% of the kernels infected with *Fusarium*. For comparison, Table 1 also includes results from 1987, which are markedly different. In that year, more than half the samples were heavily infected with *Aspergillus* and more than 63% with *Fusarium*. From discussions with breweries, it was confirmed that 1987 was the year when gushing overshadowed all other problems in the brewing industry.

In 1988 a limited carry over of the gushing problems was seen, but overall the results were similar to 1979 and 1980 (Table 1). The results from 1989 revealed that almost two thirds of the samples were heavily infected with *Aspergillus* (Table 1). Beer brewed from this harvest, however, did not show gushing, indicating that *Aspergillus* is not the primary cause of gushing. However, the need for more detailed information on the composition of the mycoflora of barley is emphasised.

Other methods

The methods we have been using appear not to be optimal although they provide some information of practical importance. However, similar considerations apply to other recommended methods as prescribed by the European Brewing Convention (EBC, 1981).

The general EBC method comprises incubation of surface disinfected barley kernels on wetted filter paper for 20 days at 20-22°C and exposure to UV light during nights. The specific method for *Fusarium* detection involves incubation of surface disinfected kernels on Potato Dextrose Agar with 0.2% pentachloronitrobenzene at 25°C for 7-10 days (European Brewing Convention, 1981). No UV light is prescribed. Storage fungi, especially *Aspergillus* and *Penicillium* are evaluated on Malt salt agar with 7.5% (w/w) NaCl.

The main disadvantages of these methods is that identification of the fungi to the species level involves subculturing onto identification media, long periods of time and very highly skilled people.

Table 1. Percentage of Danish barley samples infected with more than 30% *Aspergillus* or more than 10% *Fusarium* in selected years between 1979 and 1989, the generally accepted maximum levels of infection for malting

Year	Number of samples	>30% kernels with *Aspergillus*	>10% kernels with *Fusarium*
1979	35	30	10
1980	35	36	10
1987	43	55	63
1988	104	46	23
1989	33	60	0

We feel certain that identification of the fungi to species level is necessary in order to gain more knowledge on the correlation between the fungal flora of barley and gushing of beer.

DISCUSSION

New approaches

We are now testing a new analytical scheme, with the aim of identifying media which produce characteristic growth of the fungi present in barley. With this medium it is intended to identify the rather limited number of fungi growing in barley kernels primarily on the basis of their colony morphology.

These media have been chosen for this study. For the detection of *Fusarium*, *Alternaria*, *Cladosporium* and *Epicoccum* spp., Potato Dextrose Agar with Dichloran and Iprodione (PDAID; Thrane, *et al.*, 1992) is used. It consists of PDA (Difco) with added chloramphenicol (Sigma C-0378), 0.05 g; trace metal solution (Frisvad and Filtenborg, 1983), 1 ml; dichloran (Fluka AG 36010), 2 mg; chlorotetracycline (Sigma C-4881) 0.05 g; iprodione (Rovral 50 WP, Rhone-Poulenc Agro-chemie) 0.006 g. For detection of *Eurotium* and other xerophilic *Aspergillus* and *Penicillium* species, DG18 (Hocking and Pitt, 1980) is used, and for *Penicillium* and *Alternaria*, *Epicoccum* and *Cladosporium* spp.. Dichloran Rose bengal Yeast Extract Sucrose agar (DRYES; Frisvad, 1983).

Plates are incubated at 25°C for 7 days. Colony appearance and microscopy are used for describing the fungi. For confirmation, the TLC techniques of Frisavad *et al* (1989) are used.

It is envisaged that with time the scheme will allow the relevant fungi to be identified on the basis of colony appearance and microscopy. This will provide a valuable tool for establishing a better understanding of the relationship between fungal activity in barley and beer gushing as well as other changes of importance in brewing.

REFERENCES

AMAHA, M., KITABATAKE, K., NAKAGAWA, A., YOSHIDA, J. and HARADA, T. 1973. Gushing inducers produced by some mould strains. *Proc. Europ. Brew. Conv.* 14th Congr., Salzburg, 381-399.

BOL, J., KLOPPER, W.J., VERMEIRE, H.A. and MOTSHAGEN, M.E. 1985. Relation between the microflora of barley and malt quality. *Proc. Europ. Brew. Conv.* 20th Congr., Helsinki, 643-650.

EUROPEAN BREWING CONVENTION. 1981. Analytica Microbiologica. Part II. Analytical Methods. *J. Inst. Brew.* **87**: 303-321.

FLANNIGAN, B., OKAGBUE, R.N., KHALID, R. and TEOH, C.K. 1982. Mould flora of malt in production and storage. *Brew. Dist. Int.* 31-37.

FRISVAD, J.C. 1893. A selective and indicative medium for groups of *Penicillium viridicatum* producing different mycotoxins in cereals. *J. Appl. Microbiol.* **54**: 409-416.

FRISVAD, J.C. and FILTENBORG, O. 1983. Classification of terverticillate Penicillia based on profiles of mycotoxins and other secondary metabolites. *appl. Environ. Microbiol.* **46**: 1301-1310.

FRISVAD, J.C., FILTENBORG, O. and THRANE, U. 1989. Analysis and screening for mycotoxins and other secondary metabolites in fungal cultures by thin-layer chromatography and high-performance liquid chromatography. *Arch. Environ. Contam. Toxicol.* **18**: 331-335.

GJERTSEN, P., TROLLE, B. and ANDERSEN, K. 1963. Weathered barley as a contributory cause of gushing in beer. *Proc. Europ. Brew. Conv.* 9th Congr., Brussels, 320-341.

GYLLANG, H. and MARTINSON, E. 1976a. *Aspergillus fumigatus* and *Aspergillus amstelodami* as causes of gushing. *J. Inst. Brew.* **82**: 182-183.

GYLLANG, H. and MARTINSON, E. 1976b. Studies on the microflora of malt. *J. Inst. Brew.* **82**: 350-352.

GYLLANG, H., SÄTMARK, L. and MARTINSON, E. 1977. The influence of some fungi on malt quality. *Proc. Europ. Brew. Conv.* 16th Congr., Amsterdam, 245-254.

HAIKARA, A. 1983. Malt and beer from barley artificially contaminated with *Fusarium* in the field. *Proc. Europ. Brew. Conv.* 19th Congr., London, 401-408.

HOCKING, A.D. and PITT, J.I. 1980. Dichloran-glycerol medium for enumeration of xerophilic fungi from low moisture foods. *Appl. Environ. Microbiol.* **39**: 488-492.

PRENTICE, N. and SLOEY, W. 1960. Studies on barley microflora of possible importance to malting and brewing quality. I. The treatment of barley during malting with selected microorganisms. *Am. Soc. Brew. Chem. Proc.* 28-33.

THRANE, U., FILTENBORG, O., FRISVAD, J.C. and LUND, F. 1992. Improved methods for detecting toxigenic *Fusarium* species. *In* Modern Methods in Food Mycology, eds R.A. Samson, A.D. Hocking, J.I. Pitt and A.D. King. pp. 295-291. Amsterdam: Elsevier.

DIFFERENTIATION OF *ALTERNARIA* SPECIES ISOLATED FROM CEREALS ON DICHLORAN MALT EXTRACT AGAR

S. Andrews

School of Chemical Technology
South Australian Institute of Technology
Ingle Farm, SA 5098, Australia

SUMMARY

In a survey of Australian grains, only two species of *Alternaria*, *A. alternata* and the *Alternaria* state of *Pleospora infectoria* were isolated. These two species can be differentiated on Dichloran Chloramphenicol Peptone Agar but are more readily differentiated on a new medium Dichloran Chloramphenicol Malt Agar. The two species of *Alternaria* are distinguished by colony morphology, chain formation, ornamentation of the conidia, overall shape and size of the conidia including beak formation. This differentiation is less readily made on Potato Dextrose Agar, Carnation Leaf Agar or any one of the other traditional media used for the growth of the dematiaceous Hyphomycetes.

INTRODUCTION

Alternaria species are commonly isolated from cereal grains, tomatoes and some vegetables. Of the commonly isolated saprophytic species, *Alternaria alternata* (= *A. tenuis*), is thought to be potentially mycotoxigenic. Isolates of *Alternaria* are usually isolated and characterised on Potato Dextrose Agar (PDA), V8, Malt Agar (MA) or Potato Carrot Agar (PCA). However, few reports describe the identification of *Alternaria* isolates from foods, the topic investigated here.

The taxonomy of *Alternaria* has been described in detail by Neergaard (1945), Ellis (1971, 1976) and Simmons (1967, 1981, 1982a, 1982b), usually with an emphasis on herbarium specimens, host specificity and plant pathology.

Speciation of *Alternaria* by Neergaard (1945) was based predominantly on morphological characteristics of the conidia with pathogenicity and host specificity as additional parameters. He divided the genus into three subgroups according to the length and degree of branching of the conidial chains. Neergaard (1945) compared the characteristics of isolates on natural hosts as well as on media such as PDA and MA. Ellis (1971) in his system of classification, adopted chain formation as a taxonomic characteristic along with abruptness of the transition from spore body to beak and the length of the beak in relation to the length of the spore body. However, many descriptions by Ellis (1971) are based on herbarium specimens, limiting their usefulness. Descriptions by Simmons (1967, 1981, 1982a, 1982b) are based on both herbarium specimens and cultures, especially on PCA. Simmons has adopted the system of Neergaard (1945) and clarified some close speciation by adoption of *forma speciales*.

These schemes were regularly consulted during our investigation of *Alternaria* species isolated from Australian cereals. However the published descriptions were not always adequate for differentiation of these isolates, so alternative procedures for the isolation and characterisation of *Alternaria* species from cereals and foodstuffs were investigated.

METHODS

Isolation of *Alternaria*

In a study of Australian cereal grains that were the basis for formulation of poultry and pig feed pellets, 266 samples were examined. Media initially used for isolation of *Alternaria* were PDA and MA, which permitted luxuriant aerial growth, but minimal sporulation. Unfortunately, these media also supported luxuriant growth of *Fusarium,* dematiaceous Hyphomycetes and other fungi, making isolation of *Alternaria* very difficult and in some cases impossible. DRBC was ineffective for the isolation of *Alternaria*, as colonies were quite small, tough and velutinuous with very poor sporulation, and poor differentiation from other dematiaceous Hyphomycetes.

As the recovery of *Alternaria* from blended or homogenised samples was very low, direct plating of surface sterilised grains became the standard procedure. Dichloran Chloramphenicol Peptone Agar (DCPA) (Andrews and Pitt, 1986), developed for the isolation and characterisation of *Fusarium* species and dematiaceous Hyphomycetes, proved to be suitable for isolation of *Alternaria.* Best results were obtained after incubation at 25°C for 7 days under a light bank (12 hours of simultaneous illumination with UV and white light followed by 12 hours darkness) to optimise sporulation.

RESULTS

Characterisation of *Alternaria* on DCPA

From cereal samples, 148 *Alternaria* isolates were obtained, including at least two different species. One of these was identified as *Alternaria alternata.* On DCPA colonies were approximately 30-33 mm in diameter, of low and sparse hyaline mycelium, slightly raised centrally. Sporulation occurred over most of the colony giving it a brownish black appearance. The other major species, identified as the *Alternaria* state of *Pleospora infectoria* produced colonies of similar size (30-34 mm) with hyaline mycelium but with a distinct white grey central button (papilla) and a tendency to be more floccose than colonies of *A. alternata.* Sporulation was not as heavy and colonies were less intensely coloured, being pale to mid brown.

Although the colony morphologies were different, the differences were not distinctive. *In situ* examination revealed that both species produced a profusion of short conidiophores with both solitary and short, often branched chains of conidia. Both species produced erostrate conidiospores, with marked indentations at the septa, and both produced essentially smooth conidia measuring 30-40 μm x 11-14 μm. Conidia of the *Alternaria* state of *Pleospora infectoria* frequently contained more transverse and longitudinal septa than those observed for *A. alternata,* and also had a tendency to produce slightly longer beaks. Beaks in *A. alternata* were shorter and frequently constricted at the junction with the spore body. However the distinction between the two species was subtle.

Characterisation of *Alternaria* on DCMA

To improve differentiation of these two species, Dichloran Chloramphenicol Malt Agar

(DCMA) was formulated. This medium is composed of 1% Malt extract with 2 mg dichloran/L and 200 mg chloramphenicol/L. The concentration of malt extract was selected because it restricted development of aerial mycelium with minimal effects on sporulation. Dichloran restricted radial growth, and enhanced sporulation, while chloramphenicol was only used for primary isolation, to suppress bacterial growth.

Colonies of *Alternaria* on DCMA are larger than on DCPA, 40-45 mm in diameter of hyaline mycelium, submerged at the margins. Colonies of *A. alternata* had a distinct green black colour, sporulation was heavy and velutinous, occurring in well defined concentric rings over the entire colony. In contrast, colonies of the *Alternaria* state of *P. infectora* were green grey to dark greenish black, but were not so intensely coloured as those of *A. alternata*. Sporulation was fasciculate rather than velutinous and was heavy centrally but less dense towards the periphery.

On DCMA, *A. alternata* produced predominantly long unbranched chains of up to 11 ovoid to obclavate conidia, each possessing 3-5 transverse septa with 0-3 longitudinal septa. Beaks were generally short. The *Alternaria* state of *P. infectoria* produced short to medium length chains of conidia, which were invariably branched. Beaks were often long, up to half the length of the conidia, or occasionally longer. Beaks were usually flexuous or geniculate giving rise to secondary conidia and the appearance of "secondary conidiophores". Conidia of the *Alternaria* state of *P. infectoria* were generally roughened whereas those of *A. alternata* were generally smooth. The erostrate conidia of both species were very similar in size, however *A. alternata* rarely produced the long beaked conidia observed with the *Alternaria* state of *P. infectoria.*

An additional differential feature on DCMA was the formation of protoperithecia by most isolates of the *Alternaria* state of *P. infectoria.* The teleomorph of *P. infectoria* consists of perithecia enclosing long cylindrical asci containing eight ascospores with both transverse and longitudinal septa. At the base of the DCMA plates, the *Alternaria* state of *P. infectoria* produced highly convoluted aggregates of mycelium, but despite prolonged incubation, alternating temperature and light cycles, none of these protoperithecia matured. This was not unexpected as the culturing of anamorphs isolated from ascospores of *Pleospora herbarum,* despite prolonged and varied incubation protocols (Simmons, 1952; Leach, 1971), failed to produce the teleomorph from which they originated.

Extensive studies with 2% MA showed that the two species could be differentiated on this medium by colony and conidia morphologies. In addition, some isolates of *Pleospora infectoria* produced a yellow green water soluble pigment which helped to differentiate it from *Alternaria alternata.* However the differentiation on DCMA is more definitive and is the preferred medium for species identification of *Alternaria.*

Occurrence of the two species

Of the 148 isolates of *Alternaria* examined by the criteria described on DCMA, 68 were identified as *A. alternata* (Table 1). The *Alternaria* state of *P. infectoria* was the predominant species isolated from wheat, oats and barley, but noteably was not isolated from sorghum in this survey. Subsequently the *Alternaria* state of *P. infectoria* was isolated from sorghum but only after exhaustive searching.

Distribution of toxigenic strains of *Alternaria* on cereals

All of the 148 isolates were grown on moist rice for 21 days at 25°C. Cultures were extracted with hot ethanol and thin layer chromatograms of the extracts developed in toluene: ethyl acetate: formic acid (5:4:1 v/v). Sixty six of the *A. alternata* isolates

(97%) were found to be toxigenic. Of these, 94% produced alternariol methyl ether (AME) and alternariol (AOH) and about 60% produced tenuazonic acid (TeA) and altenuene (ALT) but not necessarily simultaneously. None of the 80 isolates of the *Alternaria* state of *P. infectoria* was toxigenic under these conditions. This finding reinforces the need to be able to readily differentiate these two species.

Table 1. Distribution of *Alternaria alternata* and *Pleospora infectoria* on cereal samples

Sample	Number of isolates of *A. alternata*	*P. infectoria*
Wheat	16	30
Oats	5	23
Barley	3	12
Triticale	14	12
Sorghum	29	0
Peas	1	2
Lupins	0	1
TOTAL	68	80

Other applications of DCMA

From grain survey isolates, DCMA has also been used to identify isolates of *A. tenuissima, A. radicinia, A. citri*, a number of species of *Curvularia, Ulocladium,* and *Drechslera*. In most cases these moulds produced moderate amounts of mycelium with good sporulation but did not grow as luxuriantly as they did on PDA, PCA or V8 agar.

Alternative identification schemes for *Alternaria*

The utility of the five plate regime used for the identification of *Penicillium* and other food-borne fungi as described by Pitt and Hocking (1985) was investigated.

On Czapek Yeast extract Agar, *A. alternata* and the *Alternaria* state of *P. infectoria* mostly produced similar size colonies (56-63 mm in diameter), that were low and slightly floccose. The aerial mycelium was mostly greyish but more greenish grey towards the periphery, with brownish black reverses. Isolates of the *Alternaria* state of *P. infectoria* that sporulated were difficult to differentiate from *A. alternata*. Many isolates of the *Alternaria* state of *P. infectoria* spored poorly, and were characterised by deeply floccose greenish grey mycelium; reverses were brownish black centrally, surrounded by a pale orange annulus. These isolates of the *Alternaria* state of *P. infectoria* were readily differentiated from *A. alternata*, but they could not be positively identified because of the absence of conidia.

On Malt Extract Agar, *A. alternata* was deeply floccose, dominated by abundant greenish grey mycelium, with limited brownish black sporulation. The growth of the *Alternaria* state of *P. infectoria* was less luxuriant, and in many cases was fasciculate with moderate sporulation, olive brown to brownish black colonies. Poorly sporulating colonies were off white to grey. Many isolates of both species were almost sterile on MEA, limiting the usefulness of this medium. When adequate sporulation did occur, characteristics similar to those described on DCMA were observed.

Both species produced undistinguished colonies of 7-12 mm on 25% Glycerol Nitrate Agar, which were olive brown to greyish brown, of low dense mycelium. The five plate regime of Pitt and Hocking (1985) was found to be unsatisfactory for distinguishing these two species of *Alternaria*: DCMA is the most appropriate medium for this purpose.

Variation within a defined species

Great variation was observed in conidial size and morphology of isolates characterised as *A. alternata* and the *Alternaria* state of *P. infectoria*. Branching of conidial chains was also highly variable. These variations were most noticeable with changes in media, but were also observed in subsequent subcultures. Therefore, identifications require caution. Future investigations will look at variation among isolates of *A. alternata* by enzyme gel electrophoresis and DNA homology, to evaluate whether they really do belong to a single species.

RECOMMENDATION

DCMA is recommended for the differentiation of *Alternaria alternata* from the *Alternaria* state of *Pleospora infectoria*. From limited experience, DCMA is also recommended for the identification of other saprophytic species of *Alternaria* and dematiaceous Hyphomycetes. DCMA can be used as an isolation medium for the recovery of dematiaceous Hyphomycetes from cereals, although competitive growth of Mucorales, *Aspergillus* and *Penicillium* may make isolations difficult. DCMA is not suitable for the isolation of most *Fusarium* species. As Mucorales, *Aspergillus* and *Penicillium* show restricted growth and sporulation on DCPA, it is recommended as the better isolation medium for dematiaceous Hyphomycetes and *Fusarium*. After isolation on DCPA, *Alternaria* and dematiaceous Hyphomycetes should be subcultured onto DCMA for identification and for culture maintenance. Cultures may be satisfactorily preserved in silica gel and freeze dried on carnation leaves aseptically removed from 14 day old Carnation Leaf Agar plates.

REFERENCES

ANDREWS, S. and PITT, J.I. 1986. Selective medium for the isolation of *Fusarium* species and dematiaceous Hyphomycetes from cereals. *Appl. Environ. Microbiol.* **51**: 1235-1238.

ELLIS, M.B. 1971. Dematiaceous Hyphomycetes. Kew, Surrey: Commonwealth Mycological Institute.

ELLIS, M.B. 1976. More Dematiaceous Hyphomycetes. Kew, Surrey: Commonwealth Mycological Institute.

LEACH, C.M. 1971. Regulation of perithecium development and maturation in *Pleospora herbarum* by light and temperature. *Trans. Br. Mycol. Soc.* **57**: 295-315.

NEERGAARD, P. 1945. Danish Species of *Alternaria* and *Stemphylium*. Copenhagen: Einar Munksgaard and London: Oxford University Press.

PITT, J.I. and HOCKING, A.D. 1985. Fungi and Food Spoilage. Sydney: Academic Press.

SIMMONS, E.G. 1952. Culture studies in the genera *Pleospora*, *Clathrospora* and *Leptosphaeria*. *Mycologia* **44**: 330-365.

SIMMONS, E.G. 1967. Typification of *Alternaria*, *Stemphylium*, and *Ulocladium*. *Mycologia* **59**: 67-92.

SIMMONS, E.G. 1981. *Alternaria* themes and variations. *Mycotaxon* **13**: 16-34.

SIMMONS, E.G. 1982a. *Alternaria* themes and variations (7-10). *Mycotaxon* **14**: 17-43.

SIMMONS, E.G. 1982b. *Alternaria* themes and variations (11-13). *Mycotaxon* **14**: 44-57.

8

RECOMMENDATIONS

RECOMMENDATIONS FROM THE CLOSING SESSION OF SMMEF II

A.D. Hocking[1], J.I. Pitt[1], R.A. Samson[2] and A.D. King[3]

[1] *CSIRO Division of Food Processing*
PO Box 52
North Ryde, NSW 2113, Australia

[2] *Centraalbureau voor Schimmelcultures*
PO Box 273
3740 AG Baarn, The Netherlands

[3] *USDA Agricultural Research Service,*
Western Regional Research Center,
Albany, CA 94710, USA

The final session of the Second International Workshop on Standardisation of Methods for the Mycological Examination of Foods considered recommendations made by various working parties on specific aspects of methodology and media for the mycological examination of foods. Recommendations were discussed in full at this final session, and, except where noted, were accepted unanimously by the Workshop participants. The Recommendations as finally accepted are set out below.

1. GENERAL PURPOSE ENUMERATION METHODS AND MEDIA

Methods

Dilution Plating. The Workshop reconfirmed the recommendations made in Boston at the First International Workshop on Standardization of Methods for the Mycological Examination of Foods (SMMEF I) (King *et al.*, 1986) on procedures for dilution plating:

1. Sample size: as large a sample as possible should be used.
2. Dilution: the initial dilution should be 1+9 in 0.1% peptone.
3. Homogenisation: use of the Stomacher is preferred over blending or shaking.
4. Homogenisation time: 2 minutes is recommended if a Stomacher is used. If a blendor is used, then the homogenisation time should be less, in the order of 30-60 sec., depending on the type and size of sample.
5. Further dilutions should be 1:10 (= 1+9) in 0.1% peptone.
   The feasibility of 1:5 dilutions for reducing "dilution error" should be further studied, preferably by means of interlaboratory trials.
6. Plating: spread plates are recommended over pour plates. Inocula should be 0.1 mL per plate.
7. Incubation: 5 days at 25°C is recommended.

Media

Dichloran Rose Bengal Chloramphenicol agar (DRBC; King *et al.*, 1979) and Dichloran 18% Glycerol agar (DG18; Hocking and Pitt, 1980) are recommended as general purpose isolation and enumeration media for foods of high water activity, i.e $a_w > 0.90$. However, the following points should be taken into consideration:

1. DG18 is less suitable for fresh fruits and vegetables, where studies have shown that DG18 yields lower counts because basidiomycetous fungi are often present.
2. Media containing rose bengal are light sensitive. Inhibitory compounds are produced after relatively short exposures to light. Prepared media should be stored away from light until used.
3. Media should be of approximately neutral pH and contain appropriate antibiotics.
4. The antibiotic of choice is chloramphenicol, used at a concentration of 100 mg/kg (ppm). Chloramphenicol is heat stable and may be added before autoclaving.
5. Where bacterial overgrowth may be a problem (e.g. in fresh meats), chloramphenicol (50 mg/kg) plus chlortetracycline (50 mg/kg) is recommended. Chlortetracycline is heat labile, and must be filter sterilised and added to the media after autoclaving. It is relatively unstable in solution and must be freshly prepared or refrigerated. Gentamycin is not recommended, as it has been reported to cause inhibition of some yeast species.

2. SELECTIVE MEDIA

Interlaboratory trials are essential if selective media are to be generally adopted by regulatory bodies for use in guidelines for enumeration of specific fungi in foods.

Media for xerophilic fungi

DG18 is the best medium currently available for enumeration of common xerophilic fungi in foods. However, growth of *Eurotium* species is rather rapid on DG18, and colonies do not have discrete margins. More work is needed on media to restrict growth of *Eurotium*.

For isolation of extreme xerophiles, e.g. *Xeromyces bisporus*, *Eremascus* species and xerophilic *Chrysosporium* species, Malt Yeast 50% Glucose agar (MY50G; Pitt and Hocking, 1985) is recommended, as outlined by the SMMEF I workshop in Boston (King *et al.*, 1986)

Media for *Fusarium* species

Interlaboratory studies are needed to compare Czapek Iprodione Dichloran agar (CZID; Abildgren *et al.*, 1987) with DRBC, with Dichloran Peptone Chloramphenicol agar (DCPA; Andrews and Pitt, 1986) as an optional addition. A check of the plant pathology literature for other useful media should be made. Work is needed on an immunoassay for *Fusarium* species, with a view to a future comparison with the chosen *Fusarium* selective medium.

Media for toxigenic *Penicillium* species

Dichloran Rose bengal Yeast Extract Sucrose agar (DRYES; Frisvad, 1983) appears to be valuable for differentiating nephrotoxigenic species in subgenus *Penicillium*. Interlaboratory studies are needed to compare counts on DRYES with those on other standard media such as DRBC, DG18 and Oxytetracycline Glucose Yeast Extract Agar (OGY; Mossel *et al.*, 1970).

Media for aflatoxigenic species

Interlaboratory studies of Aspergillus Flavus and Parasiticus Agar (AFPA; Pitt *et al.*, 1983) *vs* DRBC and Aspergillus Differential Medium plus antibiotics (ADMCC; Bothast and Fennell, 1974) are required. These should include dilution counts as well as direct plating comparisons (at least 50, preferably 100 pieces per medium per test). Dilution and direct plating should be compared on the same samples in parallel.

3. METHODS FOR YEASTS

General methods for enumeration of yeasts

For products such as beverages, where the mycoflora normally comprises mainly yeasts, a non-selective medium such as Tryptone Glucose Yeast extract agar (TGY) plus an antibiotic is recommended. Recommended antibiotics are chloramphenicol and oxytetracycline.

For products where yeasts must be enumerated in the presence of moulds, DRBC is recommended.

Preservative resistant yeasts

More work is needed to establish the validity of the methods in current use. Media containing 10% glucose appears to be better than other media. A collaborative study on the following media was recommended:

1. TGY + 0.5% acetic acid
2. MEA + 0.5% acetic acid
3. any other medium currently in use.

Diluents for yeasts

For yeasts in high a_w foods and beverages, the recommended diluent is 0.1% peptone. However, there is no evidence that 0.85-0.9% saline peptone is disadvantageous, and use may be continued.

For concentrates, syrups and other low a_w samples, a diluent containing at least 20 to 30% glucose is recommended. More work is needed on these and other glucose concentrations (e.g. 40, 50%) to determine the optimum concentration.

Detection of low numbers of yeasts

Membrane filtration is the recommended method for detection of low numbers of yeasts. MPN techniques are not recommended. Emerging technologies may eventually produce better methods.

4. PREPARATION OF DRIED SAMPLES FOR DILUTION PLATING

Dried samples intended for dilution plating where the fungi are deep seated or internal (e.g. as in grains and nuts), should be soaked for 30 minutes in 0.1% peptone before Stomaching or blending. For powders and other homogeneous samples, no soaking appears to be required prior to mixing. It was recommended that a collaborative study on soaking of powdered samples should be carried out to verify this.

5. ENUMERATION OF STRESSED FUNGAL PROPAGULES AND METHODS FOR HEAT RESISTANT FUNGI

1. Enumeration of stressed fungal propagules
A general purpose medium, such as DRBC, should be used for enumeration of foods which may contain stressed fungal propagules, with incubation for at least 5 days at 25°C. The medium may need to be modified if xerophiles are likely to be present.

Resuscitation of the initial 1+9 dilution in 0.1% peptone by soaking for 30 minutes may be useful, but more studies are needed.

More work is needed to determine the optimal conditions for recovery of fungal propagules stressed by heat, desiccation or freezing. Factors to be investigated should include:
1. Interactions between pH, organic acids and recovery of stressed cells.
2. Determination of optimal times, media and temperatures for resuscitation and maximal recovery of stressed cells.

Enumeration of heat resistant fungi
By definition, a heat resistant fungus produces ascospores which can survive heating at 75°C for 30 minutes. The raw materials most likely to contain heat resistant fungi are those that have been contaminated with soil, e.g. fruits and products derived from them, and raw milk. The procedures outlined at SMMEF I (King *et al.*, 1986) are effective.

Recommended method. At least 100 g of product should be heated at 75°C for 30 minutes and dispersed in an equal volume of double strength agar of an appropriate composition (which may contain antibiotics to prevent growth of *Bacillus*). The whole sample should be plated and incubated, and counted after at least 14 days incubation at 25°C.

Determining and reporting D values of heat resistant fungi
D values should be reported by methods which will allow comparison with other published reports. Sources of variability in D values include incubation conditions and the maturity of the ascospores.

Recommendations
1. To ensure that ascospores are mature, cultures at least 3 weeks old should be used.
2. Heat in a standard medium, such as that of Bayne and Michener (1979), which consists of 16 g glucose + 0.5 g tartaric acid per 100 mL water, pH 3.6, and also heat in the appropriate food.
3. Heat at 3 different temperatures, one of which should be 80°C, using an initial inoculum of at least 10^5 cfu/mL.
4. The time course of death should be followed over at least 3 log cycles using Czapek Yeast Extract agar or Malt Extract Agar as a plating medium.

6. BASELINE COUNTS

The workshop participants acknowledged the compilation of an impressive amount of data on baseline counts as a major achievement. With adequate data, the setting of baseline standards for levels of fungal populations in some foodstuffs is realistic and attainable.

Some areas where baseline counts may be of value include:

1. Raw materials monitoring for production of high quality end-product.
2. Monitoring of sanitation systems and processing GMP.
3. Monitoring risk levels for public health.
4. Monitoring shelf life expectations.

Currently, most criteria for foods relate to levels of bacteria, but mycological testing should be considered in specific areas such as mycotoxigenic fungi, preservative resistant yeasts, xerophilic yeasts, heat resistant fungi and extreme xerophiles.

In setting acceptable levels for fungi in foods, products tested should have been produced under GMP, sampling plans as outlined by ICMSF should be used, and standardised laboratory methods should be employed.

Recommendation

It was recommended that further surveys should be conducted to gather baseline data from the USA and Europe. Some marketing expertise may be needed in the gathering of this data, and it may be necessary to seek grant funds to carry out the work.

7. MONITORING OF MEDIA

It was recommended that all laboratories should monitor media quality and selective properties as follows:

1. A needle point inoculum of the test species should be taken from a 1-3 week old slant culture and dispersed in semi-solid agar.
2. A sterile, standardised loop should be used to inoculate the spore suspension onto 3 points of the test medium.
3. After 5 days, colony diameters of test species should be measured. Colony characteristics should also be observed.

The following test fungi were suggested:

Rhizopus stolonifer
Aspergillus niger
Eurotium chevalieri
Cladosporium cladosporioides
Zygosaccharomyces rouxii
Saccharomyces cerevisiae
Hansenula anomala
Rhodotorula glutinis

A substantial debate followed on the necessity or desirability of including bacterial test cultures. The final consensus was that if a laboratory wishes to test the antibacterial properties of a medium, then *Bacillus subtilis* is a suitable species.

These recommendations, which apply to DRBC, DG18, OGY and TGY, were accepted by the Workshop participants with one dissenter to the addition of bacteria to the list of recommended species.

An extensive interlaboratory study was recommended to examine the following points:

1. To evaluate the test species to determine their need and appropriate nature
2. To determine the minimum and maximum acceptable colony diameters for each test species on each medium.
3. To determine the expected recovery of each test species on each medium.
4. To examine colony morphology for each species.

8. GUIDELINES ON ACRONYMS FOR MYCOLOGICAL MEDIA

It was considered that a uniform approach should be adopted in the naming of new mycological media. However, the guidelines recommended and set out below do not affect acronyms in current use.

1. Ingredients should be named in the order:
   i. Basal medium;
   ii. Selective agent;
   iii. Antibiotic.
2. "A" for agar need not be included.
3. Media should not be named after those who devised them.

These recommendations were accepted by the Workshop participants, with one dissenter.

REFERENCES

ABILDGREN, M.P., LUND, F., THRANE, U. and ELMHOLT, S. 1987. Czapek-Dox agar containing iprodione and dicloran as a selective medium for the isolation of *Fusarium* species. *Lett. Appl. Microbiol.* **5**: 83-86.

ANDREWS, S. and PITT, J.I. 1986. Selective medium for isolation of *Fusarium* species and dematiaceous hyphomycetes from cereals. *Appl. Environ. Microbiol.* **51**: 1235-1238.

BAYNE H.G. and MICHENER, H.D. 1979. Heat resistance of *Byssochlamys* ascospores. *Appl. Environ. Microbiol.* **37**: 449-453.

BOTHAST, R.J. and FENNELL, D.F. 1974. A medium for rapid identification and enumeration of *Aspergillus flavus* and related organisms. *Mycologia* **66**: 365-369.

FRISVAD, J.C. 1983. A selective and indicative medium for groups of *Penicillium viridicatum* producing different mycotoxins in cereals. *J. Appl. Bacteriol.* **54**: 409-416.

HOCKING, A.D. and PITT, J.I. 1980. Dichloran-glycerol medium for enumeration of xerophilic fungi from low moisture foods. *Appl. Environ. Microbiol.* **39**: 488-492.

KING, A.D., HOCKING, A.D. and PITT, J.I. 1979. Dichloran-rose bengal medium for enumeration of molds from foods. *Appl. Environ. Microbiol.* **37**: 959-964.

KING, A.D., PITT, J.I., BEUCHAT, L.R. and CORRY, J.E.L. Eds. 1986. Methods for the Mycological Examination of Food. New York: Academic Press.

MOSSEL, D.A.A., KLEYNEN-SEMMELING, A.M.C., VINCENTIE, H.M., BEERENS, H. and CATSARAS, M. 1970. Oxytetracycline-glucose-yeast extract agar for selective enumeration of moulds and yeasts in foods and clinical material. *J. Appl. Bacteriol.* **33**: 454-457.

PITT, J.I. and HOCKING, A.D. 1985. Fungi and Food Spoilage. Sydney: Academic Press.

RECOMMENDED METHODS FOR MYCOLOGICAL EXAMINATION OF FOODS, 1992

J.I. Pitt[1], A.D. Hocking[1], R.A. Samson[2] and A.D. King[3]

[1] *CSIRO Division of Food Processing*
PO Box 52
North Ryde, NSW 2113, Australia

[2] *Centraalbureau voor Schimmelcultures*
PO Box 273
3740 AG Baarn, The Netherlands

[3] *USDA Agricultural Research Service,*
Western Regional Research Center,
Albany, CA 94710, USA

In the Editors' opinions, the following methods are the most satisfactory currently available for the mycological examination of foods. Many have been proven effective by collaborative studies. Others have been shown to be effective in a number of laboratories, but still require collaborative study. The following media and methods are recommended for use by all who study fungi in foods.

1. GENERAL PURPOSE ENUMERATION METHODS AND MEDIA

Methods

Dilution Plating. Dilution plating is recommended for liquid foods and powders, and also for particulate foods where the total mycoflora is of importance, as for example in grains intended for flour manufacture.

Samples. Samples should be as representative as possible. For suitable microbiological sampling procedures refer to the publication by the International Commission on Microbiological Specifications for Foods (ICMSF, 1986).

The sample size used should be as large as possible, consistent with the equipment used for homogenisation. If a Stomacher 400 is used, 10 to 40 g samples are suitable.

Diluents and dilution. The recommended diluent for fungi including yeasts is 0.1% peptone in water. Dilutions should be 1:10 (1+9). Some evidence exists that dilutions of 1:5 may be superior for mycological counts, but this has not been rigorously proven, and lacks the simplicity of the 10 fold dilution system.

Homogenisation. The Colworth Stomacher is the preferred homogeniser, used for 2 minutes per sample. Blending for 30-60 sec or shaking in a closed bottle with glass beads for 2-5 min are less preferable alternatives.

Plating. For mycological studies, spread plates are recommended; pour plates are less effective. Inocula should be 0.1 mL per plate, spread with a sterile, bent glass rod.

Incubation. For general purpose enumeration, 5 days incubation at 25°C is recommended. Plates should be incubated upright. A higher temperature, e.g. 30°C, is suitable in tropical regions.

Results. Results are expressed as log count/g of the original sample.

Direct plating. Direct plating is considered to be the more effective technique for mycological examination of particulate foods such as grains and nuts. In most situations, surface sterilisation before direct plating is considered essential, to permit enumeration of fungi actually invading the food. An exception is to be made for cases where surface contaminants become part of the downstream mycoflora, e.g. wheat grains to be used in flour manufacture. In such cases, grains should not be surface sterilised.

Surface sterilisation. Surface sterilise food particles by immersion in 0.4% chlorine (household bleach, diluted 1:10) for 2 minutes. A minimum of 50 particles should be sterilised and plated. The chlorine should be used once only.

Rinse. After pouring off the chlorine, rinse once in sterile distilled or deionised water. Note: this step has not been shown to be essential, but is recommended.

Surface plate. As quickly as possible, transfer food particles with sterile forceps to previously poured and set plates, at the rate of 5-10 particles per plate.

Incubation. The standard incubation regime for general purpose enumerations is 25°C for 5 days. Plates should be incubated upright.

Results. Express results as per cent of particles infected by fungi. Differential counting of a variety of genera is possible using a stereomicroscope.

Media

Dichloran Rose Bengal Chloramphenicol agar (DRBC; King *et al.*, 1979) and **Dichloran 18% Glycerol agar** (DG18; Hocking and Pitt, 1980) are recommended as general purpose isolation and enumeration media.

DRBC is recommended for fresh foods, including fruits and vegetables, meats and dairy products. Note that media containing rose bengal are light sensitive. Inhibitory compounds are produced after relatively short exposures to light. Prepared media should be stored away from light until used.

For foods of reduced water activity, i.e less than 0.95 a_w, DG18 is preferred. Although originally formulated for enumeration of xerophilic fungi, DG18 is now widely accepted as an effective general purpose medium. Its water activity (0.955) reduces interference by both bacteria and rapidly growing fungi.

Although originally chlortetracycline was used as the antibacterial agent, both media are now formulated with chloramphenicol (100 mg/kg), which is heat stable. Chloramphenicol may be added before autoclaving, for ease of preparation.

Where bacterial overgrowth may be a problem (e.g. in fresh meats), chloramphenicol (50 mg/kg) plus chlortetracycline (50 mg/kg) is recommended. Note: chlortetracycline is labile, and must be freshly prepared or refrigerated, filter sterilised and added to the media after autoclaving. Gentamycin is not recommended.

2. SELECTIVE MEDIA

Media for xerophilic fungi

Dichloran 18% Glycerol agar (DG18; Pitt and Hocking, 1980) is the recommended medium for enumeration of common xerophilic fungi in foods. Incubation at 25°C for up to 7 days is recommended. Growth of *Eurotium* species on DG18 is rather rapid, and colonies do not have discrete margins.

For isolation of extreme xerophiles, e.g. *Xeromyces bisporus*, *Eremascus* species and xerophilic *Chrysosporium* species, **Malt Yeast 50% Glucose agar** (MY50G; Pitt and Hocking, 1985) is recommended. Direct plating should be used. Incubate at 25°C and examine plates after 7 days and, if no growth occurs, after longer periods, up to 21 days.

Media for *Fusarium* species

No general agreement exists on a suitable selective medium for all common food-borne *Fusarium* species. Czapek Iprodione Dichloran agar (CZID; Abildgren *et al.*, 1987), Dichloran Chloramphenicol Peptone agar (DCPA; Andrews and Pitt, 1986) and pentachloronitrobenzene peptone agar (Burgess *et al.*, 1988) have all been recommended.

Media for toxigenic *Penicillium* species

Dichloran Rose bengal Yeast Extract Sucrose agar (DRYES; Frisvad, 1983) is of value for distinguishing *Penicillium verrucosum* and *P. viridicatum*. Collaborative studies are still needed to confirm that recoveries are satisfactory.

Media for species producing aflatoxins

Aspergillus Flavus and Parasiticus Agar (AFPA; Pitt *et al.*, 1983) is effective for detecting and enumerating *Aspergillus flavus* and *A. parasiticus*. Incubate at 30°C for 2-3 days. Collaborative studies are still needed to confirm that recoveries are satisfactory.

3. METHODS FOR YEASTS

Diluents

For the enumeration of yeasts in high a_w foods and beverages, the recommended diluent is 0.1% peptone. For concentrates, syrups and other low a_w samples, 20 to 30% glucose (w/v) in 0.1% peptone is recommended.

General purpose media

For products such as beverages, where yeasts usually predominate, nonselective media such as **Malt Extract Agar** (Pitt and Hocking, 1985) or **Tryptone Glucose Yeast Extract agar** (TGY) plus chloramphenicol or oxytetracycline (100 mg/kg) are recommended.

For products where yeasts must be enumerated in the presence of moulds, DRBC is recommended.

For products where bacteria may be encountered, or are suspected, examination of representative colonies under the microscope is important.

Detection of low numbers of yeasts

For detection of low numbers of yeasts, membrane filtration is recommended. Other methods, such as most probable number (MPN) techniques, are not recommended.

Preservative resistant yeasts

An effective medium for the detection of preservative resistant yeasts is Malt Acetic agar (MAc), which is Malt Extract Agar containing 0.5% acetic acid, added just before pouring (Pitt and Hocking, 1985). Other media incorporating 0.5% acetic acid are probably equally effective. The addition of 10 % (w/v) glucose may be helpful.

4. PREPARATION OF DRIED SAMPLES FOR DILUTION PLATING

Where dried particulate foods such as cereals are to be dilution plated, soaking for 30 minutes in 0.1% peptone before homogenising is recommended.

5. ENUMERATION OF HEAT RESISTANT FUNGI

Heat 100 g of product, preferably in 2 x 50 mL portions, at 75°C for 30 minutes, mix with equal volumes of double strength potato dextrose or other nutrient agar, and pour into plates. Large (150 mm diameter) Petri dishes can be used with advantage. Incubate at 30°C, which ensures development of *Byssochlamys*, *Neosartorya* and *Talaromyces* asci. Examine plates after 7 days incubation and, if no growth occurs, after longer periods, up to 4 weeks.

Care should be taken to avoid contamination during pouring of plates. Heavily sporing colonies of e.g. *Penicillium* or *Aspergillus* species indicate contamination. Such colonies should be ignored.

Concentrated samples (greater than 35° Brix) should be diluted 1:1 with 0.1% peptone water before heating. Very acid samples should be adjusted to pH 3.5 to 4.0 with NaOH before heating.

REFERENCES

ABILDGREN, M.P., LUND, F., THRANE, U. and ELMHOLT, S. 1987. Czapek-Dox agar containing iprodione and dicloran as a selective medium for the isolation of *Fusarium* species. *Lett. Appl. Microbiol.* **5**: 83-86.

ANDREWS, S. and PITT, J.I. 1986. Selective medium for isolation of *Fusarium* species and dematiaceous hyphomycetes from cereals. *Appl. Environ. Microbiol.* **51**: 1235-1238.

BURGESS, L.W., LIDDELL, C.M. and SUMMERELL, B.A. 1988. Laboratory Manual for *Fusarium* Research. 2nd ed. Sydney: University of Sydney.

ICMSF (1986). Microorganisms in Foods. 2. Sampling for Microbiological Analysis: Principles and Specific Applications. 2nd ed. Toronto: University of Toronto Press.

FRISVAD, J.C. 1983. A selective and indicative medium for groups of *Penicillium viridicatum* producing different mycotoxins in cereals. *J. Appl. Bacteriol.* **54**: 409-416.

HOCKING, A.D. and PITT, J.I. 1980. Dichloran-glycerol medium for enumeration of xerophilic fungi from low moisture foods. *Appl. Environ. Microbiol.* **39**: 488-492.

KING, A.D., HOCKING, A.D. and PITT, J.I. 1979. Dichloran-rose bengal medium for enumeration of molds from foods. *Appl. Environ. Microbiol.* **37**: 959-964.

PITT, J.I. and HOCKING, A.D. 1985. Fungi and Food Spoilage. Sydney: Academic Press.

PITT, J.I., HOCKING, A.D. and GLENN, D.R. 1983. An improved medium for the detection of *Aspergillus flavus* and *A. parasiticus*. *J. Appl. Bacteriol.* **54**: 109-114.

APPENDIX - MEDIA

APPENDIX - MEDIA

This list includes media commonly used to enumerate fungi in foods. Some are commercially available. For information, see Difco (1984) and Oxoid (1990) Manuals.

Unless otherwise stated, all ingredients should be combined and sterilised by autoclaving at 121°C for 15 min. Chemicals should be of high quality. Brands of yeast extract differ, and should be changed if inhibitory effects are detected. Regular bacteriological peptones are effective. Distilled water should be used. Chloramphenicol can be added to media before autoclaving, but other antibiotics should be filter sterilised, then added to sterile media just before pouring. Chlortetracycline solutions should be freshly prepared or refrigerated.

1. **AFPA: Aspergillus Flavus and Parasiticus Agar** (Pitt *et al.*, 1983)

Yeast extract	20	g
Peptone	10	g
Ferric ammonium citrate	0.5	g
Dichloran (0.2% in ethanol)	1	mL
Chloramphenicol	0.1	g
Agar	15	g
Water	1000	mL

Final pH 5.6. On this medium *Aspergillus flavus* and *A. parasiticus* are detected by production of a bright orange yellow reverse colour after 2-3 days incubation at 30°C.

2. **CREAD: Creatine Sucrose Dichloran Agar** (Frisvad, 1985)

Creatine.H_2O	3	g
Sucrose	30	g
KCl	0.5	g
$MgSO_4.7H_2O$	0.5	g
$FeSO_4.7H_2O$	0.01	g
K_2HPO_4	1.0	g
Dichloran (0.2% in ethanol)	1	mL
Bromocresol purple	0.05	g
Agar	15	g
Water	1000	mL

Final pH 8.0 (adjusted after medium is autoclaved). For differentiation of species in *Penicillium* subgenus *Penicillium*.

3. **CZID: Czapek Dox Iprodione Dichloran Agar** (Abildgren *et al.*, 1987)

Czapek-Dox Broth (Difco)	35	g
$CuSO_4.5H_2O$	0.005	g
$ZnSO_4.7H_2O$	0.01	g
Chloramphenicol	0.05	g
Dichloran (0.2% in ethanol)	1	mL
Agar	20	g
Distilled water to	1000	mL

After autoclaving and cooling to 50°C, add chlortetracycline solution, 10 mL and iprodione suspension, 1 mL. Chlortetracycline solution is 0.5% aqueous, and iprodione suspension (which should be shaken before addition to CZID) contains Roval 50WP (Rhone-Poulenc, Agro-Chemie, Lyon, France) 0.3 g in sterile water, 50 mL. CZID was developed for the selective detection of *Fusarium* species.

4. **CYA: Czapek Yeast Extract Agar** (Pitt, 1979)

Sucrose	30	g
Yeast extract	5	g
$NaNO_3$	3	g
KCl	0.5	g
$MgSO_4.7H_2O$	0.5	g
$FeSO_4.7H_2O$	0.01	g
K_2HPO_4	1.0	g
Agar	20	g
Water	1000	mL

If distilled water is used, add 0.01 g of $ZnSO_4.7H_2O$ and 0.005 g of $CuSO_4.5H_2O$ per litre. Copper ions are essential for the formation of normal colours in *Penicillium* and some *Aspergillus* conidia.

5. **CY20S: Czapek Yeast Extract 20% Sucrose Agar** (Pitt and Hocking, 1985)

Sucrose	200	g
Yeast extract	5	g
$NaNO_3$	3	g
KCl	0.5	g
$MgSO_4.7H_2O$	0.5	g
$FeSO_4.7H_2O$	0.01	g
K_2HPO_4	1.0	g
Agar	20	g
Water	1000	mL

Recommended for growth and identification (but not isolation) of *Eurotium* species.

6. **DCMA: Dichloran Chloramphenicol Malt Extract Agar** (Andrews, 1992)

Malt Extract	10.0	g
Dichloran (0.2% in ethanol)	1.0	mL
Chloramphenicol	0.1	g
Agar	15.0	g
Distilled Water	1000	mL

Recommended for identification of *Alternaria* species.

7. **DCPA: Dichloran Chloramphenicol Peptone Agar** (Andrews and Pitt, 1986)

Peptone	15.0	g
KH_2PO_4	1.0	g
$MgSO_4.7H_2O$	0.5	g
Chloramphenicol	0.1	g
Dichloran (0.2% in ethanol)	1.0	mL
Agar	15.0	g
Distilled Water	1000	mL

For detection and identification of *Fusarium* species and dematiaceous Hyphomycetes.

8. **DG18: Dichloran 18% Glycerol Agar** (Hocking and Pitt, 1980)

Glucose	10	g
Peptone	5	g
KH_2PO_4	1.0	g
$MgSO_4.7H_2O$	0.5	g
Chloramphenicol	0.1	g
Dichloran (0.2% in ethanol)	1.0	mL
Agar	15	g
Water	1000	mL

After steaming for 30 min, add 220 g of glycerol (analytical reagent grade), then sterilise by autoclaving. Final pH 6.5, a_w 0.955. For general purpose fungal enumeration, especially in reduced a_w foods.

9. **DRBC: Dichloran Rose Bengal Chloramphenicol Agar** (King *et al.*, 1979)

Glucose	10	g
Peptone	5	g
KH_2PO_4	1.0	g
$MgSO_4.7H_2O$	0.5	g
Rose bengal (5% soln., w/v)	0.5	mL
Chloramphenicol	0.1	g
Dichloran (0.2% in ethanol)	1.0	mL
Agar	15	g
Water	1000	mL

Final pH 5.6. For general purpose enumeration.

10. **DRYES: Dichloran Rose Bengal Yeast Extract Sucrose Agar** (Frisvad, 1983)

Yeast extract	20	g
Sucrose	150	g
Dichloran (0.2% in ethanol)	1.0	mL
Rose bengal (5% soln., w/v)	0.5	mL
Chloramphenicol	0.1	g
Agar	20	g
Water to	1000	mL

Final pH 5.6 (adjusted after medium is autoclaved). This medium detects *Penicillium verrucosum* and *P. viridicatum* by production of a purple reverse colour.

11. **MA: Malt Agar**

Malt extract	20	g
Agar	20	g
Water	1000	mL

This medium is recommended by Centraalbureau voor Schimmelcultures, Baarn, as a general maintenance and identification medium.

12. **MEA: Malt Extract Agar**

Formula 1 (Oxoid). Final pH 5.4.

Malt extract	30	g
Peptone	5	g
Agar	15	g
Water	1000	mL

12. MEA: Malt Extract Agar

Formula 2 (Pitt, 1979)

Malt extract	20	g
Peptone	1	g
Glucose	20	g
Agar	20	g
Water	1000	mL

For identification of *Penicillium* and *Aspergillus*.

13. OGY: Oxytetracycline Glucose Yeast Extract Agar (Mossel *et al.*, 1970)

Glucose	10	g
Yeast extract	5	g
Oxytetracycline (0.1% soln., w/v)	100	mL
Agar	15	g
Water	1000	mL

Final pH 7.0. Oxytetracycline solution should be filter sterilised and be added to sterile medium. Gentamycin (50 mg/L, final concentration) can replace or be used in combination with oxytetracycline to control the growth of Enterobacteriaceae.

14. PDA: Potato Dextrose Agar (Booth, 1971)

Potatoes, infusion from	200	g
Dextrose	15	g
Agar	20	g
Water	1000	mL

Final pH 5.6. For identification of *Fusarium* species.

15. RBC: Rose Bengal Chloramphenicol Agar (Jarvis, 1973)

Peptone	5	g
Glucose	10	g
KH_2PO_4	1.0	g
$MgSO_4.7H_2O$	0.5	g
Rose bengal (0.5% soln., w/v)	10	mL
Chloramphenicol	0.1	g
Agar	20	g
Water	1000	mL

Final pH 7.2. General purpose enumeration medium.

16. SNA: Synthetischer Nährstoffarmer Agar (Nirenberg, 1976)

KH_2PO_4	1.0	g
KNO_3	1.0	g
$MgSO_4.7H_2O$	0.5	g
KCl	0.5	g
Glucose	0.2	g
Sucrose	0.2	g
Agar	20	g
Water	1000	mL

This medium is used for cultivation of *Fusarium* species. Pieces of sterile filter paper may be placed on the agar to promote production of sporodochia.

17. TGY: Tryptone Glucose Yeast Extract Agar

Glucose	100	g
Tryptone	5	g
Yeast extract	5	g
Agar	15	g
Distilled water to	1000	mL

For enumeration of yeasts in products where moulds are not present.

18. YES: Yeast Extract Sucrose Agar

Yeast extract	20	g
Sucrose	150	g
Agar	20	g
Water	1000	mL

Primarily used for mycotoxin production.

REFERENCES

ABILDGREN, M.P., LUND, F., THRANE, U. and ELMHOLT, S. 1987. Czapek-Dox agar containing iprodione and dicloran as a selective medium for the isolation of *Fusarium* species. *Lett. Appl. Microbiol.* **5**: 83-86.

ANDREWS, S. 1992. Differentiation of *Alternaria* species isolated from cereals on dichloran malt extract agar. *In* Modern Methods in Food Mycology, eds R.A. Samson, A.D. Hocking, J.I. Pitt and A.D. King. pp. 351-355. Amsterdam: Elsevier.

ANDREWS, S. and PITT, J.I. 1986. Selective medium for the isolation of *Fusarium* species and dematiaceous Hyphomycetes from cereals. *Appl. Environ. Microbiol.* **51**: 1235-1238.

BOOTH, C. 1971. Fungal culture media. *In* Methods in Microbiology, ed. C. Booth. pp. 49-94. London: Academic Press.

DIFCO MANUAL. 1984. Dehydrated Culture Media and Reagents for Microbiology, 10th ed. 1155 pp. Detroit, Michigan: Difco Laboratories.

FRISVAD, J.C. 1983. A selective and indicative medium for groups of *Penicillium viridicatum* producing different mycotoxins in cereals. *J. Appl. Bacteriol.* **54**: 409-416.

FRISVAD, J.C. 1985. Creatine-sucrose agar, a differential medium for mycotoxin producing terverticillate Penicillium species. *Lett. Appl. Microbiol.* **1**: 109-113.

HOCKING, A.D. and PITT, J.I. 1980. Dichloran-glycerol medium for enumeration of xerophilic fungi from low-moisture foods. *Appl. Environ. Microbiol.* **39**: 488-492.

JARVIS, B. 1973. Comparison of an improved rose bengal-chlortetracycline agar with other media for the selective isolation and enumeration of moulds and yeasts in foods. *J. Appl. Bacteriol.* **36**: 723-727.

KING, A.D., HOCKING, A.D. and PITT, J.I. 1979. Dichloran-rose bengal medium for enumeration and isolation of molds from foods. *Appl. Environ. Microbiol.* **37**: 959-964.

MOSSEL, D.A.A., KLEYNEN-SEMMELING, A.M.C., VINCENTIE, H.M., BEERENS, H. and CATSARAS, M. 1970. Oxytetracycline-glucose-yeast extract agar for selective enumeration of moulds and yeasts in foods and clinical material. *J. Appl. Bacteriol.* **33**: 454-457.

NIRENBURG, H.I. 1976. Untersuchungen über die morphologiesche und biologische Differenzierung in der *Fusarium*-Sektion *Liseola*. *Mitteilung aus der Biologische Bundesanstalt für Land- und Forstwirschaft* Berlin-Dahlem **169**: 1-117.

OXOID MANUAL. 1990. Sixth Edition. Basingstoke, Hampshire: Unipath Limited.

PITT, J.I. 1979. The Genus *Penicillium* and its Teleomorphic States *Eupenicillium* and *Talaromyces*. London: Academic Press.

PITT, J.I. and HOCKING, A.D. 1985. Fungi and Food Spoilage. Sydney: Academic Press.

PITT, J.I., HOCKING, A.D. and GLENN, D.R. 1983. An improved medium for the detection of *Aspergillus flavus* and *A. parasiticus*. *J. Appl. Bacteriol.* **54**: 109-114.

LIST OF PARTICIPANTS

LIST OF PARTICIPANTS

AUSTRALIA

Dr S. Andrews, School of Chemical Technology, South Australian Institute of Technology, P.O. Box 1, INGLE FARM, SA 5098
Dr Ailsa D. Hocking, CSIRO Division of Food Processing, P.O. Box 52, NORTH RYDE, NSW 2113
Dr J.I. Pitt, CSIRO Division of Food Processing, P.O. Box 52, NORTH RYDE, NSW 2113

BELGIUM

Dr D. Stynen, Eco-Bio, Diagnostics Pasteur, Woudstraat 25, B-3600 GENK

DENMARK

Dr O. Filtenborg, Food Technology Laboratory, Technical University of Denmark, DK-2800 LYNGBY
Dr J.C. Frisvad, Food Technology Laboratory, Technical University of Denmark, DK-2800 LYNGBY
Dr U. Thrane, Food Technology Laboratory, Technical University of Denmark, DK-2800 LYNGBY
Mr P. Nielsen, Food Technology Laboratory, Technical University of Denmark, DK-2800 LYNGBY
Ms Annette Lillie, Alfred Jørgensen Laboratories, Frydendalsvej 30, DK-1809 FREDERIKSBERG

FRANCE

Mad. Helene Girardin, Unité de Mycologie, Institut Pasteur, 25 Rue de Dr. Roux, 75015 PARIS

HUNGARY

Dr T. Deák, Department of Microbiology, University of Horticulture, Somloi ut 14-16, H-1118 BUDAPEST

NETHERLANDS

Dr S. Notermans, National Institute of Public Health, P.O. Box 1, 3720 BA BILTHOVEN
Mr H. Kamphuis, National Institute of Public Health, P.O. Box 1, 3720 BA BILTHOVEN
Ir. G.A. de Ruiter, Department of Food Microbiology, Agricultural University, P.O. Box 8129, 6700 EV WAGENINGEN
Ir. Hetty Karman, Food Inspection Services, Nijennoord 6, 3552 AS UTRECHT
Ms Marjolein van der Horst, Centraalbureau voor Schimmelcultures, P.O. Box 273, 3740 AG BAARN
Drs Ellen S. van Reenen-Hoekstra, Centraalbureau voor Schimmelcultures, P.O. Box 273, 3740 AG BAARN
Dr R.A. Samson, Centraalbureau voor Schimmelcultures, P.O. Box 273, 3740 AG BAARN

SWEDEN

Ms Kirsten Åkerstrand, Statens Livsmedelsverk, P.O. Box 622, S-751 26 UPPSALA
Mr P. Adamek, Swedish Institute for Food Research, P.O. Box 5401, S-402 29 GOTHENBERG
Mrs Birgitta Bergström, Swedish Institute for Food Research, P.O. Box 5401, S-402 29 GOTHENBERG

SWITZERLAND

Mad. Nadine Braendlin, Quality Assurance Department, Nestec Limited, Av. Nestlé 55, 1800 VEVEY

TURKEY

Dr Necla Aran, TUBITAK Research Institute, P.O. Box 21, 4401 GEBZE, Kocaeli
Dr Dilek Heperkan, Hikmet sok. Baris apt., 11/2, 81090 ERENKOY-ISTABUL

UNITED KINGDOM

Dr J.H. Clarke, MAFF Central Science Laboratory, London Road, SLOUGH, Berks. SL3 7HJ
Dr Janet E.L. Corry, J. Sainsbury PLC, Stamford House, Stamford Street, LONDON SE1 9LL
Mrs Judith Kinderlerer, Department of Biological Sciences, Sheffield City Polytechnic, Pond Street, SHEFFIELD S1 1WB
Dr M.O. Moss, Department of Microbiology, University of Surrey, GUILDFORD GU2 5XH
Miss Nicola Pope, Unipath Ltd, Wade Road, BASINGSTOKE, Hants. RG24 OPW
Dr D.A.L. Seiler, Flour Milling and Baking Research Association, CHORLEYWOOD, Herts. WD3 5SH

UNITED STATES OF AMERICA

Dr L.R. Beuchat, Department of Food Science, University of Georgia Experiment Station, GRIFFIN, GA 30223
Dr L.B. Bullerman, Department of Food Science, University of Nebraska, LINCOLN, NE 68583
Dr D.E. Conner, Department of Poultry Sciences, Auburn University, AUBURN, AL 36849-5416
Dr A.D. King, USDA ARS Western Regional Research Center, 800 Buchanan Street, ALBANY, CA 94710
Dr D.F. Splittstoesser, Department of Food Science and Technology, Cornell University, GENEVA, NY 14456-0462
Dr Gwen Reynolds, Gene-Trak Systems, 31 New York Avenue, FRAMINGHAM, MA 01701

INDEX